GEOMORPHOLOGICAL STUDIES IN SOUTHERN AFRICA

PROCEEDINGS OF THE SYMPOSIUM ON THE GEOMORPHOLOGY OF
SOUTHERN AFRICA / TRANSKEI / 8-11 APRIL / 1988

Geomorphological Studies in Southern Africa

Edited by

G.F.DARDIS
Department of Geography, University of Transkei, Umtata

B.P.MOON
*Department of Geography and Environmental Studies, University of the
Witwatersrand, Johannesburg*

A.A.BALKEMA / ROTTERDAM / BROOKFIELD / 1988

The texts of the various papers in this volume were set by typists under the supervision of the editors.

Published by
A.A.Balkema, P.O.Box 1675, 3000 BR Rotterdam, Netherlands
A.A.Balkema Publishers, Old Post Road, Brookfield, VT 05036, USA

ISBN 90 6191 831 6

TABLE OF CONTENTS

Geomorphological Studies in Southern Africa, G.F.Dardis & B.P.Moon (eds)
© *1988 Balkema, Rotterdam. ISBN 90 6191 831 6*
                                                                               V

3 Soil erosion and soil movement

4 Weathering and environmental change

5 Coastal environments

6 Fluvial environments

PREFACE

The original idea of holding a meeting, to provide a forum where researchers interested in geomorphological studies in southern Africa could interact, was conceived during informal discussions during the First International Conference on Geomorphology, held at the University of Manchester in September 1985.

Much work concerned with the geomorphology of southern Africa has been carried out by researchers in southern Africa and also by earth scientists from North America, Europe and elsewhere. In order to bring together as many researchers and up-to-date studies as possible, the Symposium on the Geomorphology of Southern Africa was organised, to run between 8-11 April 1988. The response to the meeting was sufficient to merit publication of the proceedings as a special volume devoted to geomorphological studies in southern Africa.

It is hoped that this proceedings volume will provide a necessary first-step in bringing together some of the most recent geomorphological studies undertaken in the region. It is further hoped that the contributions to the proceedings volume will stimulate further meetings or workshops, and stimulate new or previously neglected avenues of research. It is clear, from the papers included here, that much remains to be investigated, much remains either unsolved or enigmatic, and many areas of research are ripe for picking.

The Symposium, and compilation of this book, has been possible only through the combined efforts of many individuals to whom the editors are most grateful. We thank Professor W. Nkuhlu, Principal of the University of Transkei, and the University Council for considerable encouragement and logistical support, without which this Symposium could not have taken place; Heinz Beckédahl, who shouldered much of the responsibility for the organisation of the Symposium and post-Symposium field excursion; Amie Mlungwana and Pauline Dardis, who provided valuable support in re-typing the contributors manuscripts onto the word-processor; and Frith Webster (Adfocus, East London), who provided valuable support in typesetting. Patricia Hanvey provided considerable help in proof-reading and setting the manuscripts.

Geomorphological Studies in Southern Africa, G.F.Dardis & B.P.Moon (eds)
© 1988 Balkema, Rotterdam. ISBN 90 6191 831 6

The METEOSAT VIS image of 13.6.1979, 13h30 on the cover, showing dust plumes off the coast of Namibia, and the Okavango River and delta, was received and processed by the Satellite Remote Sensing Centre of the CSIR, Pretoria, to whom we are grateful for allowing its reproduction. We thank Tony Watts, (Reprographics, University of Transkei) for screening the satellite image.

The editors are also grateful to the reviewers of the papers, who have provided useful and constructive criticism of the contributions, often at very short notice.

Finally, our thanks go to the individual contributors, who made considerable efforts to submit their papers on time and in the manner requested. It is hoped that these papers will be a valuable contribution to further the understanding of the southern African landscape.

GEORGE F. DARDIS
BERNARD P. MOON

INTRODUCTION

GEORGE F. DARDIS
Department of Geography, University of Transkei
BERNARD P. MOON
Department of Geography and Environmental Studies,
University of the Witwatersrand

It is apparent from many of the papers presented at the Symposium, that geomorphological research in the southern African region is being directed primarily towards three or four major areas of study.

Macrogeomorphological studies, which have traditionally dominated geomorphological research in the region through the work of L.C. King, J.H. Wellington and others, are well represented, with papers from Partridge & Maud, Klein, Jacobs & Thwaites, De Wit, and Höverman, and in an invited paper by Twidale. This area of research still remains highly controversial, with debate concentrated on the genesis of land (or planation) surfaces, and their ages. Whilst such studies have been largely abandoned outside southern Africa (or subsumed within macrogeomorphology), the nature and antiquity of the southern African landscape still demands strong adherence to, or at least strong consideration of, the erosional development of land surfaces. They are important, for example, in determining the nature and efficacy of small-scale erosional forms (e.g. soil erosion). These studies have received new impetus through plate tectonics (Klein), from a clearer understanding of Cainozoic sedimentation patterns in the main offshore basins encircling southern Africa, and studies of erosion associated with kimberlite pipes (Partridge & Maud); detailed stratigraphic and sedimentological studies of land surfaces in the southern Cape (Rogers; Jacobs & Thwaites); fission track dating of Tertiary and older surfaces in the north-western Cape (De Wit), which allows more precise dating of the formation of the west coast escarpment and poses new problems in terms of landscape evolution in southern Africa; and through studies of landform development on particular land surfaces (Twidale).

Studies of arid and semi-arid environments in southern Africa, which have been one of the most active research areas in recent years, are well represented in papers by Höverman, Wilkinson, Harmse, Thomas, De Dapper, Teller and co-workers, and Rowntree. Höverman provides a macroscale comparison of the major African deserts (Sahara, Namib, Kalahari) and draws interesting comparisons of the relationship be-

Geomorphological Studies in Southern Africa, G.F. Dardis & B.P. Moon (eds)
© 1988 Balkema, Rotterdam. ISBN 90 6191 831 6

tween macrogeomorphological and climatic changes during the Quaternary period. Wilkinson presents detailed studies of linear dune forms and their relationship to prevailing Berg Winds. Harmse presents grain-size data on eolian deposits on the west coast of South Africa, which suggests that only 9 per cent of Sandveld sand is mobile in response to the present wind regime. De Dapper describes the geomorphology of the sand-covered plateaux in southern Shaba, Zaire, where linear, micro-relief dune forms on the plateaux are presently undergoing reactivation and erosion. Detailed studies of termites and their palaeoenvironmental significance are outlined by Teller and co-workers.

Thomas examines the relationship between dunes and vegetation in the Kalahari, and presents important findings, which demonstrate that up to 35 per cent vegetation cover may be a normal component of some types of active linear dunes. The relationship between vegetation and landform development (in particular soil erosion) is the central theme of Rowntree's paper, who argues that the supposed soil (i.e. accelerated) erosion in the Karoo may result from short-term changes in vegetation cover, but reflect longer-term ecosystem stability. The soil erosion may therefore be geologically "normal" rather than "accelerated".

The theme of soil erosion cannot be readily ignored in the southern African region, particularly in eastern southern Africa, where "active badlands" are ubiquitous. Papers by Dardis and co-workers, Weaver, Beckedahl & Dardis, Dardis and Beckedahl, Marker, Rowntree, and van Rheede van Oudtshoorn are presented here, which examine various geomorphological aspects of the soil erosion problem. They consider, in particular, gully forms (Rowntree, Dardis et al.), soil erosion indices (Weaver), bedrock-incised gullies (Dardis & Beckedahl), soil pipe systems (Dardis & Beckedahl), the role of artificial drainage in soil pipe and gully development (Beckedahl & Dardis), changes in the rates and patterns of soil erosion over time (Marker), and the relationship between soil erosion and soil properties (van Rheede van Oudtshoorn). Marker's study shows that significant changes in patterns of soil erosion have accompanied resettlement in the Ciskei, associated with increased population density, removal of (woody) vegetation and degeneration of veld as a result of increased stock numbers. (Additional papers on the theme of *Soil Erosion and Parent Materials* were presented in a Workshop during the Symposium and will be published elsewhere). These contributions indicate the awareness for applied geomorphological research, but emphasise the complexity of the soil erosion problem.

Research on hillslope processes is reflected in papers on terracettes by Verster & Van Rooyen, and Watson. Verster & Van Rooyan examine soil movement over an eight and six year period at four sites in the Natal Drakensberg and the eastern Transvaal. Watson has examined the influence of soil, topographic and biomass factors on the distribution and morphological characteristics of terracettes in the Natal Drakensberg.

2

Research on weathering is reflected in papers on spheroidal weathering in basalts (Saraccino & Prasad) and a review of recent developments in weathering studies (Hall), emphasising, in particular, the need for more quantitative studies of weathering processes in southern Africa, and the need to embrace new techniques and new technologies to aid in the understanding of weathering processes.

Meadows & Sugden, and Dewey, examine evidence for long term changes in vegetation and landform development in the Karoo and Winterberg respectively. Lewis & Hanvey present new data on the sedimentology of Quaternary debris slope deposits in the Drakensberg, emphasising the high probability of Pleistocene "periglacial" conditions having prevailed in that area. Sänger, on a similar note, presents temperature data from above 2000 m a.s.l. from the Cape Fold Mountains, which, assuming a fall in MAAT of $10^{o}C$, also suggests a Pleistocene climatic regime suitable for the development of periglacial/glacial landforms.

Coastal environments are examined in papers by Illenberger, Dardis & Plumstead, Beckedahl and co-workers, Rogers, and Mulder. Illenberger examines the Holocene evolution of landforms and sediments in estuarine environments, through a case study of the Sundays estuary. Rogers presents a detailed overview of the stratigraphy and geomorphology of (predominantly) eolianite deposits in the Bredasgroup in the southern Cape. Dardis & Plumstead present observations of mudball development within the range of tidal influence on the Mbashee River, Transkei. Mulder presents new observations of stormflow response to rainfall in the Zululand coastal zone. Beckedahl and co-workers outline preliminary observations of a recurrent mass movement complex on the Transkei coast.

Many areas of research remain untouched, in particular, studies of modern fluvial systems in southern Africa (with the exception of a paper on palaeodrainage by Marshall). In view of this Baker presented an invited paper on the geomorphology and palaeohydrology of bedrock rivers. This paper reviews recent techniques, developed in North America and Australia, to examine flood events in modern and ancient fluvial systems, much of which could readily be applied to the southern African subcontinent.

THE GEOMORPHIC EVOLUTION OF SOUTHERN AFRICA: A COMPARATIVE REVIEW

T.C. PARTRIDGE
Department of Geology, University of the Witwatersrand
R.R. MAUD
P.O.Box 4122, Durban

1. INTRODUCTION

Since the early years of this century the macroscale geomorphic evolution of southern Africa has aroused much controversy and has generated a relatively voluminous literature. The observational basis of these earlier views is generally to be admired particularly in view of difficulties of transport, access, and suitable maps then available. Good evidence permitting onshore erosional events to be linked with the offshore sedimentary record has, however, until recently, been very limited. Moreover, early workers seldom applied rigorous field criteria in the correlation of denudational landsurfaces over wide areas.

In this comparative analysis, earlier views, particularly those of L.C. King, are considered with a recent evaluation based on a detailed appraisal of much newly assembled evidence (Partridge and Maud, 1987).

2. EARLIER VIEWS OF SOUTHERN AFRICAN GEOMORPHIC EVOLUTION

Apart from the early workers, Suess (1904) and Penck (1908), who both concluded that the Drakensberg escarpment on the eastern seaboard of the subcontinent was of erosional rather than tectonic origin, but that the coastline was faulted (Suess), or of monoclinal origin (Penck), much of the early geomorphic identification and correlative work was, until the late 1930s carried out in east and central Africa where some associated datable deposits are present. Workers in these areas included Wayland (1931), Beetz (1933), Veatch (1935), Jessen (1936) and Dixey (1938). These workers identified land surfaces of regional extent to which they assigned various ages. Thus a late Jurassic to early Cretaceous land surface was recognised in Uganda (Wayland 1931), in Angola by Jessen (1936), in Tanzania by Willis (1936) and in northern Malawi-Zambia by Dixey (1938).

Geomorphological Studies in Southern Africa, G.F.Dardis & B.P.Moon (eds)
© *1988 Balkema, Rotterdam. ISBN 90 6191 831 6*

Vetch (1935) described an extensive land surface of Cretaceous age which extended over much of central Africa, and which remained undisturbed until the mid-Miocene, when the cycle which produced it was terminated by upwarping of the western coastal margin.

Younger land surfaces of late Tertiary age were also recognised. Wayland recognised an early Miocene and late Pliocene surfaces in Uganda, Veatch (1935) one of late Tertiary in Angola, where Jessen (1936) also described a surface of Miocene-Pliocene age. A late Tertiary landsurface was recognised in Tanzania by Willis (1936), while in Malawi and Zambia Dixey (1938) described a replaning of the earlier surface in the early Tertiary and a lower younger surface of the end-Tertiary age, developed as a result of uplift of both the east and west coasts of the subcontinent at the end of the Pliocene and in the earliest Pleistocene.

Thus by 1940 the broad framework of the erosional land surfaces present in east and central Africa and an appreciation of their respective ages had been reasonably well established.

3. THE WORK OF KING AND DIXEY

In the early 1940's regional geomorphic investigations were extended into southern Africa.

Locally, Kent (1938) had concurred with the view of Suess (1904) that the Natal coast had a faulted origin, but King (1941) preferred to follow the view of Penck (1908) that this coastline was of monoclinal origin dating to the Jurassic, the monocline plunging gently to the south with its seaward face steeping in the same direction. The concept of coastal monoclines, first propounded in this paper, became a basic tenet, with some minor modifications, of almost all of King's later geomorphic work, both in southern Africa and world-wide (King, 1951; 1962).

In 1942 Dixey added to his suite of landsurfaces and end-Cretaceous to early Tertiary surface and a very young cycle of early Pleistocene river incision, the former being recognised around the margins of Lesotho. Importantly, he also proposed the former continuity of the main interior surface, whose development was ended in the mid-Miocene, with its coastal equivalent, their separation being ascribed to upwarping of the subcontinent along an axis generally coinciding with the Great Escarpment.

In 1944 King first entered the field of regional geomorphic studies in southern Africa, which was later extended on a global scale, leading to his international recognition in this field. He considered the summit crests of the Lesotho highlands to be a Jurassic erosion surface, but he ascribed Dixey's (1942) end-Cretaceous surface around Lesotho rather to the influence of structure. King (1944) recognised that erosion surfaces of the same age stood at different elevations above and below the Great Escarpment; this he considered as being due to the differing distances to base level on either side of this feature. Thus he equated a surface at 1200 m forming the Highveld above the escarpment. The erosion cycle which formed them was believed a lower-standing surface near Pietermaritzburg, which he considered to be of end-Tertiary age.

In 1947 King proposed that the axis of the Natal monocline was located between the escarpment and the coast, and that this feature had been initiated in the Jurassic and rejuvenated subsequently in the Tertiary. This location of the monoclinal axis of the Natal monocline was located between the escarpment and the coast, and that this feature been initiated in the Jurassic and rejuvenated subsequently in the Tertiary. This location of the monoclinal axis caused the earlier erosion surfaces to flatten west of the axis in conformity with the dip of the Karoo strata occurring there.

In 1949 King argued that the co-existence of landsurfaces of different ages is only possible if landscape evolution proceeds by means of pediplanation rather than peneplanation; he proposed further that the various erosion cycles are best dated in relation to the time of their termination by uplift. For the first time formal names were given to the various erosion surfaces, that terminated in the Miocene being called the Gondwanaland, that at the end-Tertiary the African, that in the Pleistocene the Victoria Falls and that of the present the Present. Residuals above the Gondwanaland surface were considered to be due to local upwarping and consequent scarp formation. The Gondwanaland land surface was recognised as standing at two levels, above and below the Great Escarpament.

King (1951) reiterated his view that the Lesotho Highland summits represented a surface of Jurassic age. The lower lying Gondwanaland surface, which was also regarded as pre-dating the fragmentation of Gondwanaland, was seen as occurring at two elevations in the eastern marginal regions, above and below the Drakensberg escarpment, which was regarded as having originated prior to continental disruption. The African surface was considered to have been initiated with the fragmentation of Gondwanaland, and was also seen as occurring at different levels above and below the Great Escarpment. The subsequent Victoria Falls and Congo cycles were manifested chiefly as river incision, the the former being due to updoming of the sub-continent in the Miocene. A total uplift since Miocene times of some 600 m was estimated. A group of very young 'Later' cycles related to the river terrace formation were also recognised.

Fair and King (1954) renamed the 'marginal' Gondwanaland surface below the Great Escarpment the 'Post-Gondwana' surface, considering it to have been initiated in the mid-Cretaceous and linking it with the unconformity present in the coastal sequences overlain by Senonian to Eocene strata. Subsequent erosion cycles were the African of mid-Tertiary age, the Victoria Falls of end-Tertiary age, the Congo of Pleistocene age, and the Latest cycles of Recent age.

Dixey (1955a) accepted, for the first time, the validity of the pediplanation model of landscape evolution in semi-arid and arid climates, and confirmed his preference for the dating of erosion surfaces in terms of overlying deposits and date of termination of the corresponding cycle by continental uplift. In the same year Dixey

(1955b) reiterated his 1942 position but recognised the presence of three end-Tertiary cycles. These were of upper Pliocene, Plio-Pleistocene, and Pleistocene age. These cycles were, however, produced by local warping and base level change, their significance thus not being of widespread importance. At the same time King (1955) proposed that monoclinal warping in the interior of Natal was due to crustal isostatic compensation in response to onshore erosion and concurrent offshore sedimentation.

In 1959, King abandoned his previously held view of the same erosion surfaces occurring at disparate levels above and below the Great Escarpment (King and King, 1959) and effectively embraced Dixey's (1942) view that the same surfaces were formerly continuous above and below the Great Escarpment. In this paper it was postulated that the Jurassic and end-Tertiary axes of warping in Natal occupied different positions, the African surface being carried across the Great Escarpment by the end-Tertiary uplift on an axis more or less coincident with the Great Escarpment to merge with the Highveld surface of the interior (the Gondwanaland surface of King [1951]). The Lesotho Highland summits he now considered to represent the Gondwana surface of Jurassic age, the Post-Gondwana cycle rather than the African cycle having been initiated by the fragmentation of Gondwanaland in the early Cretaceous. The prominent 'secondary' faulting of the Natal coast associated with the Jurassic monocline was, in his view, 'more or less random'. Rejuvenation of the (Jurassic) monocline, with some faulting, in the mid-Cretaceous was now considered to have initiated the African erosion cycle, the resulting erosion surface being considered to be represented by the unconformity between Cretaceous marine sediments and the overlying Miocene marine strata at Uloa on the Zululand coast. Uplift in the early Miocene on a monoclinal axis somewhat east of the Drakensberg produced renewed incision and planation in the Post-African cycle (here so named for the first time in place of the Victoria Falls cycle of 1951), the corresponding surface being correlated with an inferred unconformity below possible generations of Pliocene fossiliferous sands at Uloa in coastal Zululand. Further rejuvenation and enlargement of the inland monocline east of the Drakensberg during the Pliocene produced local planation in a Late Tertiary Phase II cycle. Finally, major upwarping along the 'Great Escarpment' monocline at the end of the Pliocene initiated major gorge incision in the sub-escarpment zone. The formation of 'treppen' (or benches) in the hinterland of Natal was ascribed to the amplitude of this end-Tertiary upwarping.

To explain the phenomenon of monoclines of continental scale, which he now perceived to be present in other continents (e.g. Brazil (King 1956) and Australia (King 1959)), King in 1961 erected a model of 'Cymatogeny'. This he defined as "...the undulating ogeny..." which involved 'a mode of crustal deformation between epeirogeny and orogeny in which the earth's surface is thrown into great waves or undulations commonly hundreds of miles across and thousands of feet high'. These undulations could be recognised by the mapping of Cainozoic denudational surfaces that are dated by reference to unconformities or hiatuses in sequences of

Cainozoic rocks deposited in relatively depressed areas. In 1962 he consolidated his views on a world-wide basis in 'The Morphology of the Earth'.

In 1972, King interposed an additional Pliocene cycle to follow that responsible for the Late Tertiary Phase II cycle. This was claimed to have produced a number of independent basins at different elevations in Natal, including the Ladysmith basin. In a subsequent paper, King (1976), proposed a new nomenclature for the succession of erosion surfaces to accord with his global perspective of their occurrence. This nomenclature was also incorporated in the second edition of 'The Natal Monocline' (1982). In it the 'Gondwana' surface retained its original name, the Post-Gondwana surface became the 'Kretacic', the African the 'Moorland', the Post African the 'Rolling', the Late Tertiary phase II cycle the Widespread', and the Quaternary cycles the 'Youngest' cycles. This nomenclature is continued and further developed in his most recent and final synthesis of global geomorphology, (King 1983).

4. THE VIEWS OF OTHER WORKERS

Obst and Kayser (1949), like Dixey, saw uplift or 'randschwellung', as proposed by Jessen (1943), along an axis coinciding with the Great Escarpment, as the means whereby the 1200 m erosion surface in the interior of Natal was carried over the escarpment into the Highveld, where it is represented at elevations between 1400 and 1700 m. This uplift was considered to be responsible for the development of the lower, end-Tertiary, erosion surface of the Natal hinterland. The earlier views of Dixey were further supported by Cahen and Lepersonne (1952), who presented evidence from the Congo and Kalahari basins for erosion surfaces current during the late Cretaceous, the Miocene and the end-Tertiary.

Wellington (1955) questioned the widespread preservation of an erosion surface of Jurassic age predating the break-up of Gondwanaland as proposed by King (1951), in the light of the continuity and extent of erosion since that time. Wellington emphasised the role of Karoo rocks in constraining the morphology of landscapes developed on them, and drew attention to the influence of the pre-Karoo surface on later landscape development, particularly in relation to inherited discordant drainage patterns. He considered the development of a Miocene erosion surface over much of southern Africa as unlikely, owing to the rapid erosion of soft Karoo strata, although he recognised a surface of this age to the north of the Limpompo River.

In the same year, Mabbutt (1955) described a series of erosion surfaces in Namaqualand, which he correlated with various surface deposits of the southwestern Kalahari. He described silicification and kaolinisation associated with a land surface to which he assigned an end-Cretaceous to earliest Tertiary age; in this he appears to have been the first worker to note this association of duricrust formation and deep weathering with a specific land surface. He concluded that the oldest cycle and land surface in the region appeared to be of upper Cretaceous or later age at the coast, and that this did not support King's concept of a 'Gondwana' surface surviving since before the breakup of Gondwanaland.

Maud (1961), as a result of investigation of the faulted structure of the coastal portion of Natal, concluded that rather than being a monocline as had been proposed by King (1941), the structure of the coast is one of mainly tilted fault blocks with related horst and graben structures, an arcuate fault pattern of major gravity faults being present (conjugate shear fractures). In a subsequent analysis Maud (1965) described laterite development and deep weathering associated with the Early Cainozoic 'African' erosion surface in coastal Natal.

In a critique of King's (1962) synthesis, De Swardt and Bennet (1974) emphasised the duality of the inland and coastal drainage of southern Africa which, in their view, precluded widespread correlation of erosion surfaces across the Great Escarpment. They recognised two polycyclic erosion surfaces in the coastal hinterland of Natal, but regarded other surfaces in the interior of the province as being of structural origin.

They dismissed evidence for a relict erosion surface of cyclic significance on the crest of the Lesotho Highlands. Progressive movement of a monoclinal arch from offshore to a position about 40 km inland was postulated in response to proceeding erosion. Extensive Pleistocene erosion in the interior of Natal was ascribed by them to glacio-eustatic sea level movements and related climatic effects, maximum planation during earlier cycles being correlated with marine transgressions. Serious problems in the dating of inland erosion surfaces were envisaged owing to claimed ambiguities in the offshore sedimentary record and a dearth of other palaeontological evidence.

In 1980 Mountain recognised the association of thick silcrete with deep kaolinisation on the 'Grahamstown Peneplain' which he considered as being possibly of early Tertiary age.

Smith (1982) sought mechanisms to explain the large Plio-Pleistocene uplifts which characterise extensive areas of the African continent. This he ascribed to phase changes in the continental mantle associated with continental plate movements across oceanic ridges.

Hendey (1983) ascribed landscape development in the southern and southwestern Cape to the effects of widespread eustatic sea level fluctuations during the Cainozoic; on this basis he tentatively correlated the largely marine-cut 150 to 200 platform of the southern Cape coast with an inferred early Eocene high sea level. Incision and the initiation of new erosion cycles were linked to rapid falls in sea level at the beginning of regressive episodes, the influence of tectonics being entirely discounted. In the same vein Dingle and Hendey (1984) suggested that river capture and changes in drainage above the Great Escarpment on the Bushmanland plain were encouraged by the major fall in sea level during late Maastrichtian to early Palaeocene time.

Ollier and Marker (1985) interpreted the Great Escarpment of southern Africa as separating two erosion surfaces only. These are high 'palaeoplain' at about 1500 m elevation above the escarpment, which is interpreted as a somewhat modified dicyclic landsurface dating to before the breakup of Gondwanaland, and a coastal plain below it. Several periods of continental uplift were documented through review and interpretation of the available evidence. This uplift was considered to have been more or less symmetrical around southern Africa, the western interior lagging, however, to form the Kalahari basin. In subsequent publications, Ollier (1985a, 1985b) has reviewed a number of alternative models which could account for this pattern of uplift as observed in Africa and other continents.

In 1985, Summerfield drew attention to the discrepancies which exist between the landscape chronology proposed by King (1962, 1972) and the most recent stratigraphic evidence from the continental margin. He placed emphasis on the complex response of passive continental margins to changes in base level as a result of varying coastal geometry and style of tectonic adjustment, reaching the conclusion that a relative fall in sea level will not always initiate a new landscape cycle and is likely to be only of local significance. The development of the Great Escarpment, and the resulting dual drainage system above and below it, he regarded as being incompatible with the formation of continent-wide erosion surfaces, and he considered that no satisfactory mechanism has been proposed that would produce the extensive uplift necessary to initiate landscape cycles of widespread significance.

In 1985 Hartnady adduced evidence for possible neotectonic activity in the Lesotho-Natal region of southeastern Africa. He considered that the uplift of this region could be related to the movement of the relevant portion of the continental plate over the former position of an oceanic spreading ridge; alternatively, the presence of a mantle hotspot beneath this region could be the cause.

5. THE ANALYSIS OF PARTRIDGE AND MAUD (1987)

In 1987 we published a paper entitled 'The Geomorphic Evolution of Southern Africa since the Mesozoic'. This publication was the result both of extensive fieldwork, over a period of some five years, which involved both field mapping and graphical correlation of erosion surface remnants, and the assessment and interpretation of the views of previous workers, most of which are summarised above. A precis of our findings in relation to those of other workers is given below.

We cannot find any conclusive evidence for the preservation of any subaerial erosion surface dating from before the breakup of Gondwanaland; indeed, modern evidence indicates that at least 300 m of material has been removed since the late Cretaceous from the summits of the Lesotho Highlands (Hawthorne 1975). In this we concur with the earlier views of Mabbutt (1955) and Wellington (1955). In addition, we find no evidence for the Post-Gondwana erosion surface of King and King (1959). Although benches do occur above the African surface, we believe them to be of structural origin. We assign such areas to a broad category of mountainous tracts above the African surface. We are, however, mindful of the hiatuses in the offshore sedimentary record in the early Cretaceous and the mid-Cretaceous, which were cited by King as the correlatives of these two surfaces. The thickness of the offshore sedimentary record in the Cretaceous indicates that the bulk of the denudation of the subcontinent occurred within a relatively short period after the completion of rifting.

We find the widespread fundamental erosion surface of the subcontinent to be the 'African' surface of King, whose earlier nomenclature we largely follow. Where it survives it is characterised by a duricrust capping of either laterite (in the east) or silcrete (in the south and west) and a deeply weathered (usually kaolinised) underlying profile. In places, this surface has not reached maturity, and we recognise a 'partly planed' phase in which considerable areas remaining above the pedimented surface display major structural control, as in the eastern Orange Free State. We also recognise a 'lowered' phase where the original surface is usually preserved on interfluves, and a 'dissected' phase where the original surface is sometimes preserved on interfluves. In the eastern Cape the surface carries marine sediments on it and here it takes the form of a marine 'platform' (Maud et al., 1987); it can be dated, in part, from associated fossiliferous deposits.

In our view the African cycle was initiated by the breakup of Gondwanaland and persisted until its termination, in the early Miocene, by marginal uplift and tilting of the subcontinent. Almost certainly it was polycyclic, as is indicated by the mid-Cretaceous break in the offshore sedimentary record, but owing to the effects of subsequent erosion and its great age, any polycyclic effects are now elided, and it survives as a single surface.

We calculate, in terms of evidence available to us, that Africa, which occupied a central position in Gondwanaland prior to continental rifting, had a high elevation, of the order of 2000 m above present sea level. After breakup this resulted in a dichotomy of base levels in the interior and marginal portions of the subcontinent; the preservation of the Great Escarpment up to the present can be attributed to this cause. Thus, erosion surfaces of the same initial age have been planed at different elevations on either side of this feature. In this regard, our conclusions accord with the earlier views of King (1951), prior to his acceptance of Dixey's invocation of upwarping of a previously continuous surface along an axis more or less co-inciding with the escarpment. We find that uplift and tilting in the eastern portion of the subcontinent has occurred along much the same axis on two occasions since the breakup of Gondwanaland; this axis is located about 80 to 100 km inland of the south eastern coastline, well seaward of the line of the Great Escarpment.

Termination of the African cycle of erosion in the early Miocene, by uplift of some 200 to 300 m in the eastern portion of the subcontinent, initiated the ensuing Post-African I erosion cycle. This continued until the late Pliocene, when it was, in turn, terminated by further major uplift of some 800 to 1000 m with associated marginal tilting. This surface was not as well planed as the African surface owing to its relatively short duration.

Uplift and marginal tilting also affected the southern and western margins of the subcontinent at both these times, but in neither case was the movement as great as in the eastern marginal regions; this explains the topographic asymmetry of the subcontinent at the present time.

Whereas in the eastern marginal areas an untitled Post-African II cycle of erosion can be recognised in places, below the level of the Post-African I surface, differentiation of Post-African cycles is not possible in the interior because of local control of base level. We recognise the following manifestations of the Post African cycles: a Post-African I surface preserved in pristine form, a dissected Post-African I surface in which the original surface is preserved on interfluves, a Post-African I marine platform, which is well developed in southern Cape, and partly planed Post African II surface in the eastern marginal regions. In the interior, we recognise an undifferentiated, predominantly partly planed, Post-African surface. As far as the Post-African cycles of erosion are concerned, our views are, therefore, more or less in accord with those of King (King and King, 1959), except insofar as the location of the axes of uplift and marginal tilting are concerned. Younger cycles of valley erosion, on the other hand, seem to us to be far more strongly controlled by local base level and structure than was recognised by King. The effects of Pleistocene glacio-eustatic sea level climatic changes can, however, be seen locally in some of these features, but in many portions of the subcontinent dissected areas of varying age are present; in these, individual landscape cycles are largely obscured by major structural controls.

Major depositional sequences identified by us include the Cainozoic sediments of the interior Kalahari Basin, and the Neogene marine and aeolian sediments of the coastal margins, some of which overlie earlier Cretaceous or Palaeogene sedimentary packages.

Our analyses to date have not   given any detailed consideration to the causes of the uplift and monoclinal tilting that has twice affected the margins of the subcontinent during the Cainozoic. These are the subject of an ongoing research programme, the findings of which should throw new light on tectonic mechanisms within passive continental margins.

6. REFERENCES

Beetze, P.F.W. 1933. The Geology of south-west Angola between Cunene and Lunda axis. Transactions of the Geological Society of South Africa, 36, 137-176.

Cahen, L., and Lepersonne, J. 1952. Equivalence entré le Systeme du Kalahari du Congo Belge et les Kalahari Beds d'Afrique australe. Mem. Soc. Belge de Geol., 8, 3-64.

Fair, T.D.J. and King, L.C. 1954. Erosional land-surfaces in the eastern marginal areas of South Africa. Transactions of the Geological Society of South Africa, 62, 19-26.

Hawthorne, J.B. 1975. Model of a kimberlite pipe. In: Ahrens, L.H., Dawson, J.B., Duncan, A.R. and Erklank, A.J. (eds.), Physics and Chem. of the Earth, Pergamon Press, Oxford, 9, 1-5.

Hartnad, C.J.H. 1985. Uplift, faulting, seismicity, thermal spring and possible incipient volcanic activity in the Lesotho-Natal Region, S.E. Africa: The Quathlamba Hotspot Hypothesis. Tectonics, 4, 371-377.

Hendey, Q.B. 1983. Cenozoic geology and palaeogeography of the fynbos region. In: Deacon, H.J., Hendey, Q.B. and Lambrechts, J.J.N. (eds.), Fynbos paleoecology: a preliminary synthesis, South African National Scientific Programmes Report, 75, CSIR, Pretoria, 35-60.

Jessen, O. 1936. Reisen und Forschungen in Angola. Reimer, Berlin. 397 pp.

Jessen, O. 1943. Der Randschwellen der Kontinente. Petermans. Geog. Mitt., 241 pp.

Kent, L.E. 1938. The geology of Victoria County, Natal, with reference to the evolution of the coastline. Transactions of the geological Society of South Africa, 41, 5-25.

King, L.C. 1941. The monoclinal coast of Natal, South Africa. Journal of Geomorphology, 3, 144-153.

King, L.C. 1944. Geomorphology of the Natal Drakensberg. Transactions of the Geological Society of South Africa, 47, 255-282.

King, L.C. 1947. Landscape study in southern Africa. Proceedings of the Royal Society of South Africa, 50, xxii-lii.

King, L.C. 1949. On the ages of African landsurfaces. Quarterly Journal of the Geological Society of London, 104, 439-453.

King, L.C. 1951. South African Scenery. 2nd Ed., Oliver and Boyd, Edinburgh, 379 pp.

King, L.C. 1955. Pediplanation and isostasy: an example from South Africa. Quarterly Journal of the Geological Society of London, 111, 353-359.

King, L.C. 1956. Rift Valleys of Brazil. Transactions of the Geological Society of South Africa, 59, 199-209.

King, L.C. 1959. Denudational and tectonic relief in south-eastern Australia. Transactions of the Geological Society of South Africa, 62, 113-138.

King, L.C. 1961. Cymatogeny. Transactions of the Geological Society of South Africa, 64, 1-22.

King, L.C. 1962. The morphology of the Earth. Oliver and Boyd, Edinburgh. 699 pp.

King, L.C. 1972. The Natal Monocline:explaining the origin and scenery of Natal, South Africa. 1st Ed., Geology Dept. University of Natal, Durban. 113 pp.

King, L.C. 1976. Planation remnants upon high lands. Zeitschrift für Geomorphologie N.F., 20, 133-148.

King,, L.C. 1982. The Natal Monocline:explaining the origin and scenery of Natal, Sout Africa. 2nd Ed. Rev., University of Natal Press, Pietermaritzburg. 134 pp.

King, L.C. 1983. Wandering Continents and Spreading Sea Floors on an Expanding Earth. John Wiley and Sons, Chichester, New York. 232 pp.

King, L.C., and King, L.A. 1959. A reappraisal of the Natal monocline. South African Geographical Journal, 41, 15-30.

Mabbutt, J.A. 1955. Erosion surfaces in Namaqualand and the ages of surface deposits in the south-western Kalahari. Transactions of the Geological Society of South Africa, 58, 13-30.

Maud, R.R. 1961. A preliminary review of the structure of coastal Natal. Transactions of the Geological Society of South Africa, 64, 247-256.

Maud, R.R. 1965. Laterite and lateritic soil in coastal Natal, South Africa. Journal of Soil Science, 16, 60-72.

Maud, R.R., Partridge, T.C. and Siesser, W.G. 1987. An early Tertiary marine deposit at Pato's Kop, Ciskei. South African Journal of Geology, 90 (3). (in press).

Mountain, E.D. 1980. The Grahamstown peneplain. Transactions of the Geological Society of South Africa, 83, 47-54.

Obst, E. and Kayser, K. 1949. Die grosse Randstufe auf der Ostseite Südafrikas und ihr Vorland: ein Betrag zür Geschichte der Jungen auslebung des subcontinents. geogr. Ges. Hanover 344 pp.

Ollier, C.D. 1985a. Morphotectonics of passive continental margins:Introduction. Zeitschrift für Geomorphologie N.F., Supplement Band, 54, 1-9.

Ollier, C.D. 1985b. Morphotectonics of continental margins with great escarpments.In: Morisawa, M. and Hack, J.T. (eds.), Tectonic Geomorphology, Allen and Unwin, Boston.

Ollier, C.D. and Marker, M.E. 1985. The great escarpment of southern Africa. Zeitschrift für Geomorphologie N.F., Supplement Band, 54, 37-56.

Partridge, T.C. and Maud, R.R. 1987. Geomorphic evolution of southern Africa since the Mesozoic. South African Journal of Geology, 90, 179-208.

Penck, A. 1908. Der Drakensberg und der Quattamba - bunch. z.d.k. Preuss. Akad. Wiss., 11, 15-30.

Smith, A.G. 1982. Late Cenozoic uplift of stable continents in a reference frame fixed to south America. Nature, 296, 400-404.

Suess, E. 1904. The Face of the Earth. Clarendon Press, Oxford, Vol. 1, 604 pp.

Summerfield, M.A. 1985. Plate tectonics and landscape development on the African continent. In: Morisawa, M. and Hack, J.T. (eds.), Tectonic Geomorphology, Allen and Unwin, Boston.

Veatch, A.C. 1935. Evolution of the Congo Basin. Geological Society of America Memoir 3, 183 pp.

Wayland, E.J. 1931. Summary of Progress for the years 1919 to 1929. Geological Survey of Uganda.

Wellington, J.H. 1955. Southern Africa: A geographical study, Vol. 1: Physical Geography. The University Press, Cambridge. 528 pp.

Willis, B. 1936. Studies in comparative seismology - East African plateaux and rift valleys. Carnegie Institution, Washington, Publ., 470 pp.

A REVIEW OF BASIC RING BLOCKS OF THE AFRICAN CRATON WITH REFERENCE TO GEOMORPHOLOGICAL ZONES AND RANGES OF RING BLOCKS OF SMALL DIAMETERS

VILKO KLEIN

Member of the Croatian Geographical Society, 41000 Zagreb, Yugoslavia

1. INTRODUCTION

Ring blocks play an important role in the development of the Earth's crust. The largest of them also influence the development of the geological structure in different areas (Brjuhanov et al., 1987). Therefore they have recently been studied with increasing care, especially in mineral prospecting (Collective of authors, 1979).

It may be said that every continent and its respective lithospheric plate are marked by the manner in which ring block units develop, and this applies to the African continent as well (Figs. 1-5, 8).

In order to point out the largest ring blocks which in some way define the geological structure of Africa and the adjacent sea-bed relief areas they were delineated on small-scale geological and topographical maps, while also making use of the results of former investigations (Brjuhanov et al., 1987; Collective of authors, 1983; Klein, 1985). The comparative maps were also prepared in this manner (Figs. 2, 3, 6, 7). This made it possible to isolate some basic characteristics of geomorphological zones resulting from these ring blocks (Fig. 13). The conclusions of this study point to the presence of several ring block ranges of comparatively small diameters (Fig. 6, 9, 10, 12).

2. WORKING METHODS AND PROBLEMS OF TERMINOLOGY

Ring blocks, or rock bodies of circular, oval or spiral delineations, were identified using small-scale geological and topographical maps. Such analysis was supported by interpretation of satellite imagery, where available.

The identification of ring blocks on the tectonic map (Fig. 1) was done first by underlining all the ring delineated geological boundaries and faults. For this sub-type of ring block the term "ring structures" ("geological ring structures") is used. Then the

Geomorphological Studies in Southern Africa, G.F.Dardis & B.P.Moon (eds)
© 1988 Balkema, Rotterdam. ISBN 90 6191 831 6

Fig. 1. Giant oval ring structures of Africa and neighbouring areas identified on the basis of an African tectonic map scale M 1:30000000 [World Atlas, 1964 p. 120]. (Units: 1. Sahara, 2. Angola-Congo [2.1. Angola, 2.2. Congo], 3. Semi oval terrestrial complex Somalia - Tanzania, 4. South Africa [Darvaro-Pilbarsk]; 1-2.2. Super-giant ring complex Africa).

Fig. 2. Comparative map of: (1) giant Sahara ring structure (according to Fig. 1) towards(2) lithospheric core (after Shapman and Pallock, 1977): (3) Central part of the Sahara nuclears (see Fig. 3).

various ring delineations of the relief, including those with a discontinuous spread, were underlined on the maps used for this purpose (Figs. 4, 8). For the sub-type of ring blocks identified in this manner the term "ring morpho-structures" (which, in fact,

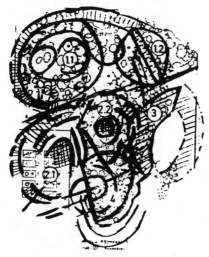

Fig. 3. Review of giant ring structures identified on the basis of an Afro-Arabian craton ring structural map (Brjuhanov et al., 1987) and corresponding bathymetric relations as presented on a tectonic map of South Africa scale 1:30000000 (World Atlas, 1964 p. 120). (Units: 1. Sahara with two symetrically arranged nuclears (1.1 West-African and 1.2 Afro-Nubian); 2. Two centre marine and terrestrial Angola-Congo (sub-units: 2.1 Angola and 2.2 Congo, which areally overlap the central-African nuclear); 3,4 One-centre marine and terrestrial units: 3. Somalia-Tanzania and 4. South-African [Darvaro-Pilbarsk] (after Brjuhanov et al., 1987).

Fig. 4. Analytic African block-relief units of morphostructural significance identified on a 1:30000000 map (Kartografija-Ucila, 1986).

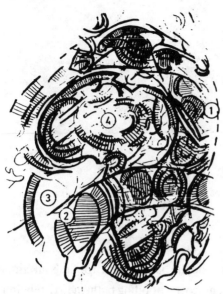

Fig. 5. Prognostic review of main African satellite ranges of relatively small ring morphostructures; (1. Sahara range; 2. South African range [singled out according to Fig. 4]; 3. Submarine ring morphostructure Angola; 4. Interpolated Central Sahara ring morphostructure [according to Figs. 8 and 9]).

Fig. 6. Generalized map of basic African ranges (according to Fig. 5).

are "relief ring structures") is used. These are developed under the combined influence of tecto-erosional factors. This terminological division has a methodological, practical and genetic meaning if such terms are used consistently in the identification of ring blocks.

The name "ring block" itself is a collective, undefined term, because, as stated before,

20

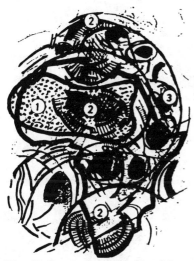

Fig. 7. Synthetic map of African ring morphostructures. Prominent elements: 1. Sahara lithospheric core (see Fig. 2); 2. Central meridional range of ring morphostructures (according to Fig. 8); 3. Ring morphostructural range of relatively smaller diameters which bends around the great Sahara oval block (see Fig. 5).

Fig. 8. Analytical relief ring units map of Africa constructed on the basis of a 1:20000000 map (World Atlas, 1964 - p. 119).

it can mean both ring structure and ring morphostructure. However, this division offers wider possibilities of describing ring blocks. This should not call into question the possibility of using the basic term "ring structure" for blocks delineated exclusively by relief, since the terminology is not yet fully defined.

On the basis of the map of ring structures of Africa prepared by Brjuhanov et al.(1987) some basic giant ring blocks of Africa are presented (see Fig. 3).

21

Fig. 9. Review of the Central Africa meridional ring morphostructural range of small diameters (1-1; according to Fig. 8).

3. RESULTS

3.1 Basic Ring Blocks of Africa

From the point of view of ring block units, the African continent can be divided into a northern and a southern part. The former is marked by the giant oval relief unit of the Sahara, which has morphostructural features (Figs. 4, 6), and the latter by three slightly smaller giant ring structures: Congo-Angola, Tanzania-Somalia, and South Africa (Fig. 1). Considerable parts of these units are linked to the adjacent seabed relief.

The Atlantic edge of the Sahara block fits into the respective Atlantic-American edge. It follows the form of the Atlantic ridge. Its morphogenesis reveals the manner in which the Pangean continent disintegrated due to the opening of the Atlantic Ocean in the early Mesozoic and the resulting horizontal movement of the continental masses (Kennet, 1987; Figs. 6-7,6-8).

The Sahara unit is structurally defined by a large lithospheric nucleus about 250 km thick (Fig. 2; Shapman and Pallock, 1977) which explains its large area of influence on the neighbouring regions of Hercynian and Alpine Europe (Fig. 4), and neighbouring parts of Asia, from which the unit derives its morphostructural importance.

The range of block relief units of small diameters extends along the eastern half of the Sahara unit (Fig. 6).

22

Fig. 10. Prognostic singled-out ring morphostructural ranges of small diameters and diagonal orientation (1.1 and 2.2); 3. Eastern part of the African continent structurally orientated towards the Indian Ocean (according to a 1:20000000 geological map of Africa; World Atlas, 1964 p. 122-124).

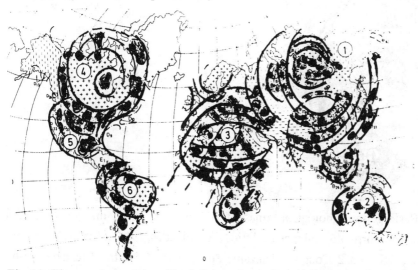

Fig. 11. Giant and supergiant Earth lithosphere plate ring blocks of morphostructural meaning constructed on the basis of a global oil and gas basins and areas (after Boskov-Steiner, 1986). (Main units: 1. Asian; 2. Australian, 3. Afro-Eurasian, 4. North American [of the Canadian shield], 5. Mexico, 6. Amazonia, 7. Indonesia).

The Sahara ring block unit is defined more precisely by its geological structure (Fig. 2). Brjuhanov et al. (1987) singled out two large nuclears in the Sahara giant unit; the West African and the Arabian-Nubian (Fig. 3). Their outlines can also be seen on the basis of relief characteristics (Fig. 4).

23

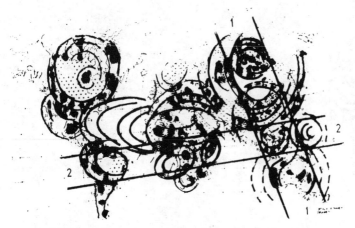

Fig. 12. Some basic planetary ring block ranges (constructed according to Fig. 11). (1.1: Submeridional Asia-Australia, 2.2: Discontinuous sub-equatorial Amazonia-Indonesia).

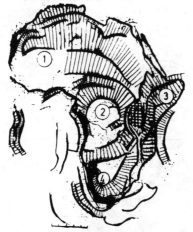

Fig. 13. Basic geomorphological zones of Africa constructed on the basis of a 1:30000000 geomorphological map of Africa (World Atlas, 1964 p. 125). (Zones connected with; 1. Sahara, 2. Congo, 3. Tanzania-Somalia and 4. South African block ring units; 1.2: Sahara-Congo geomorphic zones of continental dimensions).

Nuclears are generally defined not only as the largest but also as the oldest ring structures, the origin of which reaches far back into the Archaean Age.

Giant oval blocks, for instance the Sahara morphostructure, can be of large diameters and of more recent origin than nuclears, and the explanation of the origin should be sought in the subsequent horizontal continental drift. However, the evolution of the super-giant ring block Asia was partly the result of continental collision.

Super-giant ring blocks are singled out on a map of planetary presentation of oil and gas basins (Boskov-Steiner, 1985). By their dimensions they define the structure of the respective lithospheric plates (Figs. 11, 12).

The oval Sahara block, the longer axis of which extends in an E-W direction, is followed by the meridionally extended part of southern Africa, the equatorial part of which is built of the complex giant oval structures, Congo-Angola and Tanzania-Somalia (Fig. 1,3). Their submarine parts are shown on a bathimetrical map (Fig. 3).

The Congo-Angola unit comprises two parts. The north-eastern, continental part is defined by a lithospheric nucleus (core) about 250 km thick (Fig. 3), which is of a smaller diameter than is that of the larger Sahara unit. The south-eastern part belongs to the Angolan submarine depression.

While the longer axis of the Congo-Angola unit shows a SW-NE extension, the Tanzania-Somalia composite unit has a sub-meridional spread (Fig. 2). On its western edge the latter unit borders on the East African volcanic rift zone, which is seismically very active (Grachev, 1987; Fig. 59). The continental part of this complex contains several minor nuclears. A more detailed description of the geological structure of ring blocks of Africa is contained in a monograph by Brjuhanov et al. (1987).

The distribution of structures of nuclears within the oval Sahara block and Equatorial Africa reveals elements of symmetry in the structure of this region. Thus the two-centre oval Sahara unit is followed in the south by the nuclei of the oval blocks of Congo-Angola and Tanzania-Somalia (Fig. 3).

A wide submarine relief belt is associated with the South African ring structural complex which has morphostructural features. The relief of this belt reveals traces of Africa's northward meridional drift (Kennet, 1987; Fig. 11-12). Similar scars have been left by the rhythmic eastward migration of the sahara oval unit as part of the African craton.

The morphostructural nucleus of the negative South African ring complex is the Kalahari depression.

3.2. Geomorphological Aspects

In many cases basic ring morphostructures show corresponding zonally distributed relief characteristics. Certain departures in this sense may be due to difference in the scales of the maps used for the purpose and to differences in mapping criteria, especially when the mapping is not adapted to the manner in which ring blocks are identified.

Ring zones of Africa's relief are partly affected by climatic zones. Especially noticeable are the relief zones of the north-eastern continental part of the Anglo-Congo two-centre oval block. It reflects the influence of a lithospheric nucleus of small diameter and a thickness of about 250 km (Fig. 13). Here archaean rocks underlie the covering layers.

Ring relief zones of the Congo sub-unit are included in the much larger geomorphological zone of the oval Sahara morphostructure. Actually, they merge into wider zones of continental dimensions. However, relief features in themselves are of secondary importance when the geometry of their extension is pointed out in order to explain morphostructural and structural relations.

3.3. Ring Ranges

A special characteristics of the African craton is the meridional range of ring morphostructure of minor diameters (Fig. 9). This range can be followed far to the north. It fits the spread of the African meridional planetary structural belt of the order (Senin, 1983; Fig. 101).

In contrast to the straight-line ring ranges there are also ring ranges which bend round larger ring blocks. These are termed satellite ranges. In Figure 6 one can see a prognostically singled out semi-ring range, which borders the eastern parts of the large oval Sahara block. Another range, which lies to the south and bends in opposite direction, can also be observed.

The African nuclears are confirmed to a considerable degree by ranges of satellite ring structures of very small diameters which have developed along their inner rims (Brjuhanov et al., 1987; Fig. 15).

A special characteristic marks the Mediterranean ring morphostructure as part of the meridional range of ring blocks intersecting the African craton from south to north (Fig. 9 - upper part).

This unit connects the marginal parts of the African plate with parts of the Eurasian plate.

The Mediterranean ring morphostructure is also confirmed by an inner range of satellite ring morphostructures of smaller diameter. It covers an area of a thinner lithosphere in relation to the neighbouring regions (Levi and Lisak, 1986; Fig. 2). It also covers the respective geoidal nucleus (Gaposchkin, 1987), while areally covering the sub-maximum heat flow of 60 mVt/m2 (Cermak and Rybach, 1982; p. 58). The rift zone of East Africa also belongs to the area of increased heat flow.

Fig. 14. Review of the bending of south and central European Hercynian-Alpine belt of small ring morphostructures under the influence of the great Afro-Sahara oval block situated in the south, constructed on the basis of the geomorphological map of Europe scale 1:17 500 000 (World Atlas, 1964 p. 85).

4. CONCLUSION

Giant ring blocks are one of the basic characteristics of the geological and relief structure of the African continent.

Whereas on the one hand identified blocks integrate different continental and adjacent submarine parts of the African lithospheric plate into a major unit, on the other, those linked with the East African volcanic belt, and the spreading of the Red Sea, would indicate the beginning of disintegration processes within the plate itself.

The oval outline of the African morphostructure, i.e. Sahara, the origin of which is linked with the separation of Africa from the Americas, reveals the influence of plate tectonics on the formation of giant ring blocks.

The diameters of the oval blocks in the Sahara and Congo units are more or less proportionate to the dimensions of the respective lithospheric nuclei in their base.

Geomorphological zones reflect dramatically the influence of the structure on the character of the surface relief.

The central African meridional range of ring morphostructures reveals the influence of planetary factors on its formation.

Generally, from the results of this study it may be concluded that ring structures and morphostructures are an essential component of the geological structure of Africa, a fact that should be considered in studying geomorphological zones and in practical mineral prospecting. In this connection it should be emphasized that the basic causes of the emergence and evolution of giant ring blocks of Africa should be sought both in corresponding convective (vertical) movements of the magma in the base of the crust and in plate tectonic phenomena, which influenced the horizontal movement of the continents. Erosional processes help to uncover them.

5. SUMMARY

Giant ring blocks of the African craton and of the adjacent parts of the seabed which have structural and morphostructural characteristics are discussed. The causal and consequent connection between the diameters of the identified ring blocks and the dimensions of lithospheric nuclei is pointed out using as example the Sahara and Congo ring blocks, whereas for the disintegration processes of the African craton the instance of the Tanzania-Somalia ring complex is used. Instances are given of the influence of plate tectonics on the development of giant ring blocks and of the effect of basic single-out ring blocks on the formation of geomorphological zones of the continent.

6. REFERENCES

Boskov-Steiner, Z. 1985. Global survey of major oil and gas bearing basins and areas, Nafta 36, 4, pp. 133-143, Zagreb.

Brjuhanov, B.N., Bush, B.A., Gluhovskih, M.Z., Zverev, A.T., Katz, J.G., Makarova N.B. 1987. Ring structures of earth continents, Published in Russian by Nedra, p. 185, Moscow.

Collective of Authors, 1983. Space Information for Geology, Academy of Science of USSR, Commission of Natural Resources; Study by Space means, Published in Russian by NAUKA, p. 534, Moscow.

Collective of Authors, 1979. Analysis of cosmic photographs in tecto-magmatic and metallogenic investigations, Published in Russian by NAUKA, p. 155, Moscow.

Cermak, V. and Rybach, L. 1982. Terrestrial heat flow in Europe, Published in Russian by MIR, p. 376, Moscow. (English edn. publ. by SPRINGER-VERLAG, Berlin, 1979.

Gaposchkin, E.M. 1987. Geoid of Smithsonian astrphysical observatory reconstructed in 1973. In: Scheidegger A.E., Principles of Geodynamics, Fig. 74, Published in Russian by NEDRA, p. 384, Moscow. (English edn.publ. by SPRINGER-VERLAG, Berlin, 1982).

Grachev, A.G. 1987. Earth rift zones, Published in Russian by NEDRA, p. 235, Moscow.

Kennet, J.M. 1987. Marine Geology (two volumes), Published in Russian by MIR, p.: I.-242, II.-341, Moscow. (English edn. published by Prentice-HALL, Englewood Cliffs N.J., 1982).

Kartografija - UCILA, 1986. Geographical atlas, p. 63, Zagreb.

Klein, V. 1985. Ring-shaped and oval morphostructures of the Mediterranean and its wider hinterland, Bulletin of the INQUA Neotectonic Commission N degrees 8, pp. 28-37, Stockholm.

Levi, K.F. and Lisak, S.V. 1986. Thermal evolution and lithospheric thickness of continents. In: Basic problem of seismotectonic, Collective of authors, pp. 69-78, Published in Russian by NAUKA, p. 215, Moscow.

Senin, B.V. 1983. Global linears and their division to geologic-geomorphological, gravimetric and space survey data of high levels of generalization. In: Space Information for Geology, Collective of authors, pp. 276-287, Academy of Sciences of USSR, Commission of Natural Resources; Studies by space means, Published in Russian by NAUKA, p. 534, Moscow.

Shapman and Pallock 1977. Model showing the thickness of the lithosphere. In: Principles and Methods, Graham and Trotman, p. 384, London. (French edn. publ. by Bordas Dunod, Paris, 1983).

World Atlas 1964. Physical-Geographic, Academy of Sciences of USSR, p. 298, Moscow.

THE MISSING LINK: PLANATION SURFACES AND ETCH FORMS IN SOUTHERN AFRICA

C.R. TWIDALE
Department of Geography, University of Adelaide

1. INTRODUCTION

That many land forms, major and minor, originate not at the land surface but at the base of the weathered mantle or regolith, has been recognised for some time, though geomorphologists have been slow to appreciate either the extent of such features or the implications stemming from their genesis. Features that are shaped at the base of the regolith, at the weathering front (Mabbutt, 1961a), and later exposed as the regolith is stripped, are called etch forms or surfaces depending on their character and extent. Minor etch forms can develop on steep slopes wherever a patch of weathered rock accumulates, but major assemblages, such as those that eventually become inselberg landscapes, surely imply development over a long period beneath a stable surface of low relief. Only under such conditions can a weathering front of sensibly low relief at the regional or sub-regional scale evolve, for although groundwaters exploit every structural weakness, this tendency to diversity and the creation of a relief amplitude is more than offset, provided that there is time for various water-related alteration processes to take full effect. The preferential weathering of projections results in the production either of planate surfaces or of large-radius rounded forms that offer to external attack the least surface area in relation to volume.

2. HISTORY OF THE IDEA

In considering the origin of the etch it is necessary to distinguish between the idea and the term, as well as between major and minor forms.

Minor etch forms were recognised as such almost two centuries ago. Hassenfratz (1791) understood that the granite boulders he observed (Fig. 1a) in the Margeride district of the southern Massif Central had been more-or-less exposed (degages) as

Geomorphological Studies in Southern Africa, G.F.Dardis & B.P.Moon (eds)
© 1988 Balkema, Rotterdam. ISBN 90 6191 831 6

Fig. 1a. Exposed and partly exposed granite boulders in the Aumont district, southern Massif Central, France.

Fig. 1b. Dumonte Rock, near Wudina, northwestern Eyre Peninsula, South Australia, showing gutters that extend, in some instances converge, beneath the land surface (x-x) as a result of streams flowing along the weathering front.

Fig. 2. Pipes and prongs developed by subsurface weathering, Galong, New South Wales, Australia.

the matrix of weathered rock in which they were and are embedded was eroded. And though it has sometimes been overlooked, the two-stage concept of boulder development, the first stage involving fracture-controlled moisture weathering in the subsurface, the second the erosion of the weathered rock (in particular the stripping of the weathered granite or grus), has never since been lost (Twidale, 1978a).

Just over half a century after Hassenfratz made his perceptive observations and astute interpretations of the granite boulders of the Aumont district, the same two-stage idea was applied to the interpretation of the fluted granite surfaces exposed on Palo Ubin, an island located at the eastern end of the Jahore Strait, between Singapore and West Malaysia. Logan (1849,1851) noted that the flutings, which are developed on steep rock faces, extend beneath the soil cover and, finding no evidence to suggest burial and exhumation, but on the contrary believing the soil cover to be *in situ*, he deduced that the forms had been initiated beneath the ground. In contrast with the interpretation of granite boulders, the suggestion that minor granite forms originate at the weathering front has found support and cor-roboration only relatively recently (e.g. Twidale, 1971, 1982a; Boye and Fritsch, 1973; Twidale and Bourne, 1975a; see Fig. 1b). In other lithological environments, however, and especially in limestone terrains, the suggestion of subsoil origin goes back even beyond Hassenfratz.

Thus the subsurface provenance of *orgues geologique* or solution pipes in the Paris Basin was suspected by de Sain Fond (1784) and by Cuvier and Brogniart (1822). Several later writers, such as Eckert (1902), Lindner (1930) and more recently

Zwittkovits (1966), have distinguished between solutional forms in limestone developed at the surface and those that have evolved beneath a soil cover; and there is much field evidence (Fig. 2) to support the view that many familiar karst landforms, like their congeners developed in granite and sandstone (Klaer,1957; Twidale, 1980), originate at the weathering front and are of etch type.

On the regional scale, Falconer (1911) appreciated that the inselberg landscapes of northern Nigeria have their origins beneath a mantle of weathered rock. The concept of etch development has been clearly or concisely discussed as "..a plane surface of granite or gneiss subjected to long continued weathering at base level would be decomposed to unequal depths, mainly according to the composition and texture of the various rocks. When elevation and erosion ensued, the weathered crust would be removed, and an irregular surface would be produced from which the more resistant rocks would project..." (Falconer, 1911, p.246).

At about the same time, Jutson interpreted the New Plateau of the Yilgarn Block, in the southwest of Western Australia, as the result of stripping of a lateritic regolith formed beneath the Old Plateau (Jutson, 1914). Jutson did not use the term *etch*, but his diagrams make it clear that he fully understood the nature of the process to which the New Plateau is due (Brock and Twidale, 1984). Budel (1957) introduced a basically similar concept with the notion of double planation whereby while one epigene plain forms, another is developed simultaneously but at a lower level as the regolith is stripped.

The term *etch* was not applied to forms developed at the weathering front until 1934, when Wayland (1934, p.79) used it for certain planation surfaces in Uganda. Even so, Wayland's comments suggest very strongly that Bailey Willis (who) "... aptly names this erosive operation etching..." (Wayland, 1934; p.79) is responsible for the term used in this concept. Though Wayland's publication preceded Willis (1936) by two years it would perhaps be appropriate to attribute the term to both workers. In similar fashion, the two stage mechanism was understood for about a century and a half before Linton (1955) established the term in international literature.

3. NATURE OF ETCH FORMS

Etch forms evolve two stages. The first involves differential, structurally-controlled subsurface weathering, predominately by moisture, and through such processes as solution, hydrolysis and hydration. The second involves the differential erosion of the mass rendered heterogeneous by weathering. Wayland (1934 p. 79) has suggested that etch forms imply tectonic instability and an upward movement of the land mass, but it is alternations of standstill and relative uplift that are conducive to their development, for only then can the bedrock be weathered and then stripped. Constant uplift would not allow time for the development of a regolith.

Etch forms thus have two stages, the first denoting the period of weathering, the second that of stripping. The two may be juxtaposed in time, or they may be separated by vast aeons (Twidale, 1986a). Etch forms may also be polygenetic, in that, though weathering took place under terrestrial conditions, stripping may be due to glaciers, or waves, or, as is most commonly the case, to river action (Twidale, 1986a). Etch surfaces are due mainly to the activities of groundwater and though the precise nature of the processes at work have varied according to lithology, the closed or open character of the groundwater system, the chemistry and biotic content of the groundwaters, and the time available for interaction, the end products tend to be similar. The form of the weathering front resulting from the interplay of bedrock and groundwaters is similar regardless of the atmospheric climate. Even the nature of the bedrock is subordinate provided only that its physical characteristics (e.g. permeability, porosity, fracture density) are similar, so that comparable landform assemblages are developed as etch suites on granite, sandstone, conglomerate and limestone (Twidale, 1987a,b).

Thus groundwaters, and the etching associated with them, have significant azonal or convergent implications for landscape evolution.

4. EXAMPLES OF ETCH SURFACES

Etch surfaces of regional extent were described from the Yilgarn of Western Australia and from East Africa many years ago. As has been mentioned Wayland (1934) and Willis (1936) coined the term etch in relation to planation surfaces in Uganda, and their analysis and conclusions have been essentially corroborated (though elaborated on) by Bishop (1966). Even earlier, Falconer (1911) identified etch plains in northern Nigeria and Jutson working in Western Australia, realised that the level of the base of the regolith preserved with lateritised plateau remnants of his Old Plateau are coincident with the more extensive New Plateau, where rock platforms are numerous and where fresh bedrock is nowhere far from the surface. His interpretation has been corroborated by later workers (cf. Mabbutt, 1961a; Finkl and Churchward, 1973). Mabbutt also identified a stripped land surface related to the early-mid Tertiary duricrusted (mainly sicreted) surface in the region south of the MacDonnell Ranges in central Australia. The summit surface of the Hamersley Ranges, in the northwest of Western Australia is also of etch type, the pisolitic and ferruginous regolith that formed on it in Cretaceous times having been stripped and transferred to adjacent slopes as "canga" deposits and into river valleys where it now forms the Robe River Pisolite (Twidale et al., 1985).

Alley (1973) has plotted etch surfaces of relatively minor extent associated with the stripping of duricrust   profiles in the Mid North region of South Australia and Twidale and Bourne (1975b) have shown that the summit surface cut across sedimentary and metamorphic sequences (as well as igneous emplacements) in the eastern and southern Mt. Lofty Ranges, also in South Australia, is of etch type, resulting from the stripping of an early Mesozoic lateritic regolith (Campana, 1958; Daily et al., 1974). The summit surface of the Gawler Ranges, cut in Proterozoic volcanic rocks, is of etch type (Twidale et al., 1976) as is that of Ayers Rock (Twidale, 1978b).

In addition to etch surfaces due to the erosion of entire regoliths, which, if the regolith can be regarded as a skin, can be regarded as subcutaneous (cf. Zwittkovits, 1966), surfaces due to the stripping of only part of the weathered profile have also been recognised, termed *intracutaneous* surfaces (Twidale, 1987a). Thus Wright (1963) identified a plateau surface around Katherine, Northern Territory, as resulting from the erosion of only the upper part of the lateritic profile, and the stabilising of the land surface on a siliceous zone within the profile. Similarly, all pisolite-capped lateritic surfaces lacking the A-horizon can be regarded, if rather pedantically, as intracutaneous in type.

The Labrador Plateau, of eastern Canada, is also probably of etch type. It has been suggested (Tanner, 1941) that the present bedrock high plain, much modified by glaciation, is due to the stripping of a Tertiary weathered mantle by glaciers and ice sheets, and this seems a reasonable supposition, for little in the way of regolith remains and the glacial tills of the northeastern United States are witness to glacial erosion in the source areas to the north. Thomas and Thorp (1985) have described an extensive etch plain, the Koidu Etchplain, from Sierra Leone, and indeed, as the concept of etch planation gains credence, more examples are identified in various parts of the world.

5. PLANATION SURFACES AND ETCH FORMS

One of the key conditions conducive to etch planation is actual or comparative topographic standstill. Only then is there time for the development of essentially even bedrock surfaces at the weathering front. Planation surfaces are another expression of comparative landmass stability and it is no coincidence that well-known etch forms are associated with planation surfaces. Thus Obst (1923), Jessen (1936) and King (1949) all noted that bornhardts, most of which are etch forms (Twidale, 1982a,c), invariably occur in multicyclic landscapes. Both Wayland and Jutson, in the studies referred to previously, were working in landscapes in which are preserved flights of planation surfaces or remnants thereof. This being so, etch surfaces ought to be well developed in southern Africa, which area, thanks to the distinguished work of such as Dixey (1938, 1942) and King (e.g. 1942, 1950, 1962) is well-known for its planation surfaces (for a comprehensive bibliography of studies of planation surfaces in southern Africa see Partridge and Maud, 1987). Yet one looks in vain for any substantive mention of etch plains in the literature pertaining to that region of the subcontinent.

The topography of the subcontinent can be regarded as broadly stepped, a situation that has been interpreted by King (1962) as resulting from episodic isostatic adjustment to erosional offloading. Whatever the causation planation surfaces are well developed and preserved. There has been, and doubtless will be again, debate as to the dating of the surfaces (e.g. King, 1942, 1950; Frankel, 1960), and their complexity, in particular, whether each major surface is polyphase or whether each surface represents a separate and distinct event (e.g. King, 1976), though the latter problem is eased if it is bourne in mind that planation surfaces of regional extent are inevitably diachronous (Hills, 1955).

The latest analysis of planation surfaces is due to Partridge and Maud (1987). They identify the African Surface, of Jurassic-early Miocene age, as the dominant planation element of the landscape, and demonstrate that it has been tilted down to the west so that it stands in the east at altitudes of almost 2000 m, but is found around sea level in the western Cape. The surface is correlated with offshore sediments and sequences. The African Surface is the master surface in the sense that it is the reference plane, and it is identified by virtue of the deep weathering that affects this and no other surface. Partridge and Maud (1987) refer to '...the African Surface, with its diagnostic, kaolinized weathering profile and massive duricrusts...'. The kaolinized regolith is commonly 20-30 m thick, in contrast with the typical 2-3 metres of weathered mantle found developed on the younger erosion surfaces.

The duricrusts developed in relation to the African surface include laterite and silcrete, and the latter in particular is indicative of long standstill or only very slow topographic change, simply because of the slow rate of silica precipitation necessarily involved (e.g. Ernst and Calvert, 1969; Hutton et al., 1978). The substantial regolith developed beneath the African Surface and there was time for a regular weathering front to evolve. Could, therefore, the post African surfaces plotted by King (1962) and by Partridge and Maud (1987) be in part at least of etch type? The thin regolith typical of younger surfaces could have developed since stripping and exposure, or, alternatively, they could be in part remnant from older regoliths left by chance on the stripped surfaces. Many forms that are in other plac es construed as of etch character have been reported or observed in southern Africa. For example, flared slopes (Twidale, 1962, 1982a) are well developed on the Clarens Formation sandstone at the southern margin of the Drakensberg (Fig. 3). Gutters sculpted in the same material can clearly be seen to extend beneath the regolith (Fig. 4). Mushroom rocks, that are increasingly interpreted in terms of subsurface initiation (e.g. Centeno and Twidale, 1988), are also developed. Boulders are clearly initiated beneath the land surface (Fig. 5).

Low domes and platforms are widely developed in granitic terrains. They can be interpreted either as final remnants remaining after the disintegration of inselbergs or as incipient inselbergs, the crests of which are just exposed (Twidale, 1986b). No doubt, however, attaches to the domes uncovered in artificial excavations in several parts of southern Africa (Figs. 6 and 7; Twidale, 1982a) as well as in West Africa. Some of the pediments of the Cape Fold Belt are underlain by regoliths and the lower valley floors are surely of etch type.

Fig. 3. Flared sandstone pinnacles exposed on a cliff in the southern Drakensberg Mountains, South Africa.

Fig. 4. Gutters in sandstone and clearly extending beneath soil cover near Clarens, Southern Drakensberg Mountains, South Africa.

Turning to landscapes at the regional scale, Whitlow (1978-9) directly addresses the question of the relationship between bornhardts and planation surfaces as plotted by Lister (1976) in Zimbabwe (Fig. 8). He shows that though some

Fig. 5. Corestones or incipient boulders in basalt, southern Drakensberg Mountains, South Africa.

bornhardts stand on the African Surface (and the same is true also of parts of the Republic, see Fig. 9), and others on the lower, younger, post African surfaces, a significant and impressive majority of the forms stand on the Miocene surface immediately below the master African surface. He concludes that they may have been exhumed (sic) as the regolith formed beneath the African surface was stripped.

The Bushmanland plain of northern Namaqualand, and central Namibia is of extraordinary flatness and warrants the application of the term *ultiplain* (Twidale, 1983), yet fresh granite, gneiss, sandstone and schist are all, in different areas, exposed at the surface (Fig. 10), suggesting that it too may be of etch type.

Thus a strong prima facie case can be made suggesting that conditions conducive to etch planation have developed from time to time, and particularly from the late Mesozoic to the middle Miocene.

During that extended period the subcontinent was essentially base-levelled to produce the African Surface. That there was time for deep and intense weathering beneath that surface is widely evidenced. Etch surfaces and forms ought to be

Fig. 6. Nascent granite domes exposed in artificial excavations at Midrand, between Pretoria and Johannesburg, South Africa.

Fig. 7. Nanscent granite dome exposed in an artificial excavation at the Leeukop, near Potchefstroom, where the crest of an incipient dome (x) can be seen in the foreground.

developed in southern Africa and there is evidence that they are. A strong case can be made for a re-examination of the post-African surfaces in particular, specifically to ascertain whether they are wholly or partly of etch type: ".... seek and ye shall

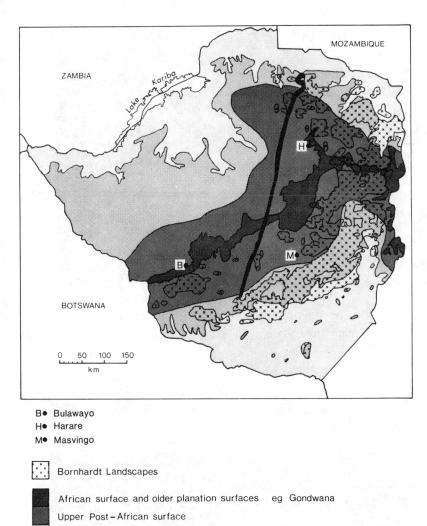

B● Bulawayo
H● Harare
M● Masvingo

Bornhardt Landscapes

African surface and older planation surfaces   eg Gondwana

Upper Post – African surface

Lower Post – African surface

Pliocene surface and more recent erosional and depositional surfaces

Great Dyke

Fig. 8. Relationship between bornhardts and planation surfaces in Zimbabwe.

find ....". Such discoveries would cause significant modification of the interpretation of the late geological history of the subcontinent.

6. SUMMARY

It is one of the geomorphological paradoxes of the century that in a subcontinent,

Fig. 9. The Bushman's Kop stands above the African Surface near White River, eastern Transvaal, but many others are partly exposed in the scarp bounding the high plain.

Fig. 10. Bushmanland Surface, cut in schist, central Namibia.

southern Africa, where the denudation chronology is so marvellously well expressed in the landscape and so superbly documented in the literature, one of the essential implications of the chronology has been so much neglected, if not ignored. A planation surface implies stability, and time for deep, intense weathering, for the formation of regolith and a weathering front. Many regoliths are weak and friable and are readily stripped to expose the front as an etch surface. Despite the evidence to the contrary, the inselbergs of southern Africa, for example, have been interpreted in terms of epigene processes not of the interplay between groundwaters and bedrock deep beneath the land surface. Inselbergs have been seen as related to scarp retreat and pedimentation, whereas in reality evidence both local and in other areas suggests that they are basically etchforms. The spectacular inselbergs of the Valley of a Thousand Hills occur in a valley incised below the African surface, and there are remnants of the weathered bedrock between closely juxtaposed domes. In Zimbabwe, Namaqualand, eastern Transvaal and in the Western Cape, inselbergs are located on plains and in valleys that stand below remnants of planation surfaces. In many parts of southern Africa (as well as in West Africa and south western Australia) nascent or incipient inselbergs can be seen exposed in quarries, road cuts etc., developed beneath the natural land surface. Some of the high plains of the interior are rolling and reflect slight stream incision into pre-existing plains of nil relief (ultiplains). In the Bushmanland Surface there is a preserved plain of virtually nil relief, cut in various lithologies and with only a few inselbergs standing above the surface of the plain. As the inselbergs are etch forms, it is possible, even likely, that the adjacent plains are of that character also. Certainly the sedimented valley floors of the Cape Fold Belt are being weathered and etched out, resulting in an increase in relief amplitude. Southern Africa is famous for its pediments and pediplains, the latter being said to result from the coalescence of the former. But in southern Africa (e.g. the Western Cape Fold Belt) as elsewhere, pediments are fringing forms that merge downslope with rolling 'peneplains'. Pediments are of various morphological types, and though some are undoubtedly due to erosion by streams and wash in specific structural and hydrological situations, others, rock pediments or platforms, are again etch forms. Many facets of the landscape of southern africa have their origins in weathering beneath planation surfaces, impressive remnants of which are preserved in the landscape but extensive sectors of which have been dissected, allowing stripping of the regolith and exposure of the weathering front in its varied shapes and forms.

7. REFERENCES

Alley, N.F. 1973. Land surface development in the mid-north of South Australia. Transactions of the Royal Society of South Australia, 97, 1-17.

Bishop, W.W. 1966. Stratigraphic geomorphology: a review of some East African landforms. In: Dury, G.H. (ed.), Essays in Geomorphology, Heinemann, London, 139-176.

Boye, M. and Fritsch P. 1973. Dégagément artificiel d'un dome crystallin au Sud-Cameroun. Trav. Doc. Geol. Trop., 8, 69-94.

Brock, E.J. and Twidale C.R. 1984. J.T.Jutson's contribution to geomorphology. Australian Journal of Earth Science, 31, 107-121.

Budel, J. 1957. Die "doppelten Einebnungsfläche" in den feuchten Tropen. Zeitschrift für Geomorphologie, 1, 201-228.

Campana, B. 1958. The Mt Lofty-Olary region and Kangaroo Island. In: Glaessner, M.F. and Parkin, L.E. (eds.), The Geology of South Australia, Melbourne University Press, Melbourne, 3-27.

Centeno, J.D. and Twidale, C.R. 1988. Rocas fungiformas, pedestales y formas associadas en Anvil Hill, Mannum, Australia del Sur. Cvardernos do Lab. Xeologico de Laxe.

Cuvier, G. and Brogniart, A. 1822. Description Geologiqué des Environs de Paris. Doufour and d'Ocagne, Paris.

Daily, B., Twidale C.R. and Milnes, A.R. 1974. The age of the lateritized summit surface in Kangaroo Island and adjacent areas of South Australia. Journal of the Geological Society of Australia, 21, 387-392.

Dixey, F. 1938. Some observations on the physiographic development of central and southern Africa. Transactions of the Geological Society of South Africa, 41, 113-172.

Dixey, F. 1942. Erosion cycles in central and southern Africa. Transactions of the Geological Society of South Africa, 45, 151-181.

Eckert, M. 1902. Das Gottesacherplateau eiene Karrenfield im Allgau Wissensch.Erganz.Z.Deutsch Osterr. Alpenvereins, 33.

Ernst, W.G. and Calvert, S.E. 1969. An experimental study of the recrystallisation of porcellantile and its bearing on the origin of some bedded cherts. American Journal of Science, 267A, 114-133.

Falconer, J.D. 1911. The Geology and Geography of Northern Nigeria. MacMillan, London.

Finkl, C.W. and Churchward, H.M. 1963. The etched land surface of Western Australia. Journal of the Geological Society of Australia, 20, 295-307.

Frankel, J.J. 1960. The geology along the Umfolozi River, South of Mt Ubatuba, Zululand. Transactions of the Geological Society of South Africa, 63, 231-252.

Hassenfratz, J-H. 1791. Sur l'arrangèment de plusieurs gros blocs de differentes pierres que l'on observe dans les montagnes. Ann. Chimie, 11, 95-107.

Hills, E.S. 1955. Die Landoberfläche Australiens. Die Erde, 7, 195-295.

Hutton, J.T., Twidale, C.R. and Milnes, A.R. 1978. Characteristics and origin of some Australian silcrete. In: Landford-Smith, T. (ed.), Silcrete in Australia, University of New England Press, Armidale, 19-39.

Jessen, O. 1936. Reisen und Forschungen in Angola. Reimer, Berlin.

Jutson, J.T. 1914. An outline of the physiographical geology (physiography) of Western Australia. Geological Survey of Western Australia, Bulletin, 61.

King, L.C. 1949. A theory of bornhardts. Geographical Journal, 112, 83-87.

King, L.C. 1950. A study of the world's plainlands. Quarterly Journal of the Geological Society of London, 106, 101-131.

King, L.C. 1962. Morphology of the Earth. Oliver and Boyd, Edinburgh.

King, L.C. 1976. Planation remnants highlands. Zeitschrift für Geomorphologie, 20, 133-145.

Klaer, W. 1957. Verkarstungserscheinungen in Silikatgestein. Abb.Geogr.Inst. Freien Univ. Berlin 5, 21-27.

Lindner, H. 1930. Das Karrenphanomen. Pet.Mitt.Erganz., 208.

Linton, D.L. 1955. The problem of tors. Geographical Journal, 121, 470-487.

Lister, L.A. 1976. The Erosion Surfaces of Rhodesia. Unpublished D.Phil. thesis, University of Rhodesia, Salisbury.

Logan, J.R. 1849. The rocks of Palo Ubin. Verh.Genootsch.Kunst. Wetenschappen (Batavia), 22, 3-43.

Logan, J.R. 1850. Notices of the geology of the Straits of Singapore. Quarterly Journal of the Geological Society of London, 7, 310-344.

Mabbutt, J.A. 1961a. 'Basal surfaces' or 'weathering front'. Proceedings of the Geologists Association, 72, 357-358.

Mabbutt, J.A. 1961b. A stripped land surface on Western Australia. Transactions and Papers,Institute of British Geographers, 29, 101-114.

Mabbutt, J.A. 1965. The weathered land surface in central Australia. Zeitschrift für Geomorphologie, 9, 82-114.

Obst, E. 1923. Das abflusslose Rumpfschollenland in nordöstlichen Deutsch-Ostafrica. Mitt.Geogr.Gesell. Hamburg, 35.

Partridge, T.C. and Maud, R.R. 1987. Geomorphic evolution of southern Africa since the Mesozoic. South African Journal of Geology, 90, 179-208.

Sain Fond B.F. de, 1784. Travels in England and Scotland in 1784. (2 vols.) (Translated by A.Geikie), Hopkins, Glasgow.

Tanner, V. 1941. Outlines of the Geography, Life and Customs of Newfoundland-Labrador. Tilgmann, Helsinki.

Thomas, M.F. and Thorp, M.B. 1985. Environmental change and episodic etch planation in the humid tropics of Sierra Leone: The Koidu etch plain. In: Douglas, I. and Spencer, T. (eds.), Environmental Change and Tropical Geomorphology, Allen and Unwin, London, 239-267.

Twidale, C.R. 1962. Steepened margins of inselbergs from north-western Eyre Peninsula, South Australia. Zeitschrift für Geomorphologie, 6, 51-69.

Twidale, C.R. 1971. Structural Landforms. Austalian National University Press, Canberra.

Twidale, C.R. 1978a. Early explanations of granite boulders. Rev. Geomorph. Dynamiqué, 27, 133-142.

Twidale, C.R. 1978b. On the origins of Ayers Rock, central Australia. Zeitschrift für Geomorphologie, Supplement Band 31, 177-206.

Twidale, C.R. 1980. Origin of minor sandstone landforms. Erdkunde, 34, 219-224.

Twidale, C.R. 1982a. Granite Landforms. Elsevier, Amsterdam.

Twidale, C.R. 1982b. Les inselbergs a gradins et leur signification: l'exemple de l'-Australie. Annals of Geography, 91, 657-678.

Twidale, C.R. 1982c. The evolution of inselbergs. American Scientist, 70, 268-276.

Twidale, C.R. 1983. Pediments, peneplains and ultiplains. Rev. Geomorph. Dynam.,

32, 1-35.

Twidale, C.R. 1986a. Granite landforms evolution: factors and implications. Geologische Rundschau, 75, 769-779.

Twidale, C.R. 1986b. Granite platforms and low domes: newly exposed compartments or degraded remnants. Geografiska Annaler, 68A, 399-411.

Twidale, C.R. 1987a. Etch and intracutaneous landforms and their implications. Australian Journal of Earth Science, 34 (in press).

Twidale, C.R. 1987b. A comparison of inselbergs developed in various rocks. Ilmu Alam.

Twidale, C.R. and Bourne, J.A. 1975a. The subsurface initiation of some minor granite landforms. Journal of the Geological Society of Austalia, 22, 477-484.

Twidale, C.R. and Bourne, J.A. 1975b. Geomorphological evolution of part of the eastern Mt Lofty Ranges, South Australia. Transactions of the Royal Society of South Austalia, 99, 197-209.

Twidale, C.R., Bourne,J.A. and Smith, D.M. 1976. Age and origin of palaeo-surfaces on Eyre Peninsula and in the Southern Gawler Ranges, South Australia. Zeitschrift für Geomorphologie, 20, 28-55.

Twidale, C.R., Horwitz, R.C. and Campbell, E.M. 1985. Hamersley landscapes of Western Australia. Rev. Geol. Dynam. Phys., 26, 173-186.

Wayland, E.J. 1934. Peneplains and some erosional landforms. Geological Survey of Uganda, Annual Report and Bulletin, 1, 77-79.

Whitlaw, J.R. 1978-9. Bornhardt terrain on granitic rocks in Zimbabwe:a preliminary assessment. Zambian Geographical Journal, 33-34, 75-93.

Willis, B. 1936. East African plateaus and rift valleys In:Studies in Comparative Seismology. Carnegie Intstitute, Washington (DC) Publication 470.

Wright,R.L. 1963. Deep weathering and erosion surfaces in the Daily River Basin, Northern Territory. Journal of the Geological Society of Australia, 10, 151-164.

Zwittkovits, F. 1966. Klimabedingte Karstformen in den Alpen, den Dinariden und in Taurus. Österreichische Geographische Gesellschaft, 108, 72-97.

EROSION SURFACES IN THE SOUTHERN CAPE, SOUTH AFRICA

E.O. JACOBS
South African Forestry Research Institute,George
R.N. THWAITES
School of Earth Sciences, Macquarie University

1. REVIEW

Among the most striking features of the southern Cape coastal scenery are the well-developed, elevated planation surfaces fringing the coastal mountain ranges and displaying various stages of dissection (Fig. 1). Philips (1931) identified two major plateaux in the Knysna area. He defined the lower ("Upland Plateau ") as stretching from the cliffed coast at 150-215 m inland up to the 325 m contour, and being extensively dissected by ravines. This gives way northwards to the "De Vlugt plateau" which lies at 370-550 m below the foothills and spurs of the coastal fold mountain range (Outeniqua Mountains). This plateau then gradually disappears south-eastwards within the Keurbooms-Bietou river basin at Plettenberg Bay. Phillips quoted Rogers and Du Toit (1909) as viewing both these surfaces as "... plains of the river erosion ...", whilst Schwarz (1906) thought at the least the "uplands plateau" is of marine origin and asserted that the feature was of "... no greater geological age than late Tertiary...". The reasons given for this marine erosion hypothesis are; (1) that the narrowness of the plateau feature, especially in the Tsitsikama region, would not allow rivers to form floodplains, and (2) the plateau deposits correlate with confirmed marine deposits on the raised platform near Port Elizabeth. The argument for terrestrial planation is based upon the morphological correlation with similar inland plateaux (beyond the coastal ranges) which are not obviously of marine origin, and the lack of explanation for the marked absence of any marine deposits or fossils on the "coastal platform". This is so in the George-Tsitsikama area, which is in contrast to the regions of Algoa Bay and Bredasdorp, where marine deposits and fossils are found on this planation surface (Truswell, 1931).

Further observations which show that the "coastal platform" surface actually passes below the partly marine coastal limestones (Alexandria and Bredasdorp Formations) led Du Toit (1926) to revert to the marine planation theory, although he qualified it with a largely fluvially eroded extension inland. He postulated that the

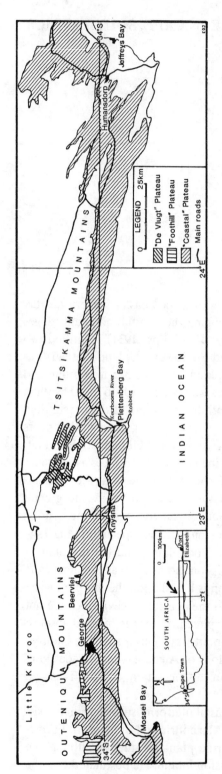

Fig. 1. Plateau features along the Southern Cape coast.

erosion event was some time during the Eocene or Oligocene, with general uplift occurring in the Pliocene (Du Toit, 1926: 366).

Haughton et al. (1937) gave a simpler explanation for the plateaux. The surface on which marine beds (the basal limestone) lie and the related, unveneered surface is the wave-cut shelf of a "pre-Tertiary" sub-aerial peneplain. The terrestrial part of this peneplain still supports silcretes and ferricretes. The two are seen to be separated by an ancient sea-cliff in the area west of George. Similarly, King (1942) referred to a single wide coastal terrace or "fringing plain". He differentiated the two features defined earlier by describing the inner as exhibiting patches of 'surface quartzite' and ferricrete, and the seaward being veneered with "...subaqueous beds containing marine fossils..." (King, 1942 p.313). The marine-abraded seaward platform is suggested to be Eocene to Mio-Pliocene in age on the basis of undefined fossil evidence.   King (1963), however, doubted this assertion and questioned whether two surfaces are involved (King, 1963, p.264). This dilemma arises over the altitude of bevelling in relation to that of the coastal limestones. In a later description, King (in Rust, 1979) suggested that the coastal plain at George contains marine deposits "... at some places ..." with Pliocene fossils (cf. Rust, 1979, p.12) implying that the now elevated coastal platform emerged sometime within the Pliocene. He also postulated that the higher (660-880 m) gravel terraces inland in the little Karoo and fringing the southern slopes of the Langeberg Range 100 km to the west are the Southern Cape expression of his "moorland" (formerly "African") surface. The terrestrial part of the coastal plain is seen as the manifestation of the "Rolling" (or "Post-Afican 1") pedimentation episode before the Pliocene upheaval (or "Active Episode E") of his global cyclic erosion paradigm (King, 1976). The coastal platform, therefore, is the marine-cut modification of the Post-African 1 surface (King, 1963).

Butzer and Helgren (1972) and Helgren and Butzer (1977) agree with King's hypothesis for the development of the two surfaces. They visualised the "coastal platform" (lower surface) to be the product of sub-aerial pedimentation periodically inundated by a shallow sea. They believed it to be at least Eocene in age owing to the correlation of this surface with partly co-eval conglomeratic deposits at Plettenberg Bay, unofficially termed the "Keurbooms Formation" (Helgren and Butzer, 1977). These deposits , however, are generally regarded as an extension of the early Cretaceous Enon Formation (SACS, 1980: 576; Dingle et al., 1983: 116), or a late Cretaceous-Palaeogene reworking of the Enon (Marker, pers. comm.). The age of these basinal conglomerates with inter-digitated sands and argillaceous sediments remains contentious. A major aspect of the argument for a Cretaceous age is the relationship of these sediments with the silicified conglomerates of the "Robberg Formation" (Rigassi and Dixon, 1972) which are thought to be the coastal terrestrial and marine facies (the "Robberg Member") of the Enon Formation (McLachlan and McMillan, 1976; SACS, 1980; Dingle et al., 1983). Butzer and Helgren (1972) maintain the Robberg sediments differ fundamentally from the downfaulted conglomerates in the Keurbooms-Bietou area. They appear to him not to exhibit classic Enon characteristics (Butzer, pers. comm.).

Schloms et al. (1983) agreed that the plateau levels are different expressions of the same feature: the flat high-level plain at approximately 200 m in the Albertinia district is associated with red apedal soils on Bokkeveld Group (Devonian) shales, and the upland gravels at 300-400 m relate to the "... highest, dissected parts of the coastal platform ...". Hendey (1983) rightly proposed that, because the erosion surfaces of the Cape coastal regions have not yet been dated independently, correlations between these features and shorelines can only be tentative. For the Western Cape Province he stated that only relative dating can be achieved by working backwards from the most recent ("Katarra") surface (44-110 m :mid-Pliocene). From this, and the work by Schloms et al. (1983), he reviewed the highest surface, the "150-200 m marine platform", to be Palaeocene to Eocene in age (Hendey, 1983: 44). This generally occurs with the evidence from the Eastern Cape (Birbury, Needs Camp) for a marine inflence at over 200 m, in the Upper Eocene (Siesser and Miles, 1979). The South African committee for Stratigraphy (SACS) appears to take an anomalous viewpoint concerning two successive periods of planation, deposition and cementation that are evident. The first level, at 210-240 m, rising to 450 m at Alexandria is suggested as being Cretaceous. The higher peneplain, at 300-380 m rises to 670 m at Grahamstown and includes silcretes, which the SACS infers as being Early Tertiary in age (SACS, 1980: 610).

The landward terrace feature of the coastal plateau is not mentioned by Marker (in Tyson, 1971), but an indistinct plateau feature is identified at 335-396 m with its maximum expression to the east of George, and is called "the foothills zone" (Tyson, 1971: 7) or "foothills plateau" (Tyson, 1971: 8, Fig. 5). This irregular surface with convex slopes is apparently not associated with the subsequent 'coastal platform' feature at 180-240 m. Following the work of Wellington (1955), Marker suggested (in Tyson, 1971) that the major marine erosion took place in the Miocene. This contrasts with the foregoing opinion that the planation occurred in the Palaeogene (e.g. Helgren and Butzer, 1977; Hendey, 1983). A recent dicussion by Maud (1986) on the development of the two surfaces adheres to a similar theme. Following the hypotheses of King (e.g. 1976) Maud believes that the 'upper plateau" silcretized surface is the manifestation of King's "African" ("Moorland") surface, while the coastal platform is the "post Afican 1" ("rolling") surface. The latter has been marine modified during the Pliocene high sea-level (Maud, 1986). To add further intrigue to this discussion Thwaites (1986) proposed using morphological analysis that in fact, a third Tertiary planation surface is identifiable north of Plettenberg Bay.

The present altitude of over 150 m for the main plateau cannot be explained by high sea levels. Glaciotectono-eustacy compounded by local uplift must have contributed to the high level to a large extent (Dingle, 1971, Tankard et al., 1982; Winter, 1984). Important periods of uplift which might effect the formation of planation levels are known at the end of the Oligocene or Early Miocene (King, 1967; 1972; Truswell, 1970) and at the end of the Pliocene (Siesser and Dingle, 1981). Davies (1971) concluded that there has been great uplift near Keurbooms River, as well as possibly along the Tsitsikamma and Robberg ridges. The Robberg ridge rose in Mid-Tertiary times (King, 1963). Davies (1972) claimed that differential warping in certain sections of the coastal platform was restricted in extent whereas Butzer and Helgren (1972) found no evidence for either local faulting or large-scale upwarping in the Tsitsikamma Forest, as suggested by Davies (1972). Vertical uplift and seaward tilting of the whole coastal platform seem to be the major tectonic events since the planation in the Eastern Cape (Ruddock, 1968; 1973).

2. DISCUSSION

Apart from the pronounced coastal platform at 150-215 m, confusion exists on recognition and nomenclature of higher erosion levels. Due to the greater age, higher levels are heavily dissected and often are nothing more than erosion remnants or "gipfel flur" features. Local tectonics make correlation tenuous.

At least two higher levels have been identified of which the altitude partly overlaps. The dating depends on the age of the coastal limestones, silcretes, lignites and local tectonics.

Coastal limestones are missing in most of the section of the Southern Cape discussed. However, they are well-developed in the Algoa Basin in the eastern Cape and situated up to 305 m a.s.l. near Bathurst and Patterson (Ruddock, 1973; King, 1972). To the west they have been mapped from   Mossel Bay and have been found at Riversdale at 153 m and Bredasdorp at 76 m a.s.l. (Dingle et al., 1981; Malan, 1986).

In the eastern Cape the limestones have been correlated with marine benches of which the highest are at a level of 240, 200 and 90 m (Marker, 1984; 1986). The limestones are undoubtedly of marine origin and the stepped underlying palaeotopography shows clearly that the coastal plateau has been formed by marine activity dating back to the Palaeogene for the 240 m bench. The 200 m bench has been dated as Late Eocene to early Oligocene (Marker, 1984).

Of younger age are the coastal limestones to the west (Bredasdorp Group) of which the oldest formation has been dated as Miocene/Pliocene (Malan, 1986). The southern Cape plateau relates to the erosion surfaces below the coastal limestones.

The lack of unequivocal evidence of the marine deposits, which may be expected in at least the Tsitsikamma, may be due either to high sea-levels never reaching the area due to regional uplift or because Pliocene sea levels lapped on to the platform; but evidence of this has been eradicated by fluvial action during the subsequent regression or intensive weathering after deposition. In the Tsitsikamma area, relict beach gravels have been found on top of the cliff at 160 m a.s.l. Therefore the planation of a part of the plateau in this district can be explained by high energy marine activity causing gravel beaches and preventing limestone sedimentation. Their extent towards the mountains is unknown due to a cover of thick silty and clayey sediments. Do these clays evidence of marine origin? As Winter (1984) mentioned: ".... clay settling would tend to occur on quiet shallow shelves of a transgressing sea ..." the clays may form a key to the reconstruction in the geomorphological development of this area. Along the plateau between Knysna and Plettenberg Bay marine benches have been established up to 250 m, but the sequence is not as clear (Marker, 1986b).

In the southern Cape most of the superficial material on the coastal platform is characteristic of extensive high and low energy fluvial and aeolian action. Miocene lignites near Knysna, situated between 260 m and 315 m, are interpreted as terrestrial deposits and therefore post-date all levels higher than 300 m (Jacobs, in prep.).

No consensus has been reached about the dating of the silcretes in the southern Cape, which are partly based on the dating of the coastal limestones. Silcretes occur between 120 and 300 m in elevation (Summerfield, 1983) and over 300 m near Beervlei between George and Knysna (pers. comm. Schafer). In contrast to the inland silcretes, the coastal formations are associated with deep weathering profiles formed in broad shallow valleys under a humic tropical or sub-tropical climate. The lack of such topography of the De Vlugt plateau may explain why silcretes have not developed here.

The first attempt has been made to date inland silcretes just north of the Langeberg outside the coastal plateau in the Little Karoo. Two datings by ESR have proved an age of 7.3 and 4.4 Ma (Hagedorn, in prep.). If this dating technique proves successful the application to the southern Cape is of extreme importance. If the coastal silcretes are of the same age as those inland, the formation of the plateau remnants on which they have been developed will provisionally support the existing ideas of a Pliocene date for the plateau.

The most recent hypothesis of the formation of planation levels forms part of a synthesis of erosion levels throughout southern Africa based on King's chronology of erosion cycles (King, 1976). Unfortunately the model seems too simplistic for the southern Cape situation.

3. CONCLUSION

Three erosion surfaces have been identified in the Southern Cape. All of them have been formed by sub-aerial processes, mainly by fluvial action, although a marine influence in the Tsitsikamma may preceed these processes. Local tectonics prevented marine modification which is well known to the west and east of the investigated area. More detailed investigations are necessary to discriminate the different levels spatially and to refine King's erosion cycles (cf. Partridge and Maud,

1987). The higher erosion surfaces must be dated as not younger than the Miocene based on the age of the Knysna lignites. The higher plateau may be as old as Mid-Cretaceous. The main coastal platform at 150-215 m level is postulated to be of Miocene age with an emergence in the Pliocene.

4. SUMMARY

At least three erosion surfaces have been identified in the Southern Cape. Their nomenclature, position, altitude, genesis and age is still a controversial subject among many authors. This is caused by the scattered occurrences of good datable deposits although in the near future dating of limestone, lignites and silcretes may reveal promising results. All of the erosion surfaces have been formed by sub-aerial processes based on occurrences of extensive fluvial and aeolian deposits, silcretes and the lack of marine sediments. On the most prominent coastal platform which ranges from 150 -215 (or 335) m these latter sediments occur extensively in the west, but are nearly lacking in the investigated area possibly because of relatively high sea levels during transgressive periods. Local Tertiary tectonics make correlation of plateau remnants difficult. They are often so dissected as to become restricted to nothing more than a "gipfel flur". This causes the altitudinal ranges of the different levels to overlap. Preliminary datings of the lignites (as Miocene) situated between the main and the next higher platform supports the hypothesis that the higher level surfaces are pre-Miocene possibly as old as Mid-Cretaceous while the main coastal platform is Miocene/Pliocene with upheaval in the Pliocene period. Morphometric analyses and future dating of the superficial should refine King's erosion cycle model.

5. REFERENCES

Butzer, K.W. and Helgren, D.M. 1972. Late Cainozoic evolution of the Cape coast between Knysna and Cape St. Francis, South Africa. Quaternary Research, 2,143-169.

Davies, O. 1971. Pleistocene shorelines in southern and south-eastern Cape Province (part 1). Annals of Natal Museum, 21(1), 183-223.

Davies, O. 1972. Pleistocene shorelines in the southern and south-eastern Cape Province (Part 2). Annals of Natal Museum, 21(2), 225-279.

Dingle, R.V. 1971. Tertiary history of the continental shelf off Southern Cape Province, South Africa. Transactions of the Geological Society of South Africa, 74, 173-186.

Dingle, R.V., Siesser, W.G. and Newton, A.R. 1983. Mesozoic and Tertiary Geology of South Africa. Balkema, Rotterdam, 375pp.

Du Toit, A.L. 1926. The Geology of South Africa. Oliver and Boyd, Edinburgh, 445pp.

Hagerdon, J. (in press) Silcretes in the Western Little Karoo and their relation to geomorphology and paleoecology. Palaeoecology of Africa, 19.

Haughton, S.H., Frommurze, H.F. and Visser, D.J.L. 1937. The Geology of the Country Around Mossel Bay, Cape Province. Explanation of sheet 201. Geological Survey, 48pp.

Helgren, D.M. and Butzer, K.W. 1977. Palaeosols of the Southern Cape coast, South africa: Implications for laterite definition, genesis and age. Geographical Review, 67, 430-445.

Hendey, Q.B. 1983. Cenozoic geology and palaeogeography of the fynbos region. In: H.J.Deacon, Q.B. Hendey and J.J.N. Lamberechts (eds.), Fynbos Palaeoecology: A Preliminary Synthesis. South African National Scientific Programmes Report 75, CSIR, Pretoria, 35-60.

King, L.C. 1942. South African Scenery. Oliver and Boyd, Edinburgh, 342pp.

King, L.C. 1963. South African Scenery. Oliver and Boyd, Edinburgh, 302pp.

King, L.C. 1967. Scenery of South Africa 9. Oliver and Boyd, Edinburgh, 308pp.

King, L.C. 1972. The Natal Monocline: Explaining the Origin and Scenery of Natal, South Africa. Department of Geology, University of Natal. 108pp.

King, L.C. 1976. Planation remnants upon high lands. Zeitschrift für Geomorphologie, 20, 133-148.

Malan, J.A. 1986. The Bredasdorp Group in the Southern Cape Province. In: H.J.Deacon (ed.), Cainozoic of the Southern Cape. SASQUA Occasional Publication, 1, Stellenbosch, 15-18.

Marker, M.E. 1973. The Tertiary limestones of the southern coastal regions of Cape Province, South Africa. In: G. Blant (ed.). Sedimentary Basins of the African Coasts, Part 2 south and east coast, Paris, Association of African Geological Survey, 79-82.

Marker, M.E. 1984. Marine Benches of the Eastern Cape, South Africa. Transactions of the Geological Society of South Africa, 87, 11-18.

Marker, M.E. 1986a. Relative age of the Coastal Limestone as determined from the intensity of Karst development. In: W.K. Illenberger and W.J. Smuts (Eds.). Tertiary to Recent coastal geology. Proceedings of a seminar, University of Port Elizabeth, 70-75.

Marker, M.E. 1986b. Marine Benches of the Southern Cape. In: H.J. Deacon (ed.), Cainozoic of the Southern Cape. SASQUA Occasional Publication, 1, Stellenbosch, 5-7.

Maud, R.R. 1986. Older deposits on the southern Cape Coastal Platform. In: H.J. Deacon (ed.), Cainozoic of the Southern Cape. SASQUA Occasional Publication, 1, Stellenbosch, 10-11.

MacLachlan, I.R. and McMillan, I.K. 1976. Review and stratigraphic significance of southern Cape Mesozoic palaeontology. Transactions of the Geological Society of South Africa, 79, 197-212.

Partridge, T. and Maud, R.R. 1987. Geomorphic evolution of southern Africa since the Mesozoic. South African Journal of Geology, 90, 179-208.

Phillips, J.F.V. 1931. Forest succession and ecology in the Knysna region. Memoir of the Botanical Survey of South Africa, 14, Pretoria, 307pp.

Rigassi, D.A. and Dixon, G.E. 1972. Cretaceous of the Cape Province, Republic of South Africa. In: T.F.J. Dessauvagie and A.J. Whitman (eds.), African Geology 1970, University of Ibadan, Nigeria, 513-527.

Rogers, A.W. and Du Toit, A.L. 1909. The Geology of the Cape Colony. Longmans, Green & Co., London, 491pp.

Ruddock, A. 1968. Cenozoic sea-levels and diastrophism in a region bordering Algoa Bay. Transactions of the Geological Society of South Africa, 71, 209-233.

Rust, I.C. (ed.) 1979. Geokongress'79: excursions guidebook. Geological Society of South Africa, Johannesburg, 180pp.

SACS (South African Committee for Stratigraphy) 1980. Stratigraphy of South Africa. Handbook 8, Part 1. Geological Survey, Dept Mineral and Energy Affairs, Pretoria, 642pp.

Scholms, B.H.A., Ellis, F. and Lamberchts, J.J.N. 1983. Soils of the Cape coastal platform. In: H.J. Deacon, Q.B. Hendey and J.J.N. Lambrechts (eds.), Fynbos Palaeoecology: A Preliminary Synthesis, South African National Scientific Programs Report 75, CSIR, Pretoria, 70-86.

Schwarz, E.H.L. 1906. Geological survey of the coastal plateau in the divisions of George, Knysna, Uniondale and Humansdorp. 10th Annual Report of the Geological Commission (Cape of Good Hope), Dept. of Agricilture Cape Town, 49-93.

Siesser, W.G. and Dingle, R.V. 1981. Tertiary sea-level movements around southern Africa. Journal of Geology, 89, 83-86.

Siesser, W.G. and Miles, G.A. 1979. Calcareous nanofossils and planktonic foraminifera in Tertiary limestones, Natal and Eastern Cape, South Africa. Annals of the South African Museum, 79, 139-158.

Summerfield, 1983. Silcrete as a palaeoclimatic indicator: evidence from southern Africa. Palaeogeography, Palaeoclimatology, Palaeoecology, 41, 65-79.

Tankard, A.J.T., Jackson, M.P.A., Erickson, K.A., Hobday, D.K., Hunter, D.R. and Minter, W.E. 1982. Crustal Evolution of Southern Africa, Springer-Verlag, New York, 523pp.

Truswell, J.P. 1970. An Introduction to the Historical Geology of South Africa. Purnell, Cape Town. 167pp.

Thwaites, R.N. 1986. Evidence for three major erosion surfaces in the southern Cape. In: H.J.Deacon (ed.), Cainozoic of the Southern Cape. SASQUA Occasional Publication, 1, Stellenbosch, 1-3.

Tyson, P.D. (ed.) 1971. Outeniqualand: The George-Knysna Area.- South African Landscape Series 2. South African Geographical Society. 23pp.

Wellington, J.H. 1955. Southern Africa, Vol. 1. Physical Geography, Cambridge University Press.

Winter, H. de la R. 1984. Tectonostratigraphy, as applied to analysis of South African Phanerozoic basins. Transactions Geological Society of South Africa, 87, 169-179.

ASPECTS OF THE GEOMORPHOLOGY OF THE NORTH-WESTERN CAPE, SOUTH AFRICA

M.C.J. DE WIT

De Beers Consolidated Mines Ltd., Kimberley

1. INTRODUCTION

The study area in the north-western Cape which comprises an area bounded by the towns of Prieska, Pofadder, Garies and Calvinia in the east, north, west and south respectively is characterised by very subdued topography. This is an expression of the mature nature of the landscape. Several stages of geomorphological evolution have been described in the past by various authors (e.g. Dixey 1942, King 1951) often based on elevation data. The present study has focused on the pedogenic characteristics of the land surface representing a non-depositional unconformity as well as sedimentary sequence associated with these. The geomorphological history of the area was supported by fission track dating, which in other parts of the world has been applied to study uplift rate and erosion levels e.g. (Wagner et al 1977, Gleadow 1978, Kohn and Eyal 1981).

2. TREND SURFACE ANALYSIS

The elevation of every point from a five kilometer square grid over the area was obtained from the 1:50000 topocadastral maps. A second degree polynomial surface was fitted resulting in a residual topographical map highlighting all major valleys and hills (Fig. 1).

Two main trends were observed. An east-north-east, west-south-west trend dominating the southern half of the area and a north-south trend in the north. The former is associated with dolerite dykes and sills intruded preferentially along specific horizons in the lower Karoo sequence and therefore follow the lithological boundaries of the Karoo sediments which strike in the same direction.

Geomorphological Studies in Southern Africa, G.F.Dardis & B.P.Moon (eds)
© *1988 Balkema, Rotterdam. ISBN 90 6191 831 6*

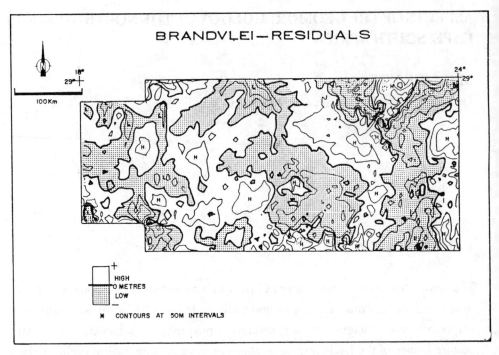

BRANDVLEI—RESIDUALS

100 Km

HIGH
O METRES
LOW

CONTOURS AT 50M INTERVALS

Fig. 1. A second degree Polynomial surface showing residuals, highlighting major valleys and hills.

In the north the latter is associated with north-south trending valleys such as the Koa, Sak and Hartebees. The underlying geology comprises metamorphic and igneous rocks of the Namaqua mobile belt which have been scoured by the Carboniferous glacial ice from a northerly direction (Visser, 1981). These glacial valleys are now exhumed as a result of the removal of a thin veneer of Karoo sediments and have subsequently been preferentially exploited by post-Karoo drainage systems. The present day geomorphology is therefore directly related to the different lithologies in the area.

3. DURICRUSTS

The interfluvial areas are characterised by remnants of mature calcrete (Fig. 2). These display the classic pedogenic profile starting with calcified rock at the base through powdery, nodular and hardpan calcrete capped by laminar rims. Thickness varies from almost one metre on permeable sediments to thirty centimetres on hard basement. Most of the duricrust area is littered with calcrete boulders and cobbles indicating that the calcrete is fossil and presently breaking down. Petrologically, this calcrete differs from the calcified Vanwyksvlei Plio-Pleistocene and Brandvlei Miocene sediments

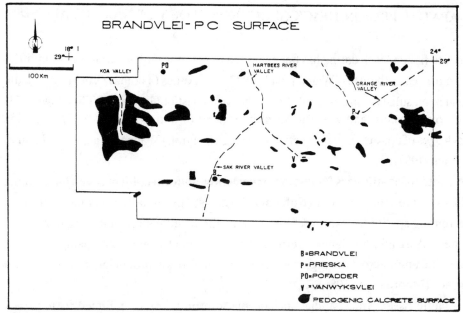

Fig. 2. Remnants of mature calcrete in the study area.

(De Wit, in prep.). Isolated outliers of this calcrete cover can be found throughout the area but also further in the northern Cape, Botswana (Halliwell, 1987) and Namibia (Ward 1984). There is no doubt that this paleosol formed under stable climatic (arid to semi-arid) and tectonic conditions which affected most of the sub-continent. At Bosluispan (De Wit, in prep.), in the Koa valley, the calcrete was cut by Miocene fluvial gravels deposited on fresh bedrock containing pebbles of calcrete which are petrologically similar to the main calcrete deposit, The upper Cretaceous ultrabasic plugs in Bushmanland have also been affected by this pedocrete and the calcrete therefore formed between the late Cretaceous and the middle Miocene.

It is suggested that the formation of the calcrete was related to the major mid-Tertiary regression when palaeoclimatic conditions changed from a tropical humid Palaeogene to a drier earlier Oligocene (Dingle et al., 1983). This more arid phase terminated during the middle Miocene with the onset of a warmer and wetter climate. This climatic change is based on the terrestrial evidence from deposits at Arris Drift (Hendey, 1978) and Bosluispan (De Wit, in prep.) and on planktonic foraminiferal studies by Shackleton and Kennet (1975). It was a major regression as it was also recognised in western Australia (Quilty, 1977) and is unlikely to have been caused by local tectonics. In addition, it probably caused a westerly shift of the climatic belt in Namibia and therefore set the scene for extensive calcrete development in the present Namib (Ward, 1984) and south western Namibia (Corbett, pers. comm.).

4. APATITE FISSION TRACK GEOCHRONOLOGY OF NAMAQUALAND

Apatite fission track dating has been applied to geomorphological studies of uplifted terrain, such as the Alps (Wagner et al., 1977), the Andes (Thomas Crough, 1983), the Sinai peninsula (Kohn and Eyal, 1981), the west coast of Greenland (Gleadow, 1978) and south eastern and western Australia (Gleadow and Lovering, 1978), as fission track ages of apatites increase with increasing elevation (Wagner et al., 1977; Thomas Crough, 1983).

A scatter of apatite ages for instance reveals non-uniform uplift between fault-bound blocks in the Red sea area (Kohn and Eyal, 1981). Similarly a regular pattern of decreasing apatite fission track ages towards the rifted continental margin in south-eastern Australia has been interpreted by a thermal event accompanied by considerable uplift across the region with subsequent erosion producing the present land surface (Moore et al., 1986).

Apatite fission track studies include the interpretation of confined fission track length distribution which is a useful source of palaeo-temperature information and is subsequently used for thermal history analysis (Gleadow et al.,1986a).

Thirteen samples were taken along an E.N.E. traverse from the Atlantic coast at Kleinzee to just south-east of Kenhardt (Fig. 3 and Table 1) from rocks of the Namaqualand mobile belt. In addition, one sample was taken of basement to the south near Vredendal (Fig. 3).

The traverse was selected on the following basis: (1) It is at right angles to the rifted margin of the African continent and based on evidence of the Sinai peninsula these type of continental margins are clearly associated with the youngest ages (Kohn and Eyal, 1981); (2) It cuts across the Great Escarpment which forms a prominent feature in the southern African geomorphology, and which has been the subject of considerable debate (De Swart and Bennet, 1974; Dixey, 1955; King, 1951; McCarthy et al., 1985); and (3) It lies in an area which is presently been subjected to a detailed geological and geomorphological investigation (De Wit, in press).

Sample treatment and analysis are described in detail by Gleadow and Duddy (1987) and the results are presented in Table 1. The most insignificant aspect of the results is that the fission track ages of all the samples fall into two distinct groups. A first and older group which is made up of a remarkably uniform suite of apatite ages falling into the narrow range of 108 + -4 and 129 + -6 Ma (Fig. 4). These are presented by samples 1-10 and 13. According to Gleadow and Duddy (1987) the ages from the coastal and the inland end are indistinguishable and no trends can be observed that relate to that of a typical rifted margin (Moore et al., 1986; Kohn and Eyal, 1981).

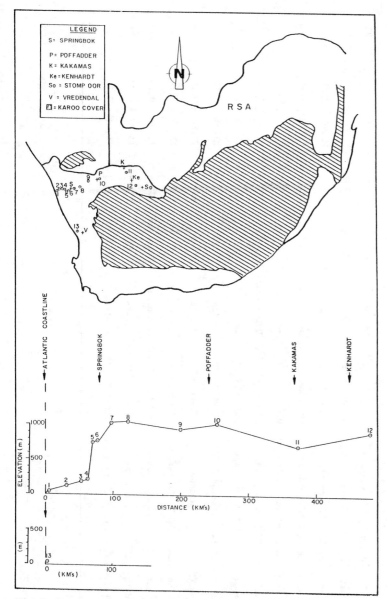

Fig. 3. Sample positions and elevations.

The second group is associated with much younger apatite ages close to 70 Ma (Fig. 4). The mean track lengths are also very homogeneous and similar in both groups and are approximately 14 microns. Additionally all track length distributions are very narrow with standard deviations of between 0.9-1.4 microns (Gleadow and Duddy, 1987).

These characteristics conform with the typical "volcanic-type" length distribution of Gleadow et al.(1986) and suggest that both groups of apatites ages have undergone

61

Table 1. Sample information and results (after Gleadow and Duddy, 1987).

Sample Number	Elevation (m)	Farm Name & Number	Magisterial District	Distance (km)	Lithology	Apatite Yield	Age Ma	Mean track length μm	Standard deviation μm
PTD001	40	Dikgat (195)	Namaqualand		Gneiss	High	119 ± 9	14.03 ± 0.12	1.07
002	120	Strydrivier (188)	Namaqualand	001-002= 26	Leuco-gneiss	High	108 ± 4	13.68 ± 0.12	1.17
003	180	Graces Puts (201)	Namaqualand	002-003= 21	Gneiss	Moderate	123 ± 7	13.69 ± 0.11	1.11
004	200	Schaap Rivier (208)	Namaqualand	003-004= 9	Gneiss gran.	Low	111 ± 12	13.33 ± 0.68	2.04
005	730	Ezelsfontein (214)	Namaqualand	004-005= 6	Gneiss gran.	Moderate	129 ± 6	14.01 ± 0.12	1.18
006	760	Modderivier (215)	Namaqualand	005-006= 9	Leuco gran.	Very low	125 ± 26	13.16 ± 0.79	2.51
007	1050	Melkboschkuil (132)	Namaqualand	006-007= 20.5	Gneiss	High	125 ± 4	14.05 ± 0.12	1.19
008	1040	Groot Kau (128)	Namaqualand	007-008= 25	Gneiss	High	126 ± 6	13.93 ± 0.11	1.13
009	940	Aroam (57)	Namaqualand	008-009= 77	Gneiss	High	111 ± 5	13.86 ± 0.11	1.10
010	1010	Pofadder East (145	Kenhardt	009-010= 54	Leuco gneiss	Moderate	107 ± 6	13.81 ± 0.14	1.37
011	685	Kakamas Suid (28)	Gordonia	010-011= 121	Leuco gneiss	Moderate	70 ± 5	13.98 ± 0.13	1.12
012	900	Moddergat (257)	Kenhardt	011-012= 107	Gneiss	High	73 ± 4	14.23 ± 0.09	0.88
013	20	Klipvlei Karookop (153)	Vredendal	011-013= 215	Gneiss	High	125 ± 6	14.01 ± 0.10	1.05

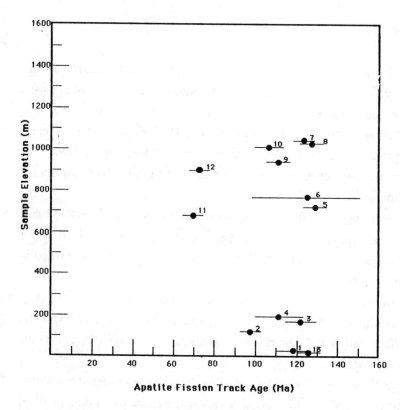

Fig. 4. Fission track ages (Ma) of Apatite samples.

an initial rapid cooling from above the partial track retention temperature (120°)

Not enough time was spent at temperature above 90 degrees to produce short tracks. This was allowed by a more gently cooling at near surface temperature (50°) and in which the apatite ages, represents the time of the cooling "event", which is significantly younger than the actual age of the crystalline basement. The cooling histories for both periods, the early and late Cretaceous, are similar.

The mean heat flow from the Namaqua Mobile belt (61 + -11 m Wm-Z) is approximately 25-50 per cent higher than in most pre-Cambian terranes (Polyak and Smirnov, 1968), including the Kaapvaal craton and confirms the generally high heat flow in this mobile belt which is regarded as a significant anomaly (Jones, 1981). According to England and Richardson (1980) the effects of uplift and erosion are not important considerations in the Namaqua Mobile belt and Jones (1981) suggests that the additional heat came from greater radiogenic crustal contribution or even an enhanced mantle heat flux. There is however no independent geophysical evidence to suggest that the aseismic Namaqua mobile belt has been subjected to lithospheric thinning or incipient rifting (Jones, 1981).

5. DISCUSSION

For the older apatite fission track date there are two possible geological explanations. The first being rapid erosion during the early Cretaceous with a removal of at least 2 to 3 km of the overlying rock. The amount of erosion would have had to be the same for those samples above and below the escarpment (Fig. 3). The alternative is an early Cretaceous overprint by a thermal event under a much shallower cover of overlying rock. The latter being a thin veneer of Karoo cover, remnants of which are for instance still present at Vioolsdrift.

The latter has been suggested the most probable by Gleadow and Duddly (1987) and explanation is also applicable to the late Cretaceous age. It is supported by the lack of intermediate apatite ages and the fact that the track lengths have formed at near-surface temperatures.

This earlier thermal event is thought to be related to the south Atlantic rifting (Fig. 5) which coincides with stage V of Dingle's (1983) igneous phases (135-125 Ma) and relates to the drift initiation and associated local eruptions of tholeiitic lavas along the new continental margin.

It is suggested that the younger thermal event coincides with a later period of alkaline volcanism (Smith et al., 1985). This period is characterised by the sediment filled diameters, "Kimberlite" pipes, olivine melilitites and olivine nephelinite plugs in Bushmanland which in the northern part of the area occurred during the late Cretaceous to early Tertiary (Moore and Verwoerd, 1985). Certainly the survival of crater lake sediments overlying melnoites near Kenhardt (Smith, 1986) and in Bushmanland (Houghton, 1931; Estes, 1977; Scholtz, 1984) is an indication of the limited amount of erosion that has taken place since emplacement of these intrusives and which is supported by the heat flow studies by Jones (1981). It is interesting to note that very similar age events were recognised along the Brazilian continental margin (Campos et al., 1974).

Figure 5 also illustrates the peak of the group 1 and group 2 kimberlite intrusions (Smith et al., 1985) and its generally associated facies, highlighting the increase of the levels with time. It should be remembered that this kimberlite data is taken from the Kaapvaal craton, an area well to the east.

One main problem remains and that is the existence of a major escarpment along the west coast of Namibia and South Africa. However both Visser (1981) and Martin (1953) recognise pre-Karoo valleys carved in the escarpment which is now being re-exposed by erosion suggesting that this escarpment was a pre-Karoo and not a rift generated feature as suggested by King (1962) and Partridge and Maud (1987). This

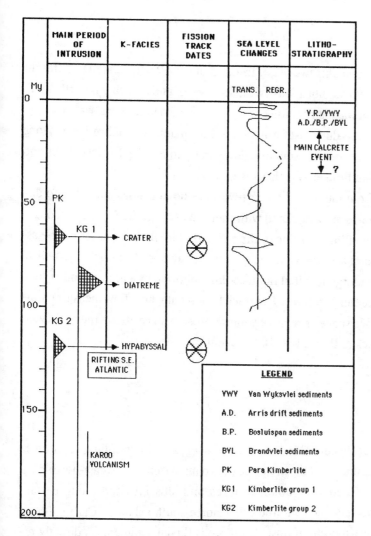

Fig. 5. Interpretation of Apatite fission track data.

is supported by the apatite fission track data which shows no linear trend away from the rifted margin and across the escarpment.

The Bushmanland plateau which stretches from the edge of the escarpment eastwards to the Doringberg fault zone near Prieska is characterized by outliers of the mature pedogenic calcrete (De Wit, in prep.) which represents a period of climatic and tectonic stability. This pedocrete affects the ultrabasic plugs in Bushmanland and can therefore be regarded as post 70 Ma in age. It is however cut by Miocene sediments at Bosluispan (De Wit, in prep.) and must therefore have formed sometime in the period between 70 and 16 Ma age (Fig. 5).

6. CONCLUSION

Trend surface anlysis highlights two major trends that are directly related to different geological events. One is east-north-east, west-south-west and the other is north-south and are instrumental in the distribution of the Tertiary palaeodrainage systems.

The extensive calcareous duricrust formation in Bushmanland is indicative of a long period of climatic and/or tectonic stability during early to middle Tertiary times and emphasises the fact that little erosion has taken place.

Fission track ages from samples in an east-west are homogenous and fall into distinct groups. Both groups are associated with thermal events. The former is thought to be related to the initiation of rifting of the S.E. Atlantic while the latter can be associated with a period of alkaline volcanicity in Bushmanland. The tracks are typical of rocks which have undergone initial rapid cooling followed by a long period at relatively low near-surface temperatures. The track length data, heat flow measurements and geological evidence suggests relatively minor erosion since 70 Ma thermal event and possibly as far back as the 120 +- 10 Ma event.

7. SUMMARY

The geomorphology of Bushmanland was studied using a multidisciplinary approach. Trend surface analysis indicated the presence of two main trends; one east-north-east, west-south-west associated with dolerite dykes and sills intruded along certain horizons in the Karoo sequence and the other north-south exhumed Carboniferous glacial valleys. The distribution of post- Cretaceous fluvial sediments is directly related to the local geomorphology associated with these trends.

Landsat studies and surface mapping of the interfluvial areas of these Tertiary deposits have outlined the extent of pre-Miocene, mature duricrusts representing a period of climatic and tectonic stability.

Differences of exposed facies of several post Karoo ultrabasic volcanic plugs of varying ages and fission track dating of basement rocks suggests an early aggressive erosional phase from the late Jurassic to early Cretaceous followed by a long intermission of stability resulting in mature pedogenic profiles. In addition the escarpment along the west coast seems to be related to a pre-Karoo period and not to the break up of Gondwanaland.

8. REFERENCES

Campos, C.W.M., Ponte, F.C. and Miura, K. 1974. Geology of the Brazilian Continental Margin. In: Burk, C.A. and Drake, C.L. (eds.), The Geology of Continental Margins, Springer-Verlag.

Cull, J.P. 1980. Climatic corrections to Australian heat-flow data. BMR Journal Australian Geolology and Geophysics, 4, 303-307.

De Swardt, A.M.J. and Bennet, G. 1974. Structural and physiographic development of Natal since the late Jurassic. Transactions of the Geological Society of South Africa, 77, 309-322.

Dingle, R.V., Siesser, W.G., Newton, A.R. 1983. Mesozoic and Tertiary Geology of Southern Africa. A.A. Balkema, Rotterdam, 375 pp.

Dixey, F. 1942. Erosion cycles in central and southern Africa. Transactions of the Geological Society of South Africa, 45, 151-181.

Dixey, F. 1955. Erosion surfaces in Africa: some consideration of age and origin. Transactions of the Geological Society of South Africa, 58, 265-280.

England, P.C. and Richardson, S.W. 1980. Erosion of the age dependence of continental heat flow. Geophysical Journal of the Royal Astronomical Society, 62, 421-437.

Estes, R. 1977. Relationships of the South African fossil frog Eoxenopoides Reuningi (Anura, Pipidae). Annals of the South African Museum, 73, 49-80.

Gleadow, A.J. and Lovering, J.F. 1978. Fission track geochronology of King Island, Bass strait, Australia: Relationship to continental rifting. Earth and Planetary Science Letters, 37, 429-437.

Gleadow, A.J.W. 1978. Fission track evidence for the evolution of rifted continental margins. In: Zartman, R.E. (ed.), Short Papers of the 4th International Conference of Geochronology, Cosmochronology and Isotope Geology, U.S. Geological Survey Open file Rep. 78-701, 146-148.

Gleadow, A.J.W. and Duddy, I.R. 1987. Fission track analysis of crystalline basement rocks from the Namaqualand mobile belt, South Africa Geotrack report No. 68, 1-4.

Gleadow, A.J.W. and Fitzgerald, P.G. 1987. Uplift history and structure of the Transantarctic Mountains: New evidence from fission track dating of basement apatites in the Dry Valleys area, southern Victoria land. Earth and Planetary Science Letters, 82, 1-4.

Gleadow, A.J.W. and Duddy, I.R., Green, P.F. and Lovering, J.F. 1986a. Confined fission track lengths in apatite: a diagnostic tool for thermal history analysis. Contributions to Mineralogy and Petrology, 94, 405-415.

Gleadow, A.J.W., Duddy, I.R.,Green, P.F. and Hegarty, K.A. 1986b. Fission track lengths in the apatite annealing zone and the interpretation of mixed ages. Earth and Planetary Science Letters, 78, 245-254.

Halliwell, G. 1987. The stratigraphy of the Kalahari formation in southern Botswana. Internal De Beers Botswana report.

Haughton, S.H. 1931. On a collection of fossil frogs trom the clays at Banke. Transactions of the Geological Society of South Africa, 19, 233-249.

Hendey, Q.B. 1978. Preliminary report on the Miocene vertebrates from Arrisdrift, South West Africa. Annals of the South African Museum, 69, 215-247.

Jones, M.Q.W. 1981. Heat flow and heat production studies in the Namaqua mobile belt and Kaapvaal Craton. Unpublished Ph.D. thesis, University of the Witwatersrand, Johannesburg.

King, L.C. 1951. South African Scenery, 2nd edn. Oliver and Boyd, Edinburgh, 379 pp.

King, L.C. 1962. The morphology of the earth. Oliver and Boyd, Edinburgh, 379 pp.

Kohn, B.P. and Eyal, M. 1981. History of uplift of the crystalline basement of Sinai and its relation to opening of the Red sea as revealed by fission track dating of apatites. Earth and Planetary Science Letters, 52, 129-141.

Martin, H. 1953. Notes on the Dwyka succession and some pre-Dwyka valleys in the South West Africa. Transactions of the Geological Society of South Africa, 56, 37-43.

McCarthy, T.S., Moon, B.P. and Levin, M. 1985. Geomorphology of the western Bushmanland plateau, Namaqualand, South Africa. South African Geographical Journal, 67, 160-178.

Moore, A.E. and Verwoerd, W.J. 1985. The olivine melilitite - "Kimberlite" carbonatite suite of Namaqualand and Bushmanland, South Africa. Transactions of the Geological Society of South Africa, 88, 281-294.

Moore, M.E., Gleadow, A.J.W. and Lovering, J.F. 1986. Thermal evolution of rifted continental margins:new evidence from fission tracks in basement apatites from south eastern Australia. Earth and Planetary Science Letters, 78, 255-270.

Partridge, T.C. and Maud, R.R. 1987. Geomorphic evolution of southern Africa since the Mesozoic. South African Journal of Geology, 90, 179-208.

Polyak, B.E. and Smirnov, Y.B. 1968. Relationship between terrestrial heat flow and tectonics of continents. Geotectonics, 4, 205-213.

Quilty, P.G. 1977. Cenozoic sedimentation cycles in western Australia. Geology, 5, 336-340.

Scholtz, A. 1984. The palynology of the upper lacustrine sediments of the Arnot pipe, Banke, Namaqualand. Annals of the South African Museum, 95 (1), 1-109.

Shackleton, J.N. and Kennett, J.P. 1975. Palaeotemperature history of the Cenozoic and the initiation of Antarctic glaciation: Oxygen and carbon isotope analysis in DSDP sites 277, 279 and 281. In: J.P. Kennett et al., Initial Rep. Deep Sea Drilling Project, 29 Washington, U.S. Government Printing Office, 743-755.

Smith, C.B., Allsopp, H.L., Kramers, J.D., Hutchinson, G. and Roddick, J.C. 1985. Emplacement ages of Jurassic - Cretaceous South African Kimberlites by the Rb-Sr method on phlogopite and whole-rock samples. Transactions of the Geological Society of South Africa, 88, 249-266.

Smith, R.M.H. 1986. Sedimentation and palaeoenvironments of Late Cretaceous crater-lake deposits in Bushmanland, South Africa. Sedimentology, 33, 369-386.

Thomas Crough, S. 1983. Apatite fission track dating of erosion in the eastern Andes, Bolivia. Earth and Planetary Science Letters, 64, 396-397.

Visser, J.N.J. 1981. Carboniferous topography and glaciation in the north-western part of the Karoo basin, South Africa. Annals of the Geological Survey of South Africa, 15, 13-24.

Viljoen, R.P., Viljoen, M.J., Grootenboer, J. and Longshaw, T.G. 1975. Ersts-1 imagery: Applications in geology and mineral exploration. Minerals Science Engineering, 7, 132-168.

Wagner, G.A., Reimer, G.M. and Jager, E. 1977. Cooling ages derived by apatite fission track, mica Rb-Sr and K-Ar dating: The uplift and cooling history of the central Alps. Padova, Societa Cooperative Tipografica, Vol. XXX, 1-27.

Ward, J.D. 1984. Aspects of the Cenozoic Geology in the Kuiseb Central Namib Desert. Unpublished Ph.D. thesis, University of Natal, Pietermaritzburg, 159 pp.

THE SAHARA, KALAHARI AND NAMIB DESERTS: A GEOMORPHOLOGICAL COMPARISON

JURGEN HÖVERMANN

Geographisches Institut, University of Göttingen, Göttingen

1. INTRODUCTION

The arid landscape of Africa, situated on both sides of the tropics of Capricorn and Cancer, are primarily characterised by common geomorphic elements.

The Namib and Sahara deserts are dominated, in the lowest and driest regions, by wind-induced landforms, characterised by large dunefields of varying shape; transverse dunes, barchans, seif dunes and wind-eroded rocks, the latter including aerodynamically-streamlined hummocks (yardangs), and inverted, forms or depressions. The middle regions, which are a little moister, comprise desert plains with less or greater sand cover (regions legerement ensablees, Sandschwemmebenen). The highest regions, which are characterised by heavy ephemeral precipitation, are strongly dissected by steep-sided gorges with steps-like longitudinal profiles.

These three different types of landscapes are strongly associated with climatic conditions, which can be expressed by mean annual precipitation, although many years may be without any rainfall. In the regions formed purely by wind erosion and accumulation the mean annual precipitation (MAP) is less than 30 mm; in the region of desert plains the mean annual precipitation is between 30 mm and 60 mm; the range of precipitation in the region of desert gorges is from 60 mm to 150 mm. These MAP values are valid not only for the Sahara and Namib but also for the Asian and North American deserts. So it appears that all the deserts in the world, the cold ones as well as the warm ones, develop the same type of landforms, dependent largely on the type and the quantity of rainfall.

The development of these different types of desert landforms effects all the structurally classified relief types (e.g. inselbergs, peneplains, table lands, cuestas etc). In general, this development transforms pre-existing generations of (Tertiary and older) planation surfaces, which may have been more or less dissected. This long term (1-10 Ma) development is superimposed on changes resulting from climatic fluctuations during the Quaternary period (on a time scale of 1-10 Ka).

As a result, changes occur not only between the three true desert types of landforms, but also between the climatic-induced desert landform types of moister climates (e.g. pediments, torrenten regions, inselbergs etc.). Smaller scale climatic

Geomorphological Studies in Southern Africa, G.F.Dardis & B.P.Moon (eds)
© *1988 Balkema, Rotterdam. ISBN 90 6191 831 6*

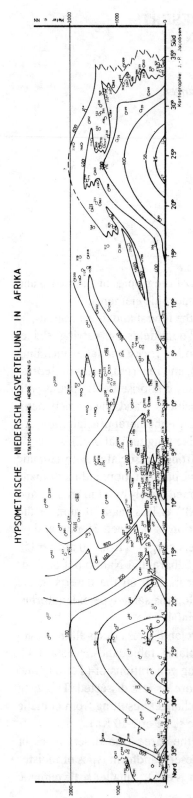

Fig. 1. Mean annual precipitation in Africa, according to latitude and elevation above sea level.

72

changes, on the time scale of 1 Ka, do not transform the whole relief of a region from one type into another one, but may result in local changes.

2. SOME GEOMORPHOLOGICAL COMPARISONS BETWEEN THE SAHARA AND NAMIB DESERTS

If we compare the Sahara and Namib deserts, significant geomorphological differences become evident.

The actual region of eolian activity in the Sahara occurs near the Tropic of Cancer, at elevations of more than 1000 m above sea level (a.s.l.). The desert plains rise up to 1400 m, while the region of desert gorges dominates the relief from 1400 m up to 2500 m in the Tibesti mountains.

In the Namib the actual wind-sculpted relief is restricted to elevations up to 800 m, the desert plains never exceed more than 1100 m, and the desert gorges occupy only a very small belt in the escarpment and in the Fish river region up to 1500 m.

These different regions in the Sahara and Namib have similar rainfall regimes, with the dunefield areas associated with the driest areas , and the desert gorge regions occupying the wettest areas (indicated above). The Sahara can be considered as a desert which rises to high altitude, while the Namib is restricted to lower elevations. The increase in MAP values with elevation above sea level is much quicker in southern Africa than in North Africa. This may be shown by two profiles showing the rainfall in northern and southern Africa according to latitude and elevation above sea level (Fig. 1).

This relationship cannot be explained by the configuration of the continent (large in the north, narrow in the south) because, in the Kalahari, the same types of climatically-induced morphogenetic landscapes, formed in that past, are now undergoing change under moister conditions. All of the peripheral areas of the Namib desert demonstrate that, in former times, the aerodynamically-formed relief (rock-sculpted forms as well as sand dunes) and the desert plains, rose up much higher than today. In the Fish river area, aerodynamic relief forms reached elevations of more than 1200 m a.s.l., while desert plains reached elevations of 1500 m a.s.l. The same relief types in the western parts of the Kalahari rise up to nearly 1300 m (dune fields) and 1500 m (desert plains).

The Sahara, in contrast, shows no indications for former drier climatic conditions than today. Aerodynamical forms and dunes never existed in elevations above 1100 m, the desert plains were never higher than 1400 m. On the other hand in the past desert gorges have been formed in the Sahara south of the tropic down to 600 m (which are today undergoing change to desert plains), whilst north of the tropic large pediments developed in that region, which is today dominated by the development of desert plains and aerodynamical landforms.

3. THE DEVELOPMENT OF CLIMATICALLY-INDUCED DESERT FEATURES

So it seems that the macroscale development of climatic-induced types of landforms is different in the arid landscapes around the tropics in the northern and southern hemisphere; today the maximum vertical extension of desert landforms in the north corresponds to a vertical shrinkage of the deserts in the south. So it seems highly likely that the former shrinkage of deserts in the north should correspond to the former enlargement of the deserts in the south.

Radiocarbon dates are useful in examining this hypothesis. The sediments dated in this way (indicating different sedimentary and erosional processes) provide evidence of climatic change, so that attempts could be made to qualify the climatic changes during the last Quaternary periods. The range of radiocarbon datas comprises a little more than the last 40000 years. Unfortunately, radically different opinions do exist concerning the patterns of climatic change, especially in southern Africa (cf. Heine, 1978, 1979,1982,1985; Cooke, 1975,1979; Grey and Cooke, 1977) to the extent that the same period in the same region has been considered either 'dry' (Heine, 1978) or 'humid' (Cooke, 1984).

To unify such different opinions and to compare systematically the climatic evolution in southern and North Africa all available radiocarbon datings for the Central and Eastern Sahara, the Kalahari, and for the Namib were amalgamated in three diagrams. Each of them concerns one actual climatically-definable region; (1) the Sahara with precipitation between 0 and 20 mm (high mountains above 2000 m up to 100 mm), (2) the Namib with precipitation between 0 and 50 mm (mountains 150 mm), and (3) the Kalahari with precipitation between 300 and 500 mm.

For the Sahara data from Pachur (1987a,b) was used, with some additions from Jakel and Schulz (1983), Molle (1971), Grunert (1983) and Skowronek (1979a,b); for the Namib the data from Heine (1983, 1984,1985), Heine and Geyh (1984),Vogel (1982), Besler (1980) and Blumel (1981) was used; and for the Kalahari the data from Cooke (1975,1979,1980,1984), Rust (1984), Rust and Schmidt (1981), Rust et al. (1984) and Heine (1978a,b,1979, 1980a,b, 1981,1982) was used, in additional to some dates from my own samples dated by Geyh. These data were quantified with respect to precipitation (see Table 1).

3.1. Precipitation values and landforms

The choice of these precipitation values needs some explanation, because they are partly new ones, partly not in good relation to values published by other authors.

The values for the aerodynamical landforms, the desert plains and the desert gor-

Table 1. Critical precipitation values for different relief forms, soils and sediments in arid landscapes.

RELIEF TYPES/ SOILS AND SEDIMENTS	PRECIPITATION VALUE (mm)
(a) Relief types	
Aerodynamic relief	< 30
Desert Plains	30-60
Desert Gorges	60-150
Pediments	150-300
Valleys with torrenten-floor	350
Inselberg landscapes	350
(b) Soils and sediments	
Salt layers	
Gypsum layers	30-100
Calcretes	100-300
Freshwater sediments (from periodic lakes	
some metres deep)	150-250
Freshwater sediments(from shallow perennial	
lakes tens of metres deep)	600-800
Cave sinter	500-700
Surface sinter	600
Freshwater sediments (from deep perennial	
lakes, hundreds of metres deep)	800-1000
Reddish soils and surface karst phenomena	350
Brown soils with calcification	300-400

ges have been derived from the work of the desert station at Barday (cf. Höverman, 1964/5). These values were tested by field work in the Sahara, in the near east, in the southern part of the USA and in the Central Asian deserts. For the active pediment forms values were derived from the Iranian highlands, from the Syrian desert and from the North American great basin. They have been confirmed by field work in the Gobi areas.

The values for the development of valleys with torrent-floors are derived from the observations in Libya, in the border regions of the Syrian desert and from the

South and North Iranian mountains. They are controlled by studies in the mountains west of Peking, in the Qilian Shan and in the Hoangho-region between Lanzhou and Hsining.

The values for the development of inselberg landscape by erosion and accumulation are derived from research work in North Ethiopia and Eritrea and from observations in the middle part of the Tchad Republic.

The conditions for the development of calcretes are very clear and defined by Abdul-Salam (1966). These findings are supported by studies in the Syrian desert itself, in the border regions of the Sahara and in South Africa. In all these regions calcretes (pedogenetic or evaporitic) are destroyed whenever the amount of precipitation exceeds 350 mm; they become destroyed also where the precipitation is less than 100 mm. So the belief, that they need a mean annual precipitation of up to 600 mm is erroneous. In southern Africa, it is easy to observe karst phenomena in calcretes and the soil formation upon the calcretes from the Etosha Pan down to Auob and Nossob.

Freshwater lakes occur in the border regions of the Taclamacan in closed depressions nourished by groundwater under very dry conditions (with mean annual precipitation less than 50 mm). The groundwater there is nourished by meltwater coming from glaciers at 4800 m a.s.l. and from snowfields occurring down to 4000 m a.s.l. in winter. Seasonal freshwater lakes are very common in the dune fields of the Gobi, nourished by precipitation of about 150 mm. Thus the value of 150 mm may be used for periodic freshwater lakes.

The value for shallow perennial freshwater lakes (tens of meters) can be derived from Etosha, from Makgadigadi (where annual precipitation of 500 mm is not sufficient) and Lake Tchad which is actually at the border of its existence. The mean annual rainfall near the lake is between 600 and 700 mm; so 700-800 mm is sufficient for perennial shallow lakes in these latitudes and surely, nearer the poles too.

The existence of sinter barred lakes is clearly controlled by precipitation of more than 600 mm/yr. This was found in Yugoslavia and in the sinter barred lakes in Afghanistan.

The samples classified in this way practically never gave contradictory results; they fit well and show a relatively detailed picture of the climatic changes and fluctuations back to 4000 years.

Some additional remarks are needed in respect to unpublished data used to construct the three profiles:

(1) Kalahari, Etos ha Pan. Well-developed raised shore-lines border the region of the pan. It is not clear if there are three or four shorelines, but the lowermost is characterised by saltbush vegetation. In the pan itself are visible at least two strandwall systems, each of them composed by a number of different walls at the same level. The outermost which seems to go round the whole southwestern part of the pan, starts west of Oshingambo and seems to be dated by Rust (1983) at Logans Island to 37900 years B.P. The inner one forms a free strandwall south of Oshingambo. Other shorelines are visible in the interior of the pan, which may correspond to the younger lake sdiments.

(2) Kalahari, Auob-Nossob. The sample Hv 9127, which I took about 20 km south of Gobabis, postdates the accumulation of calcrete-cemented higher terrace, which in the lower parts of the Nossob has accumulated into a field of red dunes, which also became calcified, giving it a brown-red colour. The sample dates a series of thin alternating layers of aeolian sand and calcrete, which were deposited in a wind blown closed depression. So between the calcified terrace, which contains rolled pieces of older calcrete, and the dated sediment (Hv 9127, 22240 +- 370 years B.P.) there was a period of strong wind erosion and deflation.

About 1 m of calcrete overlies the dated sample, which became later weathered and karstified (red soil, development of karst-towers about 2 m high). The sample Hv 9126, 30 km south of Gobabis, represents the same type of sediment also in a closed wind eroded depression. It postdates the calcrete, the karst and soil development, and predates the aeolian erosion and deflation after the soil formation. So there have been two important phases of strong wind erosion, the one before 22240 years BP, the other one, according to the dated sample Hv 9126, later than 22240 but before 15700 +- 155 years B.P.

(3) Fish river region. A sample from marl behind a bar of sinter gave the indication for a very moist period in the mountains at Voigtgrund (farm between Mariental and Maltahohe, ca. 1 200 m) A date of 15500 +- 105 years B.P. (Hv 9125) was obtained from the marl. The sinter bar and the marl developed in a strong wind blown region, so that the sample postdates a very high uprising strong wind erosion and deflation; it postdates the incision of desert gorges. Fifty km west of Maltahohe at 1500 m a.s.l., a marl has been dated to 36520 +- 1070 years B.P. (Hv 9128). The marl contains in the dated uppermost parts thin (mm) layers of red aeolian sand and is covered by a calcrete, which infiltrated also the basaltic bedrock. The sediment indicates the end of a lake, which came first into a region of desert plains and later into a region with calcrete development. So it indicates a moist climate before 36500 years B.P., followed by a very dry climate, and later still, by a moister climate.

4. RESULTS

The interpretation of the diagrams (see Fig. 2) is difficult, because the moister periods are over-represented; primarily because it is impossible to get datable samples from dunes or from erosion forms; secondly because the scale does not allow to distinguish clearly between 30, 60 or 100 mm precipitation values, so that all the values lower than 100 mm seem to be the same.

The lack of datings between 16000 and 25000 years B.P., particularly in the Sahara, seems to be caused by the fact, that it was not possible until recently to get datable samples from a pediment region. For the Namib the driest period is well represented by the Homeb silts, which indicate the same dryness at 500 m a.s.l. be-

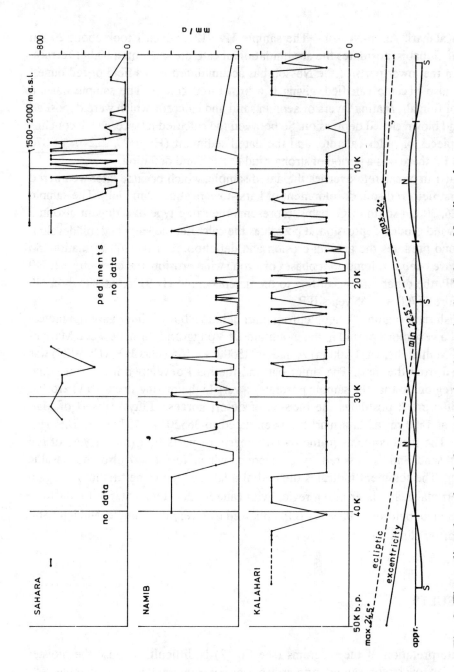

Fig. 2. The radiocarbon-dated dry and moist phases in the Sahara, Namib and Kalahari. (Cartography: Dipl.-Geogr. J.P. Jacobsen).

tween 23000 and 19000 years B.P. as exists today at sea level. But in general the Namib diagram gives no clear indication for alternating climatic conditions. The first period indicated by the impact sinter from Rossing cave was finished at 37000 years B.P..; the second indicated by the change of sinter, calcrete and aeolian sands, lasted from 37 000 to 26000 years B.P.; the third, containing the Homeb silts, never reached moister climatic conditions than pediments and calcretes. It lasted from 26000 to 16000 years B.P.; then a short interruption is indicated by the tufa from Voigtsgrund between 16000 and 15000 B.P., followed by the transition to the actual conditions. A common pattern is the lengths of the periods of about 11000 years.

More interesting is the comparison of the diagrams from the Kalahari and the Sahara, both representing large extended former deserts. The diagram from the Kalahari shows five moister and four drier phases. The drier ones are grouped in two longer periods (7000 to 8000 years) and two shorter periods (3000 years). The two shorter periods seem to be only interruptions of one very long moist period, which lasted from 29000 to 11000 years B.P. As in the Namib the first moister period of the Kalahari was finished at 37000 years B.P.; the last moist period, the actual one, started at 4000 years B.P.

From the five moist periods the three more important ones are clearly the inverse to the Saharian conditions: Before 37000 years B.P. the Sahara was dry, while the Kalahari was moist; at 22000 years B.P. the Sahara was dry, while the Kalahari was moist; since 4000 b.p. the Sahara has been dry, while the Kalahari was moist. Inversion also occurred at 30000 years B.P. (Sahara moist, Kalahari dry) and from 11000 to 7000 years B.P. (Sahara moist, Kalahari dry).

The seemingly conformable periods of dryness or wetness in the Kalahari and the Sahara are due to the fact that, the very long period of moisture in the Kalahari with its included breaks of dryness overlaps the dry period of the Sahara.

In this pattern of moisture and dryness is so regular that it cannot be purely coincidental. Considering that the climatic conditions in the tropics are depending in the first place on the position of the perihelion and the eccentricity of the orbit, a profile is presented (Fig. 3), based on the data given by Meinardus (1944), which indicates the change of the solar radiation at latitude 25 degrees north and south by the deviation from the values of today and expressed in canonical units according to Milankovitsch for the canonical summer half year.

The profile indicates that the moist periods in the Sahara correspond well with the periods of maximal solar radiation in summer, whilst the drier periods correspond with minimal solar radiation in summer. The peaks of moisture and of radiation are at 11000 and 30000 years B.P., and the minima of radiation and moisture at 22000 and 6000 years B.P. The picture for the Kalahari is more complicated, because the moist periods are longer and subdivided. But in general the moist period before 37000 years B.P. corresponds with the declining part of the radiation maximum at 47000 years B.P., the main moist period from 28000 to 11000 years B.P. comprises the first half of the ascent and the whole descent of the radiation peak at 22000 years B.P. , and the actual moist period begins also half way up the

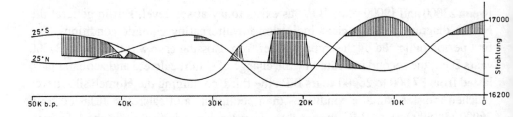

Fig. 3. The moist phases in the Kalahari and Sahara are related to the canonical summer half year radiation at latitude 25 degrees north and south during the last 50000 years. Vertical scale: canonical units according to Meinardus. (Cartography: Dipl-Geogr.J.P. Jacobsen)

ascent of the last radiation peak.

The difference between the Sahara and the Kalahari, where the Sahara occupies only the time near the peak of the radiation, while the Kalahari occupies half of the ascent and the whole of the radiation peak, may probably be explained by the different configuration of the continent, where the reaction in respect to climatic changes is generally quicker in regions far from the oceans and generally slower in more maritime areas.

5. SUMMARY

The African deserts are conformable in respect to the climatically-induced morphogenetic landscape types. This is the result of their position on the globe. But they are in inverse stages of morphogenetic development in response to the alternating climatic conditions. This alternation between pluvial and dry periods, which is generally opposite in both hemispheres, is the result of the differences in radiation caused by the changing position of the perihelion combined with the changing excentricity of the orbit.

6. REFERENCES

Abdul-Salam, A. 1966. Morphologische Studien in der Syrischen Wüste und dem Antilibanon. Berl. Geogr. Abh. 3,55S.
Adams, L.J. and Tetzalaff, G. 1985. The extension of Lake Chad at about 18 000 yr B.P. Zeitschrift Gletscherkunde und Glazialgeologie, 21, 115-123.

Agwu, C.O.C. and Beug, H.J. 1984. Palynologische Untersuchungen an marinen Sedimenten vor der westafrikanischen Küste. Palaeoecology of Africa, 16, 64; 37-52.

Besler, H. 1980. Die Dünen-Namib: Entstunhung und Dynamik eines Ergs. Stuttgarter Geogr. Studien, 96, 241S.

Blumel, W.D. 1981. Pedologische und geomorphologische Aspekte der Karlsruher Geogr. Hefte, 10, 228S.

Butzer, K.W. 1958. Quaternary stratigraphy and climate in the Near East. Bonner Geogr. Abh. 24, 158S.

Cooke, H.J. 1975. The palaeoclimatic significance of caves and adjacent landforms in Western Ngamiland, Botswana. Geographical Journal, 141, 430-444.

Cooke, H.J. 1979. K.Heine: Radiocarbon Chronology of Late Quaternary Lakes in the Kalahari, Southern Africa. A Discussion. Catena 6, 107.

Cooke, H.J. 1980. Landform evolution in the context of climatic change and neo-tectonism in the middle Kalahari of northern-central Botswana. Transactions of the Institute of British Geographers, new series 5, 80-99.

Cooke, H.J. 1984. The evidence from northern Botswana of Late Quaternary climatic change. In: Vogel, J.C. (ed.), Late Cainozoic Palaeoclimates of the Southern Hemisphere, Balkema, Rotterdam, 265-278.

Cooke, H.J. and Verstappen, H.T. 1984. The landforms of the western Makgadikgadi basin in the northern Botswana with a consideration of the chronology of the evolution of Lake Palaeo-Makgadikgadi. Zeitschrift für Geomorphologie, N.F. 28, 1-19.

Ebert, J.J. and Hitchcock, R.K. 1978. Ancient Lake Makgadikgadi, Botswana: Mapping, Measurement and palaeoclimatic significance. Palaeoecology of Africa, 10, 47-56.

Grey, D.R. and Cooke, H.J. 1977. Some problems in the Quaternary evolution of the landforms of Northern Botswana. Catena, 4, 123-135.

Grove, A.T. 1969. Landforms and climatic change in the Kalahari and Namaqualand. Geographical Journal, 135, 191-212.

Grunert, J. 1983. Geomorphologie der Schichtstufen am Westrande des Mursuk-Beckens. Relief, Boden, Paleoklima 2.

Hagedorn, H. 1971. Untersuchungen über Relieftypen arider Raume am Beispiel des Tibesti-Gebirges und seiner Umgebung. Zeitschrift für Geomorphologie N.F., Supplement Band, 11.

Heine, K. 1978a. Radiocarbon chronology of the Late Quaternary lakes in the Kalahari, South Africa. Catena, 5, 145-149.

Heine, K. 1978b. Jungquartäre Pluviale und Interglaziale in der Kalahari (südliches Afrika). Palaeoecology of Africa, 10, 31-40.

Heine, K. 1979. Reply to Cookes discussion of K. Heine: Radiocarbon Chronology of Late Quaternary Lakes in the Kalahari, Southern Africa. Catena, 6, 259-266.

Heine, K. 1980a. Studien zür jungpleistozänen Klima- und Landschafts-entwicklung in der Kalahari. Tagungsber. u. wiss. Abh. 42. Dt. Geographentag Gött. 1979, 281-283.

Heine, K. 1980b. Wann hat es der Kalahari geregnet? Umschau, 80, 8, 250-251.

Heine, K. 1981. Aride und pluviale Bedingungen während der letzen Kalt-zeit in der Süswest-Kalahari (südl. Afrika). Zeitschrift für Geomorphologie N.F. Supplement Band 38, 1-37.

Heine, K. 1982. The main stages of the Late Quaternary evolution of the Kalahari region, South Africa. Palaeoecology of Africa 15, 53-76.

Heine, K. 1983. Preliminary reconstructions of the Late Quaternary climatic History of the central Namib Desert, based on new 14 C -dates. Palaeolimnology of Lake Biwa, 11, 41-54.

Heine, K. 1984. Jungquartäre Klimascgwankungen auf der südhalbkugel. Zbl. Geol. Paläont. Teil 1, 11, 1751-1768.

Heine, K. 1985. Late Quaternary development of the Kuiseb river velley and adjacent areas, central Namib desert, South West Africa/Namibia, and palaeoclimatic implications. Zeitschrift Gletscherkunde und Glazialgeologie. 21, 151-157.

Heine, K. and Geyh, M.A. 1984. Radiocarbon dating of Speleotherms from the Rossing Cave (Namib Desert) and palaeoclimatic implications. In: Vogel, J.C. (ed.), Late Cainozoic Palaeoclimates of the Southern Hemisphere, Balkema, Rotterdam.

Hövermann, J. 1964/65. Die wiss. Arbeiten der Station Bardai im ersten Arbeitsjahr (1964/65). Berliner Geogr. Abh. 5, 7-10.

Hövermann, J. 1967. Hangformen und Hangentwicklung zwischen Syrte und Tschad. Les congré et coll. de l'universite de Liège 40, 140-156.

Hövermann, J. 1972. Die periglaziale Region des Tibetsi und ihr Verhaltnis zu angrenzenden Formungsregionen. Göttinger Geogr. Abh. 60 (Hans-Poser-Festschrift), 261-283.

Hövermann, J. 1978. Formen und Formung in der Paränamib (Flächen-Namib). Zeitschrift für Geomorphologie N.F. Supplement Band, 30, 65-68.

Hövermann, J. 1985. Das System der klimatischen Morphologie auf landschafts-kundlicher Grundlage. Zeitschrift für Geomorphologie N.F. Supplement Band, 56, 143-153.

Hövermann, J. and Hagedorn H. 1983. Klimatisch-geomorphologische Landschaftstypen. 44. dt. Geographentag Münster, Tagungs-bericht u. wiss. Abh., 460-466.

Jakel D. and Schulz, E. 1972. Spezielle Untersuchungen an der Mittel-terrasse im Enneri Tabi, Tibetsi-Gebirge. Zeitschrift für Geomorphologie N.F. Supplement Band, 15, 129-143.

Lancaster, N. 1979. Evidence for a widespread Late Pleistocene humid period in the Kalahari. Nature, 279, 145-146.

Lancaster, N. 1982. Spatial variations in linear dune morphology and sediments in the Namib Sand Sea. Palaeoecology of Africa, 15, 173-182.

Lancaster, N. 1983. Linear dunes of the Namib Sand Sea. Zeitschrift fur Geomorphologie N.F. Supplement Band, 45, 27-49.

Meinardus, W. 1944. Zum Kanon der Erdbestrahlung. Geol. Rund-schau 34, 7/8 (Klimaheft Geologie und Klima), 748-762.

Molle, H.G. 1971. Giederung und Aufbau fluvialer Terrassenakkumulationen im Gebiet des Ennerl Zourmri (Tibesti-Gebirge). Berliner Geogr. Abh. 13, 59S.

Pachur, H.J. 1987a. Gerinnenetze, Seen und Ergs der östlichen Sahara als Indikatoren quartärer Formungsdynamik. Verh. Dt. Geographentages 45, 167-173.

Pachur, H.J. 1978b. Vergessene Flüsse und Seen der Ostsahara. Geowissenschaften in unserer Zeit 5, 2, 55-64.

Pachur, H.J. and Roper, H.P. 1984a. Die Bedeutung palaoklimatscher Befunde aus den Flachbereichen der östlichen Sahara und des nördlichen Sudan. Zeitschrift für Geomorphologie N.F. Supplement Band, 50, 59-78.

Pachur, H.J. and Roper, H.P. 1984b. The Libyan (Western) Desert and northern Sudan during the Late Pleistocene and Holocene. Berl. geowise. Abh. (A) 50, 249-284.

Rust, U. 1984. Geomorphic evidence of Quaternary environmental changes in Etosha, S.W.A. /Namibia. In: Vogel, J.C. (ed), Late Cainozoic Palaeoclimates of the Southern Hemisphere, Balkema, Rotterdam, 279-286.

Rust, U. and Schmidt, H.H. 1981. Der Fragenkreis jungquartärer Klimaschwankungen im südwestafrikanischen Sektor des heute ariden südlichen Afrikas. Mitt. Geogr. Gesell. Münschen 66, 141-174.

Rust, U., Schmidt, H.H. and Dietz, K.R. 1984. Palaeoenvironments of the present day arid south western Afica 30000 - 5000 B.P. Results and problems. Paleoecology of Africa, 16, 109-148.

Sandlowsky, B.H. 1977. Mirabib - an archaeological study in the Namib. Madoqua, 10, 221-283.

Shaw, P. 1986. The palaeohydrology of the Okavango delta. Some preliminary results. Palaeoecology of Africa 17, 51-58.

Skowronek, A. 1979. Paläoböden und Vorzeitklimate in der zentralen Sahara. Mitt. dt. bodenk. Ges. 29, 821-826.

Skowronek, A. 1979b. Paläoböden in der zentralen Sahara. Würburger Geogr. Arb. 49, 163-182.

Van Zinderen Bakker, E.M. 1975. The origin and palaeoenvironments of the Namib Desert biome. Journal of Biogeography, 2, 65-73.

Van Zinderen Bakker, E.M. 1976. The evolution of Late Quaternary Palaeoclimates of southern Africa. Palaeoecology of Africa, 9, 160-202.

LINEAR DUNES IN THE CENTRAL NAMIB DESERT: THEORETICAL AND CHRONOLOGICAL PERSPECTIVES FROM WIND STREAKS

M. JUSTIN WILKINSON

Department of Geography and Environmental Studies,
University of the Witwatersrand

1. INTRODUCTION

The term linear and longitudinal dunes are used interchangeably although the former is descriptive and the latter genetic, suggesting that the formative wind is known. Greeley and Iversen (1985, p. 164) define longitudinal dunes as "...those..oriented parallel to the prevailing wind direction or to the vector of multiple wind directions...they are symmetrical in cross-section and have two slip faces which commonly meet at a sharp crest...". Crests often undulate both horizontally and vertically to form peaks and saddles (Tsoar, 1978b). Linear dunes in the Namib sand sea are of simple, compound and complex types (Lancaster, 1983a, 1983b). The simple type is restricted to a mere 2 % of the area of the sand sea in the south and east. Compound types, 25-50 m high, are commonest and comprise 2-5 crests surmounting a low-angle plinth, the crests either roughly parallel, or anastomosing or reticular (Lancaster, 1983b, p.29). Complex types reach heights of 80-170 m, with crests that are sinuous and stellate at intervals. Smaller barchanoid types occur on their flanks (Lancaster, 1983b).

The origin and growth of linear dunes has occupied geomorphologists for decades (see reviews in Cooke and Warren, 1973; Mabbutt, 1977; Tsoar, 1978; Lancaster, 1982a; Greeley and Iversen, 1985). Greeley and Iversen (1985) distinguished two sets of theories concerning the development of linear dunes, one which invokes unidirectional winds and another which invokes multiple, especially bi-directional, winds. The complexity of the issues is evident when it is considered that controls such as resultant wind directions may be important, that wind regimes are characterised by stronger and weaker winds, that dunes themselves set up secondary flows of arguably crucial importance, that subtle topographic effects seem to operate in some areas but not in others (Cooke and Warren, 1973), and that orientations of palaeo-wind fields persist in many deserts.

Implications concerning dune-forming winds from the evidence of aeolian features of the duneless flats surrounding the sand sea are explored. Aeolian phenomena are consistent with modern morphogenesis in the duneless areas of the Namib and these are usually mentioned in general descriptions of the desert.

Geomorphological Studies in Southern Africa, G.F.Dardis & B.P.Moon (eds)
© *1988 Balkema, Rotterdam. ISBN 90 6191 831 6*

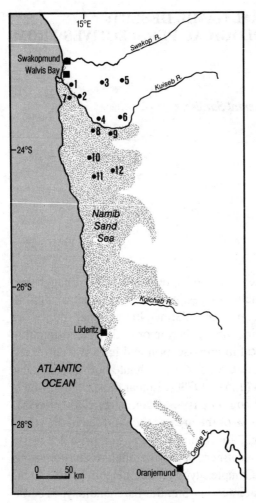

Fig. 1. Namib Desert sand sea. Meteorological stations (1-12) from Lancaster (1985).

Specific studies are, however, few (e.g. Kaiser, 1923, 1926a, 1926b; Selby, 1977; Sweeting and Lancaster, 1982; Lancaster, 1984; Wilkinson, 1987). The more pertinent aeolian features of the duneless flats are reviewed here. Wind streak features have received least attention, but (1) indicate with little doubt which winds have been formative of certain linear dunes, (2) contribute to theorization concerning the problem of linear dune formation, and (3) suggest chronological perspectives on dune-formation in the sand sea.

2. SETTING: WIND AND SAND ROSES

The Namib desert stretches along the southwestern littoral of Africa from 10° to

32° S (Meigs, 1966). The hyper-arid core of the desert is centred along the coastline. Most of the surfaces in this narrow desert are not sand-covered, being characterised instead by plains where rocky outcrops, alluvial spreads and duricrusted surfaces dominate.

The largest sand-covered area is that of the Namib sand sea which stretches 340 km between Luederitz and Walvis Bay, extending 100-120 km inland to cover an area of 34000 km^2 (Barnard, 1973) (Fig. 1). Three broad groups of dune patterns have been recognized in the northern half of the sand sea (Barnard, 1973; Besler, 1980, 1984; Lancaster, 1983a, 1983b); (1) The major central tract, of specific interest in this study, is dominated by large, N-S-aligned linear dunes (termed *N-S dunes* hereafter), averaging 50-150 m in height, and spaced 1.2 to 2.8 km apart (Lancaster, 1983a, 1983b), are of specific interest to this study; 2) a 5-30 km-wide coastal tract of transverse dune types, flanking group 1; and (3) a discontinuous eastern tract dominated by large star dunes. Several dune types have been identified within these groups (Besler, 1980, 1984; McKee, 1982; Lancaster, 1983a, 1983b), many with regular patterns and orientations (Besler, 1980, 1984; Lancaster, 1982a, 1983a, 1983b).

Lancaster (1983a, 1983b, 1985) has drawn attention to the spatial variability in surface wind energy and direction in the central and southern areas of the Namib desert because of their undoubted geomorphic importance. The following summary of Namib winds is taken from various reviews (Royal Navy and South African Air Force, 1944; Lancaster et al., 1984; Tyson and Seely, 1980). Dominant winds are recorded from three main sectors, SSW-W, NW-NNE, and NE-E, and comprise general and thermo-topographic components. SW-quadrant winds are dominant in early summer (30-40 % of winds in Sept.-Nov.) when the offshore anticyclone is better developed and as sea-breeze activity increases with rise in thermal gradients. The SW winds often reach gale force at coastal stations in the southern Namib, but gales are rare further north, winds usually attaining maxima of 50 km/hr. at Walvis Bay. The velocity of these winds also falls off rapidly inland and wind regime directionality is concomitly more varied. Breed et al. (1979) characterised wind regimes as high energy unimodal in the coastal tract, and intermediate- to low-energy bimodal in the central tract.

Weaker, northerly quadrant winds dominate from December to February (40-60 % in these months) as plain-mountain winds, often associated with the progress of coastal low pressure systems. Under conditions of poorly organised offshore high pressure patterns, northerly quadrant winds can cause sea surface temperatures to rise dramatically.

Winds from the east and northeast dominate year-round inland (40-60 %) and appear to be mountain and mountain-plain winds, but are best developed in winter. Winds from this sector become very strong for one to three days on an average of ten times per year (Wilkinson, Maclear and Bayman, in prep.) when they are experienced as warm to hot, strong, gusty "Berg Winds". These winds occur optimally when the regional pressure gradient is directed normal to the coastline. In a five-year period, Berg Winds generated the highest hourly wind speeds in the central

Namib (61 km/hr.). Interestingly, they are responsible for the highest absolute annual temperatures which thus occur in winter.

Analyses of wind frequency and direction, and recent computations by several workers of the sand-moving potential of these winds have permitted more sophisticated discussion of dune and dune pattern origin than was possible heretofore. Southerly and easterly quadrant winds are apparently the most effective winds in geomorphic terms, i.e. as sand-moving winds of 16 km/hr (4.4 m/sec.).

Computations of the sand-moving potential of winds from eight directions gives the following picture (Breed et al., 1979; Harmes, 1982; Lancaster, 1985; Ward and von Brunn, 1985).

The pattern of resultant sand movement flux accords broadly with the wind fields; sand-moving potentials generated by southerly winds in the southern Namib at Luederitz (80-90 % of sand flow) decline both northwards (55-65 % of sand flow in the northern sand sea) and inland (35-40 % of sand movement on the eastern margin of the sand sea) (Lancaster, 1985). As might be expected, easterly quadrant winds are responsible for less than 10 % of sand flux on the coast, but 30-55 % on the eastern edge of the Dune Namib (Lancaster, 1985). A major disjunction occurs at the north end of the sand sea, however, where, even close to the coast, 60-65 % of potential sand flux is generated by the Berg Wind. Resultants of sand movement are thus opposed on either side of the Kuiseb River (Fig. 9), a situation discussed below. N-NW winds account for small percentages of sand-moving winds only.

Sand flow data from the non-coastal areas reflect the fact that southerly quadrant winds attain intermediate energy status (in terms of world energy regimes set up by Breed et al. (1979). Berg Winds at full strength fall in the high energy bracket, equivalent to the very energetic coastal southerlies. Annual potential sand flow data for the non-coastal areas of the northern dune sea and neighbouring plains (Lancaster, 1985) show that flows are more than twice as high today under Berg Wind conditions as they are under SSW wind conditions, if the plains station RB (Rooibank) (total 278 tonnes/m/yr, resultant 129 tonnes/m/yr) is compared with stations in the northern and central dune field (totals 23-119 tonnes/m/yr, resultants 6-63 tonnes/m/yr) (see note 1).

3. RECENT THEORIES OF LINEAR DUNE FORMATION IN THE NAMIB SAND SEA

Two partly opposing theories of linear dune development have been propounded for the dominant dune patterns in the Namib sand sea. These may be termed the oblique propagation theory of Lancaster, and the helical roller theory of Besler.

Support for Bagnold's (1941) bi-directional theory has come from the world-wide survey of dunes and wind regimes by Fryberger (1979). Breed et al. (1979) and McKee (1982) saw particular relevance of these ideas to patterns of dune develop-

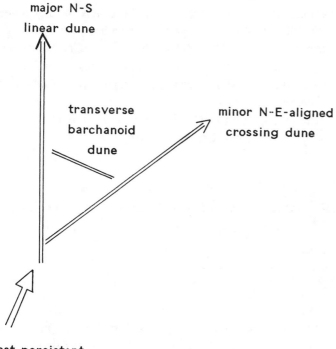

major N-S
linear dune

transverse
barchanoid
dune

minor N-E-aligned
crossing dune

most persistent
winds (SSW)

Fig. 2. Lancaster's (1983a, 1983b) model of relationship between most persistent wind (SSW) and three dune orientations in north-central Namib sand sea: oblique alignment of two sets of linear dunes (large N-S chains and small NE-SW crossing dunes) and transverse alignment of barchanoid dunes in the angle of the two (after Lancaster, 1983a).

ment in the Namib sand sea. Modifying this idea in which dune growth is seen to occur parallel to wind resultants, Lancaster has proposed a "general model" (Lancaster, 1982a, 1983a) of linear dune development, in which linear dunes are developed *obliquely* to the dominant wind. The theory of dune propagation at angles oblique ($< 40°$ degrees) to resultant winds, in a bimodal regime, has been derived from processual studies by Tsoar (1978, 1983) (see also reviews in McKee, 1982, and Lancaster, 1982a). These theories have been applied in the central tract of the Namib dune sea by Lancaster (1982a, 1983a, 1983b), who ascribed three sets of dunes to the action of the SSW wind, namely (1) the large, N-S linear dunes, and (2) minor linear dunes, aligned northeast-southwest (termed "corridor crossing dunes" by Lancaster, 1983a, 1983b, and *crossing dunes* hereafter) and (3) small, transverse dunes in the angle between the linear sets (Figs. 2, 10, 11, 12).

Besler's (1976, 1977, 1980, 1984) theory relies on Taylor-Goertler boundary layer modelling, by which spacing of the N-S dunes in the Namib demands higher-than-present wind velocities to generate horizontal helical vortices of appropriate diameter. Such vortices are regarded as formative for the chains of large linear dunes. Besler (1976, 1977, 1980, 1984) has proposed that higher wind speeds are best explained as part of more vigorous, Pleistocene atmospheric circulation. The existence of ventifacts in interdune areas and on the rocky plains is regarded as evidence of long-continued aridity (Besler, 1976) and comparative immobility of the large (draa-sized) N-S dunes.

These theorists thus differ in respect of mechanisms invoked and in terms of the currency of major dune forms in relation to present wind regimes.

4. AEOLIAN FEATURES OF THE DUNELESS FLATS AND NORTHERN SAND SEA

4.1 Northeast Winds and Wind Streaks of the Duneless Flats

Since aeolian features of the duneless flats of the central Namib desert are important to the present argument, they are briefly described, especially those in the central Namib where the writer is most familiar with them and where wind data is most complete. The plains display several aeolian features apparently forming under the influence of present-day winds. Except for a narrow coastal tract which is undoubtedly dominated by the activity of southerly winds, the northeast Berg Wind is the formative wind inland on the plains. Features such as small coppice dunes (nebkha) tied to individual scrubs, thin and thick sand veneers on windward mountain slopes, offshore dust plumes, deflation hollows, oriented scarplets, ventifacts, wind-facetted boulders and wind streaks all show the imprint of this wind. Discontinuous sand veneers blanket the windward eastern faces of Roessing and Khan Mountains (Goudie, 1972), the Chuos Mountains (Wilkinson, 1976, 1987). Large, rising dunes lie on the eastern slopes of may hills and ridges in the southern and central Namib (e.g. Saagberg) (Hueser, 1976). These dunes are now fixed by vegetation, duricrusted by caliche and fluvially incised. Sand-blast facetting, grooving and polishing of boulders is heavily dominated by northeast winds (Selby, 1977; Sweeting and Lancaster, 1982; Lancaster, 1984). Several deflation hollows of significant dimensions (up to tens of square kilometers in area) occur near Rooibank and show preferred northeast-southwest orientations. Duricrusted scarplets in the Tumas River valley, on the plains north of the sand sea, overwhelmingly face north and northeast as a result of undercutting by NE winds. Ground control confirms ephemeral dune development during Berg Wind events, the only time sand can be seen to move on the plains. Sizable coppice dunes and dust plumes are generated by the Berg Winds (Jaeger, 1965) (see note 2); the latter can extend several hundred kilometers out to sea in a southwesterly direction (Fig. 3). Analysis of available METEOSAT imagery and synoptic records shows that dust storms are re-

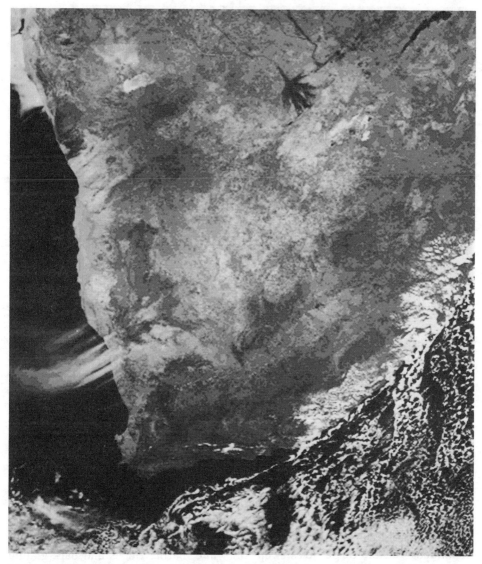

Fig. 3. Dust plumes off Namib Desert coast generated by strong NE "Berg Winds", June 1979 (METEOSAT VIS image, 13.6.1979, 13h30).

lated to Berg Wind events only (Wilkinson, Maclear and Bayman, in prep.).

Of particular significance to the present argument, however, are wind streaks, features defined by Greeley and Iversen (1985: 209) as "...patterns of contrasting albedo [which form] as a result of various aeolian processes...". The streaks appear on aerial photographs as straight, narrow, markedly parallel, straight features. Wind streaks are usually associated with topographic obstacles to wind flow. They are strongly elongated in the direction of the formative wind, and may result from grain size differences, grain mineral (colour) composition, or aeolian bedforms (Greeley

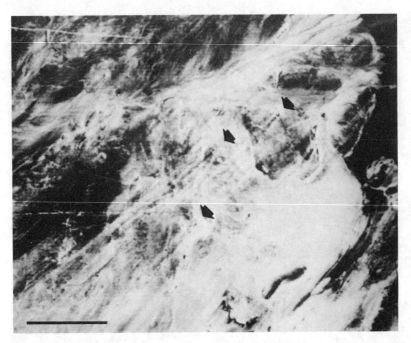

Fig. 4. Light and dark wind streak bands (arrows) trailing southwest from marble ridges (white linear features), Tumas Flats, central Namib Desert. (a): enhanced LANDSAT image (scale bar ca. 10 km). (b): aerial photograph of LANDSAT image centre -- granite dome centre (middle arrow on space image) (scale bar ca. 1 km).

and Iversen, 1985) (see note 3). Analysis of streak orientation has provided basic data in the understanding of planetary atmospheric circulations, by which, for example, Thomas and Veverka (1979) have demonstrated the asymmetry of atmospheric hemispheres on Mars. Streaks of bedform type appear to be absent in noncoastal parts of the Namib (see note 4), where they comprise instead; (1) thin, noncontinuous veneers of particles swept off white marble ridges and dark granite domes, and (2) patterns of deflation of preferred grain sizes from rock surfaces. Fine grains produce optically brighter, and coarser grains darker, streaks (Greeley and Iversen, 1985). Wind shear stress is much increased downwind of topographic obstacles in a narrow zone in which horizontal vortices are generated. Stresses can locally reach as much as an order of magnitude greater than that exerted by the ambient wind (Greeley and Iversen, 1985).

In passing reference Fryberger (1979) ascribed streaks in the Swartbank and Vogelfederberg areas (stations 2 and 3, Fig. 1) to northeasterly winds, whereas Breed et al.(1979), ascribed the same set to SSW winds. Streak phenomena have since been documented along the length of the Namib Desert from Namaqualand to southern Angola on aerial photographs (Wilkinson and Barbafiera, in prep). Two groups of differing orientation can be identified; (1) those aligned N-S (mean

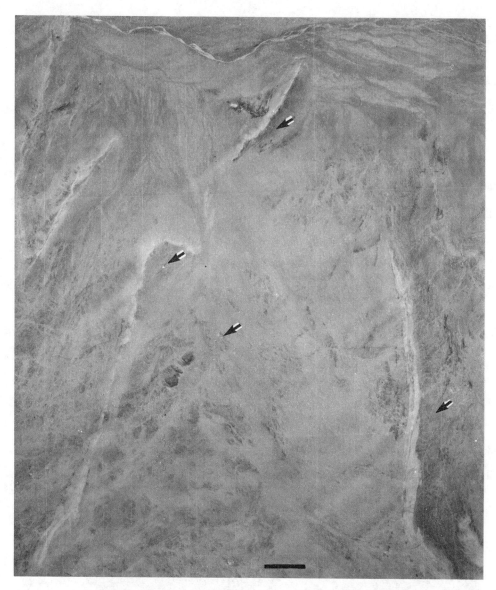

Fig. 4 b.

orientation 177°) occupy a narrow, discontinuous, coastal zone up to 10 km wide, and (2) a larger group aligned NE-SW (mean orientation 059°) in a zone up to 80 km wide inland of the first group. Formative winds in both cases are easily ascertained because points of streak origination can be identified at specific topographic obstacles, be these barchan dunes, dolerite or marble ridges, granite domes or even man-made features.

Streak orientations on the flats north of the dune sea provide a particularly instructive sample because of their number as visible on aerial photograph and space images (Figs. 4-8). Furthermore, the quantity of wind rose, and especially sand

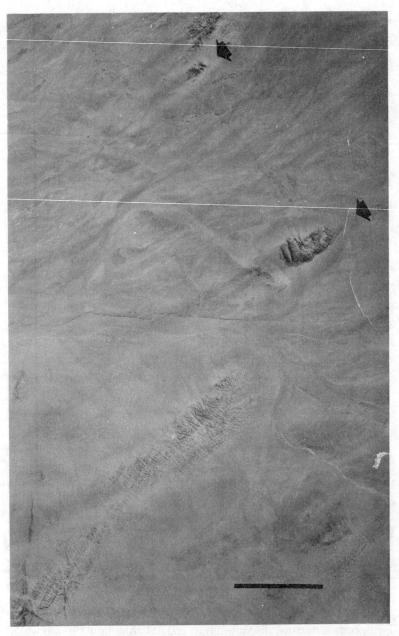

Fig. 5. Wind streaks (darker bands -- arrows) trailing southwest of granite dome (centre right) and other obstacles, Tumas Flats, central Namib Desert (scale bar ca. 1 km).

rose, data available locally from a good network of stations, allows the connection between aeolian feature and formative wind to be established. A mean wind streak alignment of 055° has been assessed from satellite imagery (n = 28; values range between 047°-063°). Mean alignment of a largely different set of streaks taken

94

Fig. 6. Wind streaks (mainly lighter bands), trailing southwest of marble ridge, Marmor Pforte, Welwitschia Flats, central Namib Desert. Note alignment independent of geological structure (scale bar ca. 1 km).

from aerial photographs was similar ($058°$, n = 19), values falling within a similarly narrow range (053-$066°$). Within the west-trending Tsondab River Valley,

Fig. 7. Broad NE-SW-aligned streaks (darker bands, best developed at top), 30-40 km north of Swakopmund (scale bar ca. 1 km).

topographic control and Ekman spiral effects exert a progressive east-west orientation on streaks with nearness to the valley centre.

Wind and sand rose resultants show the influence of Berg Winds in the central Namib. On a small sample Breed et al. (1979) showed that wind resultants were dominated by the NE wind. Data collected over five years confirmed a preponderance of strongest winds from the northeast and ENE sectors (between 034-079°) for all six plains stations (1 - 6, Fig. 1) (Lancaster et al., 1984), and although winds blow from other sectors, especially the southwest, only the strongest winds appear capable of generating an imprint on the plains of the duneless Namib. Resultant sand flow directions are accordingly towards the southwest, being dominated by northeasterly sand-moving winds (resultants from 025° and 051°;

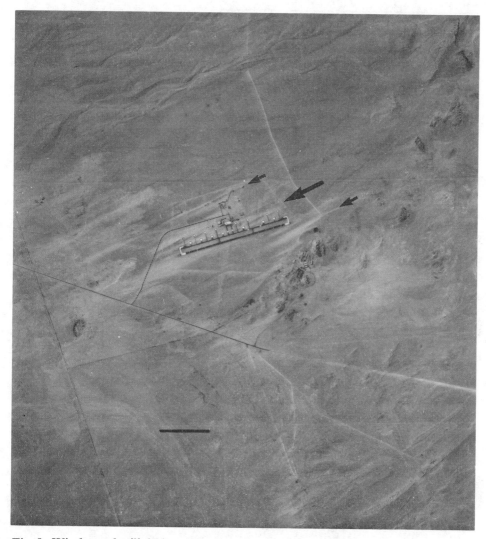

Fig. 8. Wind streaks (light bands -- arrows) to southwest of man-made installations, east of Walvis Bay, central Namib Desert (scale bar ca. 1 km).

Harmes, 1982; and 065° and 069°; Lancaster, 1985) (Fig. 9).

In an attempt to gauge how long the Berg Winds may have acted as formative winds in the central Namib, it seems relevant that ventifaction, which is strongly dominated on the plains by these winds, probably takes thousands of years to accomplish (Higgins, 1956). Besler (1976) and Selby (1977) have suggested that the Berg Wind has been facetting boulders for much of the Holocene. Streaks forming downwind (i.e. to the southwest) of man-made installations (Fig. 8) support the idea that these are features developed currently with modern wind regimes.

The correspondence of streak orientation on the duneless flats of the central Namib with the resultants of strong present-day Berg Wind flow suggests, with little

97

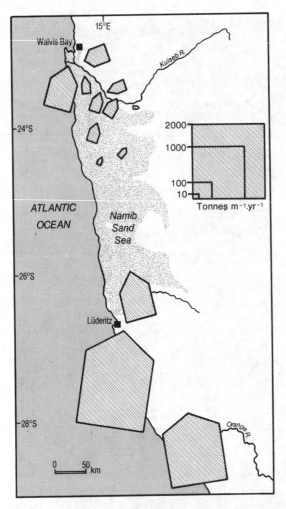

Fig. 9. Sand flow resultants in southern and central Namib (from Lancaster, 1985).

doubt, that wind streaks are features generated by the Berg Wind and that they are modern features.

4.2 Northeast Winds and Crossing Dunes of the Northern Sand Sea

Those features within the sand sea most securely related to the Berg Wind are the abovementioned linear crossing dunes. These dunes are generally small (6-8 m high, Lancaster, 1980) (Fig. 10) and usually do not extend beyond one interdune corridor. That is, they are markedly shorter features than the large linear dunes

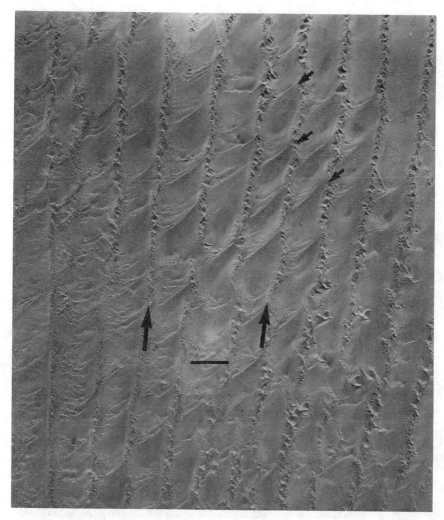

Fig. 10. Dune patterns in central tract of the Namib sand sea, dominated by large N-S linear dunes of complex type (larger arrows); smaller crossing dunes of simple type, aligned NE-SW (smaller arrows) cross interdune areas (scale bar ca. 2 km).

(Figs. 10, 11, 12). Locally, individual crossing dunes span three or four corridors. Besler (1984) has termed the resulting pattern a network complex.

The northeast-southwest orientation of the crossing dunes is notably consistent. Yet the precise correspondence of crossing dune alignments with those of the wind streaks, and hence of maximum Berg Wind flow, has not attracted attention, probably because of the compelling correspondence between very strong, well-documented southerly winds at the coast, and the striking N-S alignment of the dominant dunes in the dune sea.

That the streak alignments coincide with those of the crossing dunes is evident.

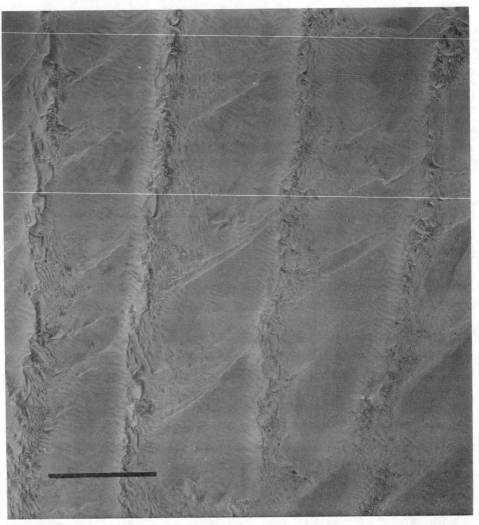

Fig. 11. Detail of Fig. 10.: dominant N-S linear dunes with transverse dunes aligned NW-SE on their east flanks; crossing dunes aligned NE-SW on interdune areas (scale bar ca. 2 km).

The mean azimuth of a population of 149 crossing dunes in the northern half of the dune sea is 058 degrees, ranging from 045° and 074°. Compared with local wind streak means of 055° and 058°, there can be little doubt that the Berg Wind is the formative wind for these features (compare Figs. 4-8 with 10-12). No statistical difference was found between streak and dune population orientations.

Since wind streaks usually display clear points of origin at topographic irregularities, southwesterly extension under the influence of Berg Winds is not in question for the streaks, and seems highly likely therefore, by inference, for the crossing dunes (see note 5). A primary difficulty in ascertaining the direction of

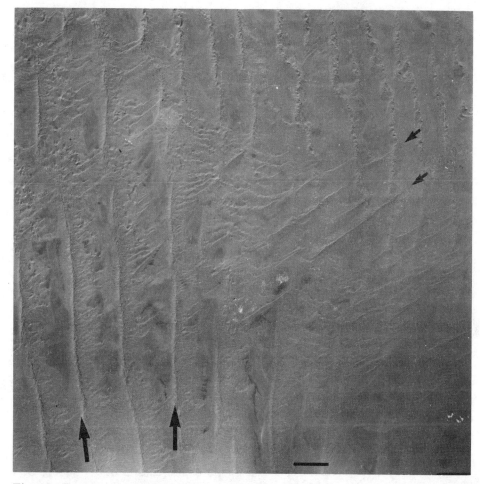

Fig. 12. Dune patterns in north central tract of Namib sand sea: N-S linear dunes (larger arrows), crossing dunes aligned NE-SW (smaller arrows), and transverse types aligned E-W to WNW-ESE (scale bar ca. 2 km).

crossing dune elongation derives from the fact that points of origin are usually impossible to determine. Some dunes peter out at their northeast extremities, some at their southwestern, and may connect at both ends with the large N-S dunes.

Several details of dune patterns in the sand sea are best ascribed to the activity of the northeast Berg Wind (Wilkinson and de Wet, in prep.); (1) Lancaster (1983a) noted that secondary arms of star dunes in the eastern tract of the dune sea are elongated northeast-southwest; (2) Dune alignments in the coastal dune cordon between Walvis Bay and Swakopmund are best interpreted as elements developed transverse to both strong SSW sea breezes and NE Berg Winds; (3) Disrupted dune patterns west of Sossus Vlei in particular, sometimes ascribed to the action of water downstream of end-point vleis (e.g. Wieneke and Rust, 1972), seem better explained as the effects of topographic channelling and velocity increase of both katabatic and Berg winds induced by valley morphology; (4) The related phenomenon of strong westward curvature of major linear dunes on the south banks of both the Kuiseb and Tsondab River valleys (the former near its mouth and inland near the Gaub drainage) seems related to the effects of more westerly resultants produced by the interaction of southerly and E-NE quadrant winds in areas most exposed to the latter winds.

5. DISCUSSION

5.1 Theoretical Considerations

The existence of sand-flow resultants of opposing direction on either side of the Kuiseb River led Lancaster (1985, p.618) to suggest that "...mesoscale roughness of the dunes acts to reduce the velocity of northeasterly winds...". But the situation seems "...anomalous..." (Lancaster, 1985 p.618) indeed, since the winds involved are not simply high localised, topographically-induced winds. They are part of deeper land-sea, mountain-plain and geostrophic systems (Jackson and Tyson, 1971; Tyson and Seely, 1980; Tyson, in prep.) (see note 6). As larger-scale flows, it is reasonable to suppose that both opposing sets of winds produce aeolian effects on both sides of the Kuiseb River. Lack of sand and dominance of the Berg Wind negate any geomorphic effects of the SSW wind on the flats north of the Kuiseb River.

The opposite is not the case south of the Kuiseb River, however, where a plentiful supply of moveable sand exists well inland of the coastal tract. Indeed, Berg Winds appear to generate a variety of effects across the entire width of the dune sea. These effects are probably both permanent and seasonal, except for the narrow coastal strip of transverse dunes where exceptionally strong southerly winds obliterate any Berg Wind features, even though Berg Winds are felt on the coast, as records at coastal locations show (Logan, 1960; Jackson and Tyson, 1971; Ward and von Brunn, 1985).

Lancaster (1982a) has persuasively applied his general model of linear dune development (by oblique propagation) to explain not one but three sets of dunes in the north-central Namib (Fig. 2). The data presented above suggest, however, that the linear element represented by crossing dunes has developed parallel to the strongest (Berg Wind) flows. Less vigorous Berg Wind flow, which is more easterly, and N-NNW winds, may provide the bimodality which seems necessary for the development of linear dunes. Indeed, may researchers have favoured the model of linear extension parallel to the resultant wind (Bagnold, 1953; Cooper, 1958; Wopfner and Twidale, 1967; McKee and Tibbitts, 1964). Greeley and Iversen (1985) suggested that the oblique propagation model may not be applicable in all deserts.

The fact that dominant sand-moving winds in the northern Dune Namib are SSW and northeasterly raises questions concerning the geometries of Lancaster's (1983a, 1983b) model as applied to the northern sand sea (Fig. 2). It seems unlikely that a bimodal regime, comprising these vector maxima, could generate linear dunes extending due north (or even only NNW) against what is the stronger of the two sand-moving winds in the northern parts of the sand sea. It has been noted that resultant sand flows are two to ten times higher than at stations within the dune sea (Lancaster, 1985).

These considerations suggest that; (1) the oblique propagation model explains evolution of the crossing dunes less well than the model of dune extension parallel to a formative wind; and (2) that computation of sand flow resultants (e.g. Fig. 9) may therefore, in some circumstances, obscure rather than illuminate the work of winds as geologic agents in deserts.

This analysis suggests that not only must the distinction be drawn between those winds which move sand and those which do not, but also, within the sand-moving group, winds capable of the construction of specific dune forms need to be isolated from those which are not. The Namib crossing dunes seem to provide a prime illustration of the distinction, since southerly winds which determine long-term sand flux resultants in much of the sand sea appear to be comparatively unimportant in the construction of crossing dune elements. Fryberger (1979) has indeed noted that actual sand movement may differ from computed (i.e. potential) sand flux.

Lancaster (1980) reported the existence of a field of small barchans on the Tsondab Flats with slip faces oriented northeast. Some barchans display strongly elongated southern horns, which Lancaster (1980) ascribed to the effect of bimodal winds from the SSW quadrant. The interpretation offered here, however, is that the elongated arms are crossing dunes since they are aligned precisely with other crossing dunes. It appears therefore, that SSW winds have subsequently caused recurving of the southwest end of crossing dunes, imparting a barchanoid form to crossing dunes in the lower topography of the Tsondab Flats protected from easterly flow. Although this explanation is not entirely satisfactory, the alternative scenario (Lancaster, 1980) does not counter the weight of evidence in favour of the present efficacy of the Berg Winds.

5.2 Diachronic and Synchronic Models

The orientation of linear dunes in particular, has been employed widely in recent years as a palimpsest of formative wind patterns, especially in the reconstruction of past wind regimes. From the arrangement of three very large, discrete sets of linear dunes in central southern Africa (Kalahari), Mallick et al. (1981) and Lancaster (1981b) have proposed that the African subcontinental anticyclone occupied three different positions during the late Cenozoic. If the Namib dunes are current forms, then linear dune alignments hold little interest for reconstructing past wind flow patterns. The opposite is true in the case of Besler's diachronic reconstruction. In a series of studies, Lancaster (1981a, 1982a, 1982b, 1983a, 1983b, 1985) has applied modern wind data to Namib aeolian bedforms, implying that dunes and dune patterns are equilibrium forms related to *present* circulation of the atmosphere.

Besler (1977, p.52), however, regarded as mobile under present conditions only comparatively thin, surficial layers of sand comprising dune crests and small "foredunes" "... Im innern Erg bei Gobabeb (wandern) nur die Kammsande der grossen Duenen und bilden Klein Vorlaeuferdunen, waehrend der Grundriss der Duenenzuege unveraendert bleibt...". Besler (1977, p.52) wrote specifically of the "...weitgehende Stabilitaet der unteren Duenenpartien...". Besler (1977, 1984) also regarded the sand as relatively immobile on the grounds that five sedimentological-ly discrete "sand provinces" of the Sossus Sand Formation (Besler, 1984) remain substantially unmixed. Lancaster (1982a, 1983a, 1983b) criticised Besler's diachronic theory as implausible because of the non-existence of helical vortices of appropriate diameter.

It is possible to envisage a synchronic situation whereby winds during Berg Wind events move sand southwest, thereby fashioning the crossing dunes as morphologic entities. Under these conditions sand will also be blown obliquely southwards along the lee (west) side of the major N-S linear dunes, reversing the orientation of the crest slipface on both these and the transverse barchanoid dunes which commonly appear on the east side of the N-S dunes (Fig. 2) (M. Seely, pers. com.). During the rest of the year time-dominant SSW winds move sand along the crossing dunes in the opposite direction, and reverse slipface directions on the major linear dunes and the associated east-side transverse dunes. Major N-S dunes are asymmetric with steeper west-side faces suggesting a morphology of dunes roughly transverse to Berg Winds, as Goudie (1972) noted. Lancaster (1985) has demonstrated the seasonal alternation of sand flow vectors at five stations within the sand sea and near its northern edge. Even at the coastal station of Wortel (Station 1, Fig. 1), Berg Winds briefly dominate sand flow regimes in June and this effect becomes progressively more important with distance inland. Ward and von Brunn (1985) have documented winter immobility of the north-moving dunes in the Kuiseb delta as an effect of the Berg Wind. Present patterns thus may be a product of complex interactions between both sets of winds.

Models related to such multiple sets of formative winds have been propsed occasionally for some dunefields (Cooke and Warren, 1973), but a diachronic model appears simpler for dunes outside the high energy coastal tract. In a diachronic model, Berg Winds are seen to be actively realigning fossil dune forms, to conform with arguably the strongest modern flows. Though not mentioning crossing dunes, Besler (1975, 1980, 1984) has suggested that the east flank barchanoids, linear northwest-oriented silk dunes and small foredunes are manifestations of such remodelling, implying by the small size of these types (in comparison with the major linear dunes on which they have formed) a younger age.

Various observations can be made in support of the diachronic theory. Wilson

(1972) considered it axiomatic, and built theories of dune development on the proposition, that bedform size and age are related. Thus, ripple bedforms are younger than larger forms such as dunes, and dunes are younger than draa-sized features. Similarly, crossing dunes in the Namib sand sea may be younger forms than the major, N-S dunes since they are on average an order of magnitude smaller than the latter. The existence of the crossing dunes is thus suggestive of set of younger dunes forming by resculpture of older dunes.

Inferences from dune spacing support the contention that stronger winds were operative in building the N-S dunes. Wilson's (1972) opinion that stronger wind speeds may relate to greater linear dune spacing lends prima facie support to Besler's arguments. It is noteworthy that the spacing between crossing dunes, where sets of crossing dunes have developed, is the same as that of the major dunes. Since present Berg Winds are stronger than, and presently move more sand than, the SSW winds, it may be true that winds with higher velocities (of the order of present Berg Wind maxima) may have been necessary in establishing the spacing of the N-S dunes.

Evidence appears to be accumulating that this was indeed the case during the Last Glacial Maximum. Nicholson and Flohn (1980), Newell et al.(1981) and Flohn (1984), among others, have argued that Glacial Maximum wind speeds were higher than present on the grounds of steeper hemispheric temperature gradients. By Besler's theory, stronger winds are required to account for the spacing of the large linear dunes. Indeed, consideration of modern resultants and trend surface analysis of dunesand sedimentology led Harmse (1980, p.iv) to conclude that "...the northeaster" displacement of Namib dune sand ..is.. a fossil phenomenon...".

The diachronic view receives support from the dating of hearth charcoal at an archaeological site on a dune 2 km west of Gobabeb (Sandelowsky, 1977). The date of 12800 years B.P. led Vogel and Visser (1981) to remark on the unexpected stability of the dune surface on which the site is situated. On a different scale, most authors agree that the sand accumulations of the dune masses in the Namib sand sea (Sossus Sand Formation) are probably some few million years old (see for example Ward et al., 1983; Wilkinson, 1987). If this is true, some aspects of the imprint of past environments might be expected to persist, as they do not further east in the Kalahari desert, particularly features such as larger dune forms and more extensive dunal patterns such as those typified by the large N-S dune chains.

The diachronic view helps explain another aspect of dune patterning, namely the regularity of spacing of the crossing dunes. Smaller aerodynamic forms (such as the crossing dunes) are known to achieve regular spacing when air flow crosses an elongated obstacle. Under such circumstances, transverse, cellularly-organized vortices develop along the lee side of the obstacle (Maull and East, 1963). Since the cells tend to develop to constant size, erosional and depositional forms associated with the cells are regularly spaced. Iversen (1979) has noted the application of this phenomenon of fluid dynamics to snow drifts downwind of ditches aligned transverse to prevailing winds (Fig. 13). Furthermore, not only are the drifts

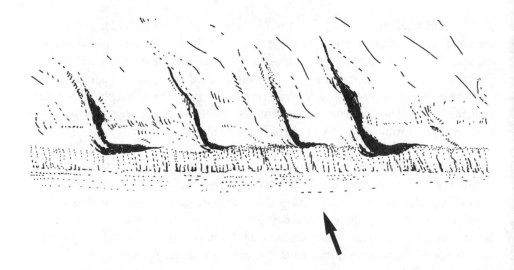

Fig. 13. Linear snow drifts generated by horizontal vortices in elongated obstacle (ditch) transverse to wind (arrow): snow drifts are periodic features related to cellular vortices which develop to constant size (from Iversen, 1979). Regularity of drift spacing and linearity of form are suggestive of crossing dunes (= snow drifts) and major N-S dunes (transverse, elongated obstacle) of Namib sand sea pattern (Figs. 2, 10, 11, 12).

developed parallel to the formative wind, but they are constructed as linear features. The correspondence between drifts and Namib crossing dunes, developed transverse to the N-S dune chains, is striking. This argument implies prior existence of the N-S dunes, even by Lancaster's (1983a, 1983b) model of crossing dune formation dominated by SSW winds (Fig. 2) (see note 7). Without the requisite linear obstacle in the form of N-S dune chains, the regularity in spacing of crossing dunes is less easy to explain.

In the arid core of a desert where sands are actively moved, the problem thus remains of knowing whether present wind fields pertain to existing dune patterns, some of which may be fossil. Whereas it is true that modern winds move sands in ways described by the synchronic models above, modern dune-forming winds form no significant part of modern sand-moving winds which are depicted by the calculated, long-term sand flux resultants.

Diachronic scenarios such as that suggested by Besler (1977, 1984) imply shifts in atmospheric circulation which are as yet unexplored. The issue of the existence of helical rollers in the boundary layer (proposed by Besler, 1977, 1984) is not central to the arguments posed here, but appears open (Barbafiera, 1986).

6. CONCLUSIONS

Three issues are addressed, one concrete and two more theoretical. The first concerns the formative winds of a host of small, linear crossing dunes in the north-central sand sea of the Namib desert; the second concerns applicability of the theory of oblique propagation to these linear elements; the third investigates the possibility that the major N-S dune pattern of the Namib sand sea may be fossil.

Wind streaks and several other aeolian features are a permanent aspect of the geomorphology of the duneless flats surrounding the sand sea of the Namib Desert. These features are expressions of present rather than past wind systems. Two groups of wind streaks occur, those aligned N-S in a narrow coastal strip, related to strong southerly winds along the coast, and those generally found inland of the coastal strip which are ubiquitously aligned NE-SW. As known from other parts of the world and from other planetary surfaces, streaks indicate the direction of formative flows in the atmosphere because their points of origin are easily ascertained. Thus, wind streak alignments have been shown to cluster closely around azimuths of the strongest expression of the Berg Wind, in this region of southern Africa, regularly a northeast wind. In turn, streaks align closely with the mean orientation of one particular set of dunes in the sand sea, namely the smaller, linear dunes termed by Lancaster (1983a, 1983b) crossing dunes. It can be concluded therefore, that strong Berg Wind flows are causally related to the existence of the crossing dunes.

The crossing dunes have been seen as the product of southerly quadrant winds, by the theory of dune propagation oblique to formative winds. The close correspondence between Berg Wind, streak and crossing dune orientations suggests that theories of parallel propagation are more appropriate for the crossing dunes, at least. Northerly winds and weaker (more easterly) Berg Wind and katabatic flow allow consideration of these linear dunes as the product of bimodal wind regimes in conformity with theory, but aligned with the stronger wind direction. Sand roses which render resultant directions of sand movement, a commonly-used method of encapsulating some dominant characteristics of wind regimes. Resultants may mean less in terms of dune morphology than analyses of strong winds of possibly short duration. Strong winds appear to be closely related to alignment and type of dune, as this analysis has shown. Future research needs to distinguish between winds of morphologic importance (dune-forming winds) which determine dune type and orientation, and those which move sand but do not control such basic characteristics (sand-moving winds).

Various arguments are presented which suggest that the major, N-S linear dune pattern of the Namib may be largely fossil in terms of present wind regimes. Firstly, the size of crossing dunes with respect to larger dunes suggests the former are younger; secondly, stronger, southerly winds, more like present Berg Winds at their strongest (rather than the present medium energy southerly winds), arguably may have been responsible for the large dunes; thirdly, regularity of spacing of the crossing dunes implies the necessary pre-existence of the larger dune chains since transverse wind cells generated in the lee of linear obstructions are known to give rise to longitudinal bedforms which are regularly spaced, both of which characteristics describe the crossing dunes; lastly, since the sand sea is reliably considered to be some millions of years old, palaeo patterns may be preserved. The syn- versus diachronic issue appears to be important for the future work on dune morphology and pattern in the sand sea.

7. SUMMARY

Small, linear dunes aligned NE-SW ("crossing dunes") in the Namib sand sea are oriented parallel with the direction of maximum flow of strong northeasterly winds (Berg Winds). This relationship is established from the alignments of wind streaks (linear, wind-induced, mainly erosional features without aerodynamic form) on the duneless plains surrounding the sand sea. The fact that points of origin of streaks are easily ascertained, where those of linear dunes are not, lends streaks methodological importance in this study.

It is concluded that crossing dunes are the product of infrequent, strong, northeast winds rather than of year-round, but less vigorous, southerly quadrant winds. Demonstrable parallelism of these dunes with northeast winds suggests that the dunes are longitudinal forms, with dune growth parallel to formative winds. This finding supports theories of linear dune development parallel to resultant winds, rather than obliquely as suggested in a recent general model of longitudinal dune development. Furthermore, it stresses the importance of strong winds of possibly short duration which may be obscured by long-term wind- and sand-rose resultants. A body of evidence, including that of the relative smallness of the crossing dunes, their association with strong winds and their regular spacing, suggests that the major dune pattern of the sand sea, of large dunes aligned N-S, is fossil in its gross morphology.

8. ACKNOWLEDGEMENTS

I thank the Anglo American Corporation for logistical support in the form of transport and camp facilities, and the Chief Director, Surveys and Mapping, Department of Public Works and Land Affairs, for permission to publish aerial photographs. D. Noli kindly provided aerial photographs on several occasions. The Satellite Remote Sensing Centre of the Council for Scientific and Industrial Research (CSIR) received and enhanced space images of the central Namib desert and dust plume. A. Lamb (CSIR) generously performed several manipulations of the former. G.J. Wilkinson and R. Linz helped with preparation of the statistical material. C.S. Breed, M.K. Seely and J.D. Ward provided welcome commentary on verbal presentations of these ideas. B.P. Moon read an interim draft, P.J. Stickler drew the maps and W. Job drew the final diagram.

9. NOTES

1. Data for Gobabeb station are avoided on the grounds of the much-quoted effects of the Kuiseb valley on wind and sand rose patterns (Tyson and Seely, 1980; Lancaster, 1985).

2. "...in dem...Swakoptal, scheint der Ostwind besonders heftig zu sein. Er hat hier...viele 1-2 m hohe Duenen um die Buesche herum angeweht...1917 was der Staubsturm einmal so stark, dass man nicht ueber die Strasse gehen konnte..." (Jaeger, 1965, p/208).

3. The term sand streak is more specific (Breed and Grow, 1979; Greeley and Iversen, 1985) where a streak is known to be of bedform type.

4. Tankard and Rogers (1978) and Lancaster (1982b) have mentioned sand streaks on the Namaqualand and Skeleton coasts of the Namib Desert respectively. Wilkinson and Barbafiera (in prep.) noted their existence at various other points on the coastline.

5. One small crossing dune gives direct support for the hypothesis presented here. Located immediately south of Sossus Vlei, this dune is attached at its northeastern end to the pointed summit of a larger dune; it disappears 500 m to the southwest on the smooth flank of the larger dune. There is little doubt that a crossing dune has extended from the northeast under the influence of the Berg Wind.

6. It is argued (Wilkinson and de Wet, in prep.) that a positive feedback relationship obtains between the large linear dunes of the sand sea and local wind regimes whereby southerly quadrant air flow, of sand-moving competence, is maintained well inland of where it otherwise occurs. This effect results partly or wholly from the protection afforded by the large linear N-S dunes. Decoupling of surface SSW winds and overlying easterly air streams is strongly suggested by various kinds of evidence, giving complex present-day patterns of mesoscale flow in this part of the Namib desert.

7. The question of the origin of regular spacing of the large N-S dunes remains moot, as is the case with so many linear dune fields worldwide (Cooke and Warren, 1972).

10. REFERENCES

Angell, J.K., Pack, D.H. and Dickson, C.R. 1968. A Lagrangian study of helical circulations in the planetary boundary layer. Journal of Atmospheric Science, 25, 707-717.

Bagnold, R.A. 1941. The Physics of Blown Sand and Desert Dunes. Methuen, London, 265 pp.

Bagnold, R.A. 1953. The surface movement of blown sand in relation to meteorology. Special Publication, Research Council of Israel, No. 2, 89-93.

Barnard, W.S. 1973. Duinformasies in die Sentrale Namib. Tegnikon (Pretoria), December 1973, 2-13.

Barbafiera, M. 1986. The Namib erg: an investigation of dune dynamics and morphology. Research report, Department of Geography and Environmental Studies, University of the Witwatersrand, Johannesburg, 25 pp. (Unpublished).

Besler, H. 1975. Messungen zür Mobilitaete von Duenensanden am Nordrand der Duenen-Namib (Suedwestafrika). Wuerzburger Geographische Arbeiten, 43, 135-147.

Besler, H. 1976. Wasserueberformte Duenen als Glied in der Landschaftsgenese der Namib. Mitteilungen, Basler Afrika Bibliographien, 15, 83-106.

Besler, H. 1977. Untersuchungen in der Duenen-Namib (Suedwestafrika). Journal, South West Africa Scientific Society, 31, 33-64.

Besler, H. 1980. Die Duenen-Namib: Enstehung und Dynamik eines Ergs. Stuttgarter Geographische Studien, 96, 241 pp.

Besler, H. 1984. The development of the Namib dune field according to sedimentological and geomorphological evidence. In: Vogel, J.C. (ed.), Late Cainozoic Palaeoclimates of the Southern Hemisphere, Balkema, Rotterdam, 445-454.

Brain, C.K. and Brain, V. 1977. Microfaunal remains from Mirabib: some evidence of palaeo-ecological changes in the Namib. Madoqua, 10, 285-293.

Breed, C.S., Fryberger, S.C., Andrews, S., McCauley, C., Lennartz, R., Gebel, D. and Horstman, K. 1979. Regional studies of sand seas using Landsat (ERTS) imagery. In: McKee, E.D. (ed.), A Study of Global Sand Seas, Professional Paper, United States Geological Survey, 1052, 305-397.

Breed, C.S. and Grow, T. 1979. Morphology and distribution of dunes in sand seas observed by remote sensing. In: McKee, E.D. (ed.), A Study of Global Sand Seas, Professional Paper, United States Geological Survey, 1052, 253-302.

Cooke, R.U. and Warren, A. 1973. Geomorphology in Deserts. Batsford, London, 394 pp.

Cooper, W.S. 1958. Coastal sand dunes of Oregon and Washington. Memoir, Geological Society of America, 72, 169 p.

Flohn, H. 1984. Climate evolution in the southern hemisphere and the equatorial region during the late Cenozoic. In: Vogel, J.C. (ed.), Late Cainozoic Palaeoclimates of the Southern Hemisphere, Balkema, Rotterdam, 5-20.

Fryberger, S.G. 1979. Dune forms and wind regime. In: McKee, E.D. (ed.), A Study of

Global Sand Seas, Professional Paper, United States Geological Survey, 1052, 137-169.

Goudie, A.S. 1972. Climate, weathering, crust formation, dunes and fluvial features of the Central Namib Desert, near Gobabeb, South West Africa. Madoqua, 1, 15-31.

Greeley, R. and Iversen, J.D. 1985. Wind as a Geological Process. Cambridge University Press, London, 333 pp.

Harmse, J.T. 1980. Die Noortwaartse begrensing van die duinsee van die sentrale Namib langs die benede-Kuiseb. Unpublished Masters thesis, University of Stellenbosch.

Harmse, J.T. 1982. Geomorphologically effective winds in the northern part of the Namib sand desert. South African Geographer, 10, 45-52.

Higgins, C.G. 1956. Formation of small ventifacts. Journal of Geology, 64, 506-517.

Hueser, K. 1976. Kalkkrusten im Namib-Randbereich des mittleren Suedwestafrika. Mitteilungen, Basler Afrika Bibliographien, 15, 51-77.

Iversen, J.D. 1979. Drifting snow similitude. Journal of the Hydraulics Division, American Society of Civil Engineering, 105, 737-753.

Jackson, S.P. and Tyson, P.D. 1971. Aspects of weather and climate over Southern Africa. Occasional Paper, University of the Witwatersrand, Johannesburg, Department of Geography and Environmental Studies, 6, 13 pp.

Jaeger, F. 1965. Geographische Landschaften Suedwestafrikas. South West Africa Scientific Society, Windhoek, 251 pp.

Kaiser, E. 1923. Abtragung und Auflagerung in der Namib, der SWA Kuestenwueste. Geologische Charakterbilder, Parts 27/28. 40 pp.

Kaiser, E. 1926a. Die Diamantenwueste Suedwestafrikas. Dietrich Reimer, Berlin, 2 vols. I, 321 pp.; II, 535 pp.

Kaiser, E. 1926b. Hoehenshichtenkarte der Deflationslandschaft in der Namib Suedwestafrikas und ihrer Umgebung. Mitteilungen, Geographisches Gesellschaft Muenchen, 19, 38-75.

Lancaster, J., Lancaster, N. and Seely, M.K. 1984. Climate of the central Namib desert. Madoqua, 14, 5-61.

Lancaster, N. 1980. The formation of seif dunes from barchans -- supporting evidence for Bagnold's model from the Namib Desert. Zeitschrift für Geomorphologie, 24, 160-167.

Lancaster, N. 1981a. Aspects of the morphometry of linear dunes of the Namib desert. South African Journal of Science, 77, 366-368.

Lancaster, N. 1981b. Palaeoenvironmental implications of fixed dune systems in Southern Africa. Palaeogeography, Palaeoclimatology, Palaeoecology, 33, 327-346.

Lancaster, N. 1982a. Linear dunes. Progress in Physical Geography, 6, 475-504.

Lancaster, N. 1982b. Dunes on the Skeleton Coast, Namibia (South West Africa): geomorphology and grain size relationships. Earth Surface Processes and Landforms, 7, 575-587.

Lancaster, N. 1983a. Controls of dune morphology in the Namib sand sea. In: Brookfield, M.E. and Ahlbrandt, T.S. (eds.), Eolian Sediments and Processes, Elsevier, Amsterdam, 261-289.

Lancaster, N. 1983b. Linear dunes of the Namib sand sea. Zeitschrift für Geomorphologie, Supplementband, 45, 27-49.

Lancaster, N. 1984. Characteristics and occurrence of wind erosion features in the Namib Desert. Earth Surface Processes and Landforms, 9, 469-478.

Lancaster, N. 1985. Winds and sand movements in the Namib sand sea. Earth Surface Processes and Landforms, 10, 607-619.

Logan, R.F. 1960. The Central Namib Desert, South West Africa. Publication, National Academy of Sciences/National Research Council, Washington, D.C., No. 758, 162 pp. (ONR Field Research Program, Report 9).

Mallick, D.I.J., Habgood, F. and Skinner, A.C. 1981. A geological interpretation of LANDSAT imagery and air photography of Botswana. Overseas Geology and Mineral Resources, 56, 1-35.

Maull, D.J. and East, L.F. 1963. Three-dimensional flow in cavities. Journal of Fluid Mechanics, 16, 620-632.

McKee, E.D. 1982. Sedimentary structures in dunes of the Namib Desert, South West Africa. Special Paper, Geological Society of America, No. 188, 64 pp.

McKee, E.D. and Tibbitts, G.C. 1964. Primary structures of a seif dune and associated deposits in Libya. Journal of Sedimentary Petrology, 34, 5-17.

Meigs, P. 1966. Geography of coastal deserts. UNESCO Arid Zone Research, 28, 140 pp.

Newell, R.E., Gould-Stewart, S. and Chung, J.C. 1981. A possible interpretation of palaeoclimatic reconstructions for 18,000 BP for the region 60 degrees N to 60° S, 60° W to 100° E. Palaeoecology of Africa, 13, 1-19.

Nicholson, S.e. and Flohn, F. 1980. African environmental and climatic changes and the general atmospheric circulation in the late Pleistocene and Holocene. Climatic Change, 2, 313-348.

Royal Navy and South African Air Force. 1944. Weather on the coasts of Southern Africa, volume 2, part 1. Meteorological Services of the Royal Navy and the South African Air Force, Cape Town, 1-61.

Sandelowsky, B.H. 1977. Mirabib - an archaeological study in the Namib. Madoqua, 10, 221-283.

Selby, M.J. 1977. Paleowind directions in the central Namib Desert, as indicated by ventifacts. Madoqua, 10, 195-198.

Sweeting, M.M. and Lancaster, N. 1982. Solutional and wind erosion forms on limestone in the Central Namib Desert. Zeitschrift für Geomorphologie, N. F., 26, 197-207.

Thomas, P. and Veverka, J. 1979. Seasonal and secular variation of wind streaks on Mars: an analysis of Mariner 9 and Viking data. Journal of Geophysical Research, 84, 8131-8146.

Tsoar, H. 1978. The dynamics of longitudinal dunes. Final Technical Report, European Research Office, United States Army, 171 pp.

Tsoar, H. 1983. Dynamic processes acting on a longitudinal (seif) sand dune. Sedimentology, 30, 567-578.

Tyson, P.D. and Seely, M.K. 1980. Local winds over the central Namib desert. South African Geographical Journal, 62, 136-150.

Vogel, J.C. and Visser, E. 1981. Pretoria radiocarbon dates II. Radiocarbon, 23, 43-80.

Ward, J.D., Seely, M.K. and Lancaster, N. 1983. On the antiquity of the Namib. South African Journal of Science, 79, 175-183.

Ward, J.D. and von Brunn, V. 1985. Sand dynamics along the lower Kuiseb River. In: Huntley, B.J. (ed.), The Kuiseb environment: the development of a monitoring baseline, Report, South African National Scientific Programmes, Council for Scientific and Industrial Research, No. 106, 51-73.

Wieneke, F. and Rust, U. 1972. Das Satellitenbild als Hilfsmittel zur Formulierung geomorphologischer Arbeidshypothesen. Wissenschaftliche Forschung in Suedwestafrika, 11, 16 pp.

Wilkinson, M.J. 1976. Preliminary report on aspects of the geomorphology of the lower Tumas Basin, South West Africa. Internal Report, Anglo American Corporation of South Africa Limited, Johannesburg, 36 pp. (Unpublished).

Wilkinson, M.J. 1987. A Late Cenozoic geomorphic history of the Tumas drainage basin in the Central Namib Desert. Unpublished Ph.D. dissertation, The University of Chicago. 294 pp.

Wilkinson, M.J. and Barbafiera, K. (in prep.). Aeolian streaks on the arid west coast of southern Africa.

Wilkinson, M.J. and de Wet, B. (in prep.). Topographic effects on aeolian features in the Namib sand sea.

Wilkinson, M.J., Maclear, P.B. and Bayman, B.A. (in prep.). Dust plumes on the Namib Desert coast.

Wilson, I.G. 1972. Aeolian bedforms - their development and origins. Sedimentology, 19, 173-210.

Wopfner, H. and Twidale, C.R. 1967. Geomorphological history of the Lake Eyre basin. In: Jennings, J.N. and Mabbutt, J.A. (eds.), Landform Studies from Australia and New Guinea, Cambridge University Press, London, 119-143.

GEOMORPHOLOGY OF THE SAND-COVERED PLATEAUX IN SOUTHERN SHABA, ZAIRE

MORGAN DE DAPPER
Geological Institute of the State University, Gent, Belgium

1. INTRODUCTION

Large areas of southern Shaba Province in Zaire are covered with loose sandy deposits, which have been reworked mainly from sand bodies of Neogene ("Kalahari superieur : Serie des sables ocre") to Plio-Pleistocene age. In situations where vegetation cover is sparse, varying from grass steppe to wooded steppe (Malaisse, 1975), an extensive and varied relief can be detected on remote sensing images. This is especially the case for isolated features such as the Kundelungu, Marungu, Kibara, Biano and Manika plateaux (Fig. 1). The latter, situated near the town of Kolweni, in the eastern part of the study area (Fig. 1), is a typical example of the plateaux in southern Shaba and is moreover reasonably accessible. It was therefore chosen for a detailed investigation combining airphoto interpretation and a dense network of field observations (De Dapper, 1981a).

2. THE SAND-COVERED PLATEAUX IN THE KOLWEZI AREA

2.1. Environmental setting

The major landforms in the vicinity of Kolwezi form a complex of plateaux situated between the peripheral plateaux that surround the Central Zaire Basin and the high-plateaux that form the SW-extension of the western horst of the East African Rift system. The relief is developed in folded schists and tillites of the Kanga-system (Upper Precambrian). In a W-E topographic profile, elevations rise from 1075m to 1515m m above sea level (a.s.l.) over a distance of 60 km. It shows a step-like form, resulting from three major series of fault scarps which transgress the area (Fig. 2).

Geomorphological Studies in Southern Africa, G.F.Dardis & B.P.Moon (eds)
© *1988 Balkema, Rotterdam. ISBN 90 6191 831 6*

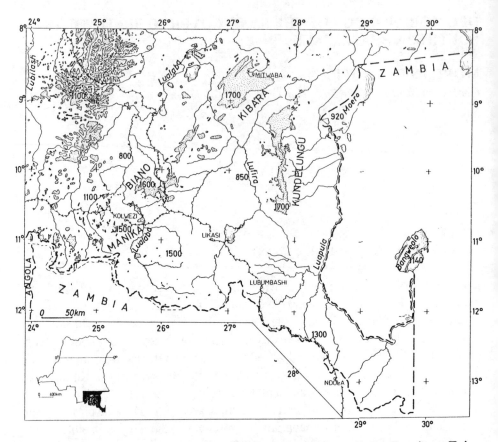

Fig. 1. Zones (dotted) with dune-like features in southern Shaba Province, Zaire as detected on controlled airphoto mosaics on a scale 1:100 000 (Based on the 1957 airphoto coverage on a scale of 1:40 000 - 1:45 000 of the former Comite Special du Katanga concession). Altitudes are in metres a.m.s.l. The major plateaux are indicated (Kundelungu ...).

The climate is characterized by an alternation of a wet and a dry season which exceeds five months. The mean monthly air temperature is 20 degrees C, with the coldest month in June (below 18 degrees C). Diurnal temperature variations are important and during the dry season occasional night frost may occur. The mean annual precipitation reaches 1200 mm. The onset of the rainy season often involves heavy storms, precipitation intensities of 50 mm/h being common.

The principal vegetation formation is the miombo, a woodland of the dry type eventually degraded to a savanna. Great parts of the plateaux are covered by the *dilungu*, a steppic grassland formation. Strips of the rainforest occur along the permanent rivers (Malaisse, 1975).

116

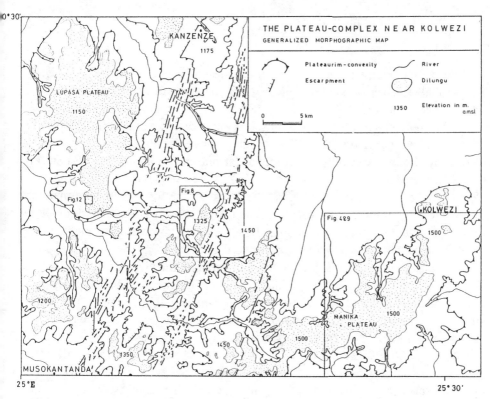

Fig. 2. Generalised morphological map of the plateaux complex near Kolwezi, Shaba, Zaire

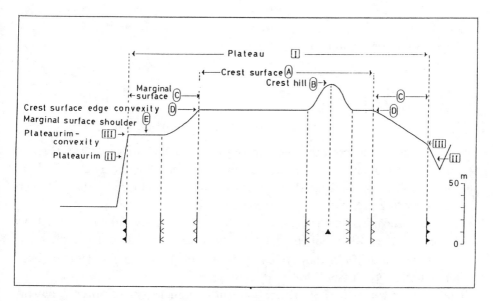

Fig. 3. The morphotype of the plateau in the Kolwezi area.

117

2.2. Morphography of the plateaux

The morphotype of the plateau in the Klowezi-regioï is built up as follows (Fig. 3). The *plateau* (I) is surrounded by a *plateau rim* (II), 25 m to 125 m high. The transition is sharply marked by a first major slope change line, the *plateau rim-convexity* (III). The greatest part of the plateau itself is made up of an almost flat (slopes 1.5 degrees) *crest surface* (A), rising 20 m to 30 m above the plateau rim-convexity. A few gently sloping crest hills (B) rise 25m (at maximum) above the crest surface that is surrounded by a marginal surface (C). A second major break of slope, the *crest surface edge convexity* (D), marks the transition between crest- and marginal surface, two major morphological units. The marginal surface shows predominantly very gently inclined, rectilinear slopes. In some cases however a distinct concave break of slope delimits a *marginal surface shoulder* (E).

The *dilungu* (see note 1), is the dominant element on the plateaux. It forms an extensive flat, covered with fine, Kalahari-type sands and is covered by a steppe vegetation of grasses and herbs. The term *crest dilungu* will be reserved for a dilungu that only covers the crest surface of a plateau. The term *over-all dilungu* will be used for a dilungu that covers both, crest and marginal surfaces.

In the region of Kolwezi the sandy cover of the dilungu is generally rather thin (0.5-4 m). Locally and especially around important valley heads, the thickness of the sandy layer may reach up to a few tens of metres. The sandy sediments lie discordantly over the weathered Precambrian substratum which sustains a perched groundwater table, flooding great parts of the malungu (i.e. plural of dilungu) during the rainy season.

2.3. Ancient erg remnants on the crest dilungu

2.3.1. Morphography

The microrelief on the crest dilungu is mainly composed of elongate forms (Fig. 4); others features that are associated with them have a more limited extent (De Dapper, 1979, 1981b, 1985). The elongate forms can be subdivided into linear and sinuous microridges, and in linear microdepressions.

The linear *microdepressions* are very shallow (depths up to 30 cm) and narrow (maximum width of 40 m) but very long (between 1 km and 3.6 km). Although difficult to survey in the field, they are easily discernible on aerial photographs. They show a remarkably constant E by SW by N direction.

The *linear microridges* (Fig. 5) are low (maximum height of 50 cm), narrow (widths between 50 m and 100 m) and also very long (lengths between 1 km and 5

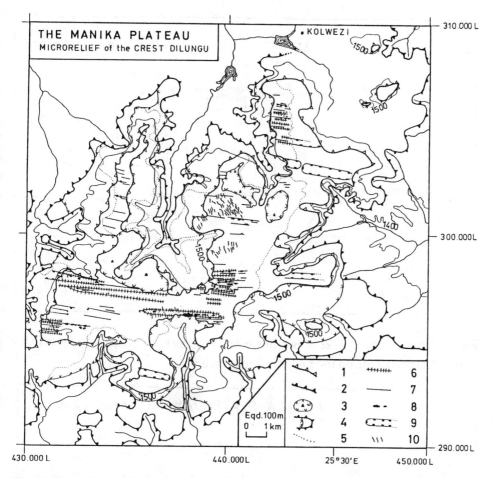

Fig. 4. The micro-relief of the crest dilungu on the Manika plateau. (1. Crest surface edge convexity 2. Plateaurim convexity 3. Crest-hill 4. Marginal surface shoulder 5. Extension of the dilungu 6. Linear micro-ridge 7. Linear micro-depression 8. Depression-pan 9. Dry trough-shaped valley 10. Sinuous micro-ridge.

km). They are always associated with the linear microridges, running parallel and subparallel with them. In some cases ridges join and then form a Y-shaped fork with two long prongs and a short stem always pointed to the W by N. It should be noted that the linear microridges are not always evenly spaced. The microridges are very low (average height of 20 cm), narrow (maximum width of 50 m) and long (200-1200 m). The direction of their long axis ranges between SSE-NNW and SE-NW. It is difficult to survey them in the field but they are also easily seen on aerial photographs. Field measurement in the Kahilu test zone on the Lupasa plateau show a net asymmetrical cross-sectional form, with a more gentle slope facing ENE or NE. Drainage differences at the onset of the dry season are well translated in the airphoto image by tonality differences (Fig. 6).

Some pans occur in the linear microdepressions. These closed microforms are

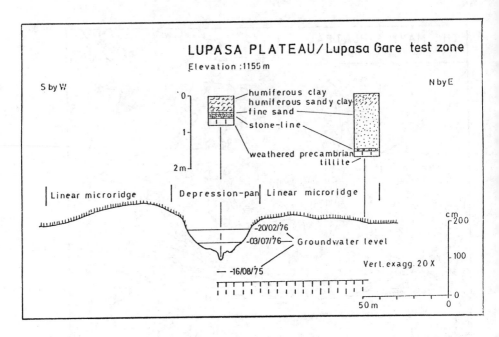

Fig. 5. Cross-section in a set of two linear microridges and a depression pan in the Lupasa Care test zone on the Lupasa plateau (see Fig. 2).

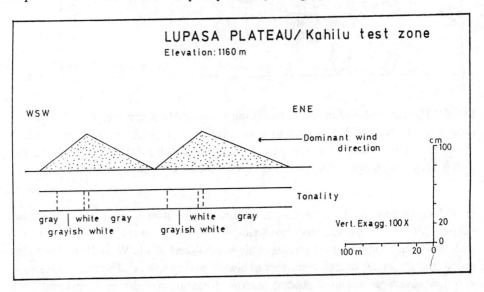

Fig. 6. Asymmetrical cross-section of sinuous microridges translated by tonality differences on the aerial photo images; as observed in the Kahilu test zone on the Lupasa plateau (see Fig. 2).

shallow (depths between 1 m and 3 m) and show a circular, elliptic or oval-shaped in plan form. Their diameter or axis varies in length from 50 m to 200 m.

120

2.3.2. Morphogenesis

The linear microdepressions, linear microridges and depression pans constitute a landform association always situated on the peripheral zones of the crest dilungu (Fig. 4). On the contrary, sinuous microridges chiefly occur on the central zone where the sand cover is slightly thicker. In some cases sinuous microridges are developed on and affect the set of linear microdepressions and microridges.

Taking account of their remarkable constant direction and their important geographical extent, eolian landforms are the most obvious origin of the linear microdepressions and microridges. Identical forms were observed on the Biano and Kundelungu plateaux by Alexandre-Pyre (1971) and by De Dapper et al. (1987) respectively. We consider the linear microridges as fixed remnants of extensive longitudinal dunes. This interpretation particularly augments the studies on dune sequences made by Verstappen (1968, 1972) showing how longitudinal dunes, many kilometres long, irreversibly can be formed out of parabolic dunes. The same author insists on the fact that the great longitudinal dune ridges are not always evenly spaced and that they often join to form a fork whose stem is always directed leeward. The depicted spatial arrangement perfectly fits with our observations in the Kolwezi area.

An eolian origin supposes an arid climatic phase with very sparse or even non-existent vegetation cover offering conditions in which the dilungu sand can easily be eroded and easily transported by the wind. Following the morphology of the ridges the dominant direction of the winds was E by S.

We suppose the linear microdepressions and the depression pans are indirectly derived from the seif landscape. In our hypothesis (Fig. 7) the evolution on an arid phase to a semi-arid or steppic climatic phase involving some amount of precipitation led to a shrub vegetation able to fix the eolian forms. Run-off concentrated in the "straats", the parallel depressions between the dunes (Goudie, 1973).

In the case of a very thin sand cover as observed in the peripheral zones (Fig. 5) this water concentration initiated a drainage pattern running parallel to the longitudinal dunes. On the other hand, in the central zones where the sand cover was more important, concentrated water soaked entirely through the permeable sands impeding stream development.

During this first stage of evolution, dune ridges were already partially degraded by the attack of sheetwash erosion. In the peripheral zone the supplied sediments were evacuated by the streams. In the central zones, in contrast, sheetwash led to aggradation in the straats.

In the long run higher precipitation amounts led to a slow rise of the average water table perched on the relatively impervious Precambrian bedrock, involving a growing seasonal hydromorphy in the peripheral zones. Under this condition it is highly probable that the shrub vegetation was gradually replaced by a steppe vegetation of suffrutex and geofrutex, similar to the one that occupies the malungu

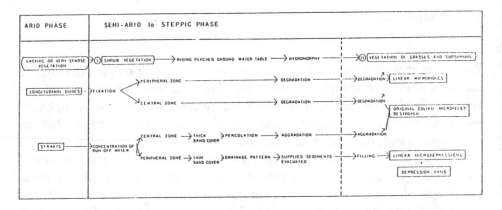

ARID PHASE	SEMI-ARID to STEPPIC PHASE

(i) LACKING OR VERY SPARSE VEGETATION → (i) SHRUB VEGETATION → RISING PERCHED GROUND WATER TABLE → HYDROMORPHY → (ii) VEGETATION OF GRASSES AND SUBSHRUBS

LONGITUDINAL DUNES → FIXATION < PERIPHERAL ZONE — → DEGRADATION — → DEGRADATION → LINEAR MICRORIDGES

CENTRAL ZONE — → DEGRADATION — → DEGRADATION → ORIGINAL EOLIAN MICRORELIEF DESTROYED

STRAATS → CONCENTRATION OF RUN-OFF WATER < CENTRAL ZONE → THICK SAND COVER → PERCOLATION → AGGRADATION — → AGGRADATION

PERIPHERAL ZONE → THIN SAND COVER → DRAINAGE PATTERN → SUPPLIED SEDIMENTS EVACUATED — → FILLING → LINEAR MICRODEPRESSIONS

DEPRESSION-PANS

Fig. 7. Scheme of the evolution of the original longitudinal dune landscape.

nowadays (see note 2). In this second stage of the evolution, topsoil being less protected by vegetation, sheet- and rill-erosion became more important and accelerated the degradation of the dune ridges. Taking account of the very gentle sloping surface it is very possible that the growing sediment supply was not sufficiently evacuated any more, envolving a filling of the drainage network in the straats. The sketched evolution generated the linear microdepressions and depression pans that we consider as regression pans. Several authors (De Ploey, 1965; Flint and Bond, 1968; Verboom and Brunt, 1970; Sterckx, 1974) consider the pans as original blowouts. Nevertheless, it must be stated that their interpretation relies mainly on the sole morphological aspect of this feature.

The shrub vegetation probably lasted longer in the central zones, the average water table laying deeper in this part of the crest dilungu. In this zone the combination of degradation of dune ridges and aggradation of straats obliterated the original eolian microrelief almost completely. There also the evolution finally ended in hydromorphic conditions forcing the shrubs to be replaced by a less protective steppe vegetation and leading to further general degradation.

In the scope of the hypothesis on the evolution of the most important part of the microrelief on the crest dilungu, as sketched above, only the peripheral zones show remains of the original longitudinal dunes and straats under the form of linear microridges, linear microdepressions and depression pans. On the central zones the original eolian landforms were completely destroyed.

Several field observations indicate the existence of an obliterated drainage pattern on the origin of the linear microdepressions. Where the crest surface convexity cuts a linear microdepression frequently water seep zone can be observed, often forming the source of small intermittent streams on the marginal surface. Possibly the straats drainage pattern here and there cut into the weathered Precambrian bedrock so that these former thalwegs now serve as collecting channels for the perched groundwater. The occurrence of linear, dry, trough-shaped valleys (Fig. 4) parallel to the linear microrelief, on the parts of the crest surface no longer covered by a

122

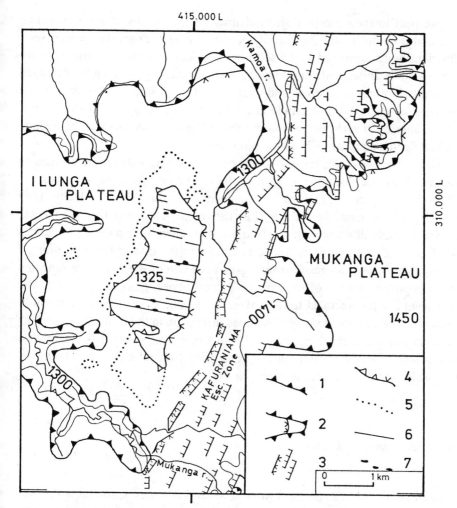

Fig. 8. Simplified morphological map of the Ilunga plateau. The linear microridges hit the basal concavity of the Kafuraniama escarpment zone (see Fig. 2). (1. Plateaurim convexity 2. Marginal surface shoulder 3. Escarpment zone 4. Crest surface edge convexity and basal concavity 5. Extension of the dilungu 6. Linear microridge 7. Depression-pan).

dilungu, corroborates this view. Though the drainage network is very sparse on the sand-covered crest surfaces, it shows a remarkable preferential E by S-W by N direction emphasizing the axes of the crest dilunu microrelief.

In some cases (e.g. on the Ilunga plateau ;Fig. 8), the set of linear landforms hits the basal concavity of the fault escarpments (Fig. 2). With respect to these morphological observations one can conclude that the formation of the original longitudinal dunes must be situated before the formation of the escarpment. If not, the very regular longitudinal aspect would be perturbed at the proximity of the escarpments, the more so as the derived dominant wind direction was approximately perpendicular to the escarpment lines.

123

With respect to their morphological characteristics and distribution an eolian origin can also be postulated for the sinuous microridges. From their rather scatterd distribution as compared to the seif remnants, one can conclude that they are not formed solely under arid conditions but they are merely seasonal forms issued from a climatic phase with long and accentuated dry seasons. The asymmetric cross-section as observed in the field and on aerial photographs and the sinuous wavy crest lines fit in with the morphography of fixed transverse dunes stretching approximately to a dominant ENE wind direction.

The origin of transverse dunes has usually been associated with low or decreasing wind speed. However, the crucial factor in the formation of transverse dunes, according to Verstappen (1972), is rather the decrease in dune speed velocity. They can be formed from parabolic and from barchans alike. If either of these two dunes types starts to move at a slower rate, the upwind dunes can catch up with the ones further downwind and thus form tranverse ridges. In the case of the crest dilungu decreasing dune speed can result from a growth in size of the dunes approaching the more important source of sand in the central zones. Since the speed of the dunes is inversely proportional to their size, the upwind dunes under such conditions will more readily catch up with the ones further on in the central zones which increased in size when reaching the area with ticker sand cover. Eventually also shrub vegetation on the central zones under drier climatic conditions, could be responsible for the decrease of the dune velocity.

2.4. Microrelief on the overall dilungu

The over-all microrelief can be subdivided in several complexes, spread in distinctive zones with very regular spatial organisation. From the edge to the center of the plateau the following sequence is observed (Fig. 9); (1) a belt with sandy microfans, (2) a belt with forms due to active sheetwash and rill erosion, and (3) a zone with mena-relief.

2.4.1. Belts with active forms

A narrow, 40m to 200m wide, belt forms the transition between the dilungu and the surrounding areas without sandy cover, generally occupied by miombo and savanna. This outermost belt is characterized by a typical vegetation of small trees, dominated by *Phillippia benguelensis* and *Uapaca robijnsii* (Malaisse, 1975). The general surface gradient varies between 1 and 1.5 degrees.

From the morphological point of view, this belt corresponds to a transit zone

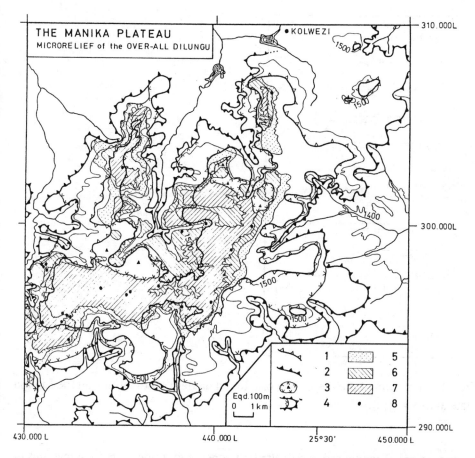

Fig. 9. The microrelief of the over-all dilungu on the Manika plateau (see Fig. 2).
(1. Crest surface edge convexity 2. Plateaurim convexity 3. Crest-hill 4. Marginal
surface shoulder 5. Belt with micro-fans 6. Belt with spurs of sheetwash- and rill-
erosion 7. Zone of mena-relief 8. Closed depression [pan]). Fig. 10. Plan form and
cross sections of a wina, based on field observations on the Lupasa plateau (see Fig.
2).

wherein fine sands supplied from the upward dilungu is temporally stocked in
microfans to be redistributed afterwards over the downward slope facets. The
microfans are a few centimeters thick and have a planform up to 1 sq. m, covering
fallen leaves and branches, blades of grass etc.

The adjoining microrelief belt has a width of 100m to 400m, going up to 1500 m
around valley heads extending on the plateaux. The surface gradient averages 0.7
degrees so that the transition to the micro-fan belt is marked by a convex slope
change line. Although slope values are small, this belt is dissected by a great number of rills. The rill-interfluves are subjected to intense sheetwash during the rainy

125

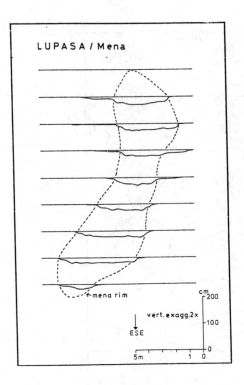

LUPASA / Mena

mena rim

vert. exagg. 2x

ESE

cm
200

100

0

5m   1   0

Fig. 11. Plan form of the mena network, based on field observations on the Lupasa plateau (see Fig. 2).

season storms. The upward end of the belt consists of a micro-scarp, a few tens of centimeters high, retreating parallel to itself, thus showing a case of micro-pedimentation. This process is very active during the rainy season as shows the denudation of roots and subsoil stems of grasses and subshrubs. As a result, part of the sediments covering the stone-line (see note 3) are affected and sorted. The fine fractions are transported in the rills to form the micro-fans of the outer belt. The soft nuclei of iron oxide accumulations in the podzol-like soil are washed out and harden as soon as they are exposeed to the air. They form angular concentrations with a rough surface that are concentrated and covered with fine material supplied by subsequent micropedimentation processes, thus forming a new embryonic stone-line.

2.4.2. Zone with mena relief

2.4.2.1. Morphography of the mena

Mena (see note 4) are small shallow closed depressions, with the rim at lower eleva-

126

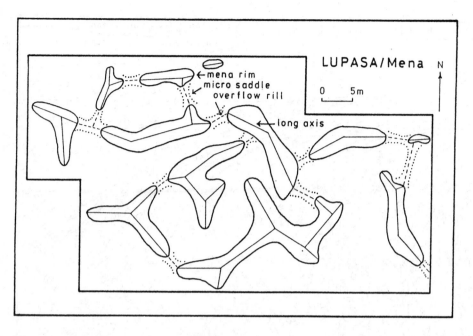

Fig. 12. The different microrelief complexes as seen on oblique low-altitude aerial photograph s (taken on the Lupasa plateau during the dry season in July after a bushfire).

tion where two mena are joined by a saddle. Mostly they are elongated and ramified. Their length varies between 3 m and 10 m, their width between 2 m and 5 m and their depth reaches 15 cm to 30 cm (Fig. 10). They never appear separately but in a dense pattern with long axes running parallel. In the Lupasa test zone the axial length averages 1180 m/ha (Figs. 2, 11 and 12).

The mena rims always form subvertical microscarps. The sandy microsaddles between the mena are often crossed by short and narrow overflow rills. The rims show an undisturbed A1 horizon, incised by the mena themselves, whose bottoms have only slight and probably more recent A1 development.

The mena interfluves consists of mainly sandy material which continues over at least 2m depth. In the mena bottoms, by contrast, occurs a superficial layer of more clayey and silty texture, but resting upon the same dilungu sands as the interfluves.

The mena bottoms and mena interfluves show definite contrasts in vegetation and termite activity. The mena bottoms are covered with grass species, mainly hemicryptophytic Gramineae and Cyperaceae. The vegetation on the mena interfluves, on the other hand, besides a few grass species (mainly Gramineae) is dominated by subshrubs, especially *Syzygium guineense* subsp. *huillense* and *Parinari capensis* subsp. *latifolia* (see note 5). Grass height varies between 10 cm and 250 cm; subshrub height averages 20 cm.

The malungu are colonised by humivorous termite species (see note 6) constructing

127

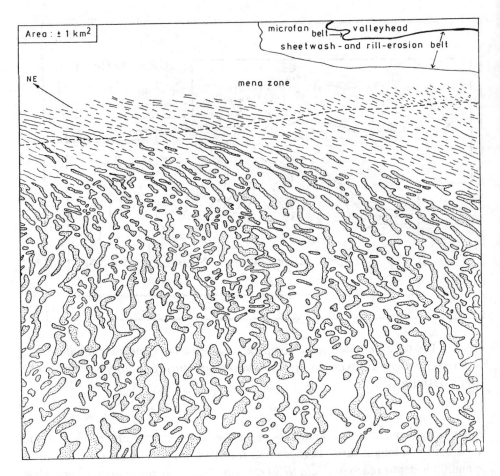

Fig. 13. Chronosequences of landforms on the sand-covered plateaux of the Kolwezi area and a tentative correlation to the climatic and environmental evolution during the late Quaternary of southern Shaba. (Climate and vegetation after Roche [1979], Roche and Mbenza [1980] and Mbenza et al. [1984]).

candle-shaped calies. Despite intense termite activity the amount of reworked substratum is quite restricted. Termite activity is distinctly higher on the mena bottoms as compared to the interfluves (Table 1).

The calies consists mainly of silt and clay as the humus on which the termites feed is attached to these fractions. There are however few differences between interfluves and bottom calies as to their clay and organic matter content. Clay content in both cases averages about 42 per cent and organic carbon content about 9.2 per cent.

2.4.2.2. Morphological processes in the mena

Fluctuations of the perched groundwater table mainly control present-day morphological processes. After a slow rise during the rainy season, a maximum level of water table is reached at the end of February. All mena bottoms are then flooded but interfluves remain dry. These contrasts in drainage account for differences in vegetation and an even more complex termite activity. In the mena bottoms some of the termite calies become completely flooded and therefore abandoned; the others all have termite activity concentrated towards the top.

As soon as the mena are flooded, cohesion of material at the depressions rims and at the base of the termite calies is lost. This decrease starts a general microslumping, a slow retreat of the rims and some aggradation on the mena bottoms. Collapse of callies by narrowing at their base promotes their breakdown. Meanwhile the interfluves show little erosion, as runoff transport is greatly impeded by the subshrubs. The transported material consists only of silt, clay and organic matter and it also settles on the mena bottoms. This reconstruction corroborates the colluvial nature of the material on the mena bottoms.

As the dilungu surface slopes slightly towards the plateau edge, a distinct water overflow starts between successive aligned mena as soon as the groundwater level rises sufficiently and floods the bottoms. The overflow transports silt and clay in suspension while some fine sand is dragged for a short distance along the overflow rills.

At the end of the rainy season the groundwater level drops and overflow stops. The mena transform into pans wherein silt, clay and organic matter settle from suspension. Later desiccation by evaporation and drop of the groundwater level occurs. At the end of the dry season, this sediment layer forms a hard skin with a thickness of several millimeters, covering most of the mena bottoms. The bottom grasses wither very quickly. Shortly afterwards they are completely burned by early bush fires; by contrast the green leaves of the subshrubs do not suffer. Towards the end of the dry season all vegetation becomes withered and late bush fires also destroy the leaves of the subshrubs.

At the onset of the rainy season heavy storm-rains fall, but erosion only affects the interfluves. The interfluve surface shows intense sheetwash onto the wina (see note 4) bottoms where sedimentation in micro-fans occurs but neither flooding nor runoff is initiated. Moreover the clayey skin and the weak slope protect the bottoms against the splash erosion.

2.4.2.3. Morphogenesis of the mena

Field observations and airphoto interpretation clearly show the mena long axes directed to the dilungu edges. Profile pits crossing the mena do show any collapse structure in the sands, showing that the formation of the depressions cannot be im-

Table 1. Occurrence of candle-shaped termite calies.

	INTERFLUVE	BOTTOM
Mean density	1.2calies/sq m	2.4 calies/sq m
Minimum height	6.0 cm	8.0 cm
Maximum height	25.0 cm	25.0 cm
Average height	13.0 cm	17.0 cm
Minimum diameter	1.3 cm	3.0 cm
Maximum diameter	7.5 cm	8.5 cm
Average diameter	3.4 cm	5.3 cm
Mean weight of calies material	98.0 g/sq m	506.0 g/sq m

puted to solution mechanisms. These arguments make it obvious that the mena corresponds to a phase of degeneration of a once-extensive open rill system, dating from a period of severe soil erosion.

Three phases can be discerned in the morphogenesis of the mena; (1) a rill phase, (2) a transition-phase, and (3) a mena-phase.

Concerning the rill-phase, objections arise as to the overall slope which is mostly below 0.5 degrees in the mena zone. According to several authors, slopes below 2 degrees are too weak to allow sufficient runoff velocity to induce severe soil erosion. However threshold slope values ought not to be generalized. Indeed, soil roughness, vegetation type and cover also condition the erosive power of the runoff. Moreover changes in these factors may at the same time affect soil erodibility.

The generalized rills could be generated during a climatic period with precipitation characteristics somewhat different from those occurring now; a lower mean annual precipitation amount, a longer dry season and a higher rainfall variability. Decrease of the annual precipitation amount results in a thinning of the vegetation cover. Increase of the length of the dry season intensifies withering of the vegetation and provokes the development of a denser network of dessication cracks affecting the structure of the A1 horizon. Increase of rainfall variability induces higher runoff aggressiveness especially by higher precipitation intensities at the onset of the rainy season and shifting of the onset itself.

The onset of the generalized bush firing practised in the region can also be imputed for the rill-phase generation. These practices can lead to a small-scale pseudo-rhexistasic situation resulting in increased soil erodibility and subsequent rill formation. Bushfires consume the protective litter cover that weakens the raindrop impact, favours percolation and attenuates runoff concentration. Besides, vegetation has to adapt to the new situation leading to temporarily thinning of the cover.

Several causes can be imputed to explain for the transition of the rills into mena.

(1) A slight climatic change to the present-day can provoke a generalized rising of the groundwater table so that the seasonal fluctuations can play a morphodynamical role on the dilungu surface. Liquefaction of the sandy material at the rims of the rills at more or less regular distances and subsequent colmatation induces the transition of rills to mena.

(2) The establishment of the bushfires as an ever-recurring practice can result in an increased sediment supply from the rill-interfluves to the rill-bottoms. Because of the very slope values, a situation can be reached whereby sediment evacuation cannot follow supply, resulting in colmatation of the rills and transition to mena.

(3) The transition mechanism can also be explained by the vegetation-hydrology relationship. The development of an extensive rill system can provoke a generalized drop of the groundwater table, resulting in an extension of the subshrubs that occur exclusively on the dry parts of the mena zone at present day. As field observations prove, soil roughness of mena bottoms and mena interfluves shows clearcut differences at the end of the dry season. The overall surface consists of a mosaic of clumbs, 2-6 cm high and of an almost circular shape, alternating with barren soil patches. In both cases clumbs cover about 22 % of the surface. In the mena bottoms their number averages 56 sq. m., on the mena interfluves only 32 sq. m. (test plots of 4 sq. m.). This clearly shows that the average diameter of the clumbs is much higher on the interfluves, increasing concentration and aggressiveness of runoff. That process can clearly be observed at the outer rim of the mena zone, where part of the mena are opened by the micropedimentation processes already mentioned above. Because of the slow adaption, mena vegetation still remains unchanged here despite lack of flooding. Because of the lack of pan formation, no skin of silt-clay organic matter can develop. Here rill erosion affects the bottoms as well as the interfluves, but is distinctly more severe on the latter. In this case too, increased sediment supply can result in a transition to the mena-phase. Once the mena form , flooding and widening by microslumping can take place, leading to the extension of the grasses to the detriment of the subshrubs. Because of the vulnerability of the grasses to bush fires, a new rill-phase can eventually develop at a critical level of their extension ushering in a new cycle.

3. CHRONOLOGY

Direct absolute datings on the micro-landforms of the sand-covered plateaux near Kolwezi are not available. Despite this inconvenience it is however possible to establish a relative geomorphological chronosequence and to frame that sequence in a general climatological evolution scheme for the area (Fig. 13). Such a scheme is worked out for the Late Quaternary of Southern Shaba by Roche (1979), Roche and Mbenza (1980) and Mbenza et al. (1984). It is based on palynological evidence for several sites of the area.

The relatively oldest recognisable generation of landforms is the major complex of longitudinal dunes. They probably belong to the important fixed eolian landform remnants surveyed in the former arid zones of Zimbabwe, Zambia and Angola by Thomas (1984, 1985). Although no dates have yet been obtained for the formation of those ergs, Lancaster (1981) and Heine (1982) suggest periods of stronger anticyclonic circulation coinciding with glacial maxima during the late Pleistocene. In the Shaba context the formation of the longitudinal dunes may be situated during the arid maximum around 20000 years BP. The transverse dunes form a second generation as they are developed on the remnants of the longitudinal dunes. Their formation can also be situated during the same major arid phase. The fact that the longitudinal dunes were already degraded before the formation of the transverse dunes may imply that the curve based on palynological evidence is more complex. A second possibility is that the longitudinal dunes predate the more humid Kalambo interstadial, that they were degraded during that humid transition and subsequently affected by the transverse dunes then formed during the arid Mount Kenya hypothermal.

The probable formation of fault scarps after the erg formation could lead to river incision that had a repercussion on the degradation of hillslopes extending into the plateau margins.

Stripping of bedrock saprilite from the plateaurim convexity (Fig. 3) could split the plateau surface into a marginal and a crest surface. Hereby the crest dilungu conserved the original micro-landforms that were destroyed on the marginal surface. The expansion of the forests during the warm-humid maximum around 7000 years BP could also activate river incision and its morphological consequences.

The mena rill-phase may be linked to the cold phase around 3000 years BP. Sedimentological and palynological (see note 7) investigations in the correlative valley bottom sediments of the Upper-Luilu (Fig. 2) indicate that the transition-phase took place not only before 2000 years BP (see note 8) De Dapper, 1981a). It can most probably be linked to the warm period around 2000 years BP.

Finally, the belts with sandy micro-fans and forms related to sheetwash and rill-erosion are active phenomena.

4. SUMMARY

Plateaux are major elements in the macro-geomorphology of Southern Shaba (Zaire). Great parts of them are covered by a dilungu, a thin sandy layer, reworked from neogene (Kalahari system) to Plio-Pleistocene sands and covered with a steppic vegetation. A detailed study of a type-area, the complex of plateaux near Kolwezi, showed that the dilungu is marked by an extensive and varied microrelief that can be subdivided into a chronosequence of at least four generations. The oldest generation is composed of remnants of important longitudinal dunes that can be linked to the late Pleistocene ergs described for Zimbabwe, Zambia and Angola. There is evidence that after the modelling of the original eolian landforms important radial tectonic movements took place that originated or accentuated the stepwise upbuilding of the complex of the plateaux. The next generation consists of remnants of more local transverse dunes developed on the seif remnants. Both microrelief generations exclusively occur on the crest surfaces of the dilungu. A third generation, the mena, occurs on the over-all dilungu and obliterates the two former ones. Mena form a very extensive network of small and shallow closed depressions. They are interpreted as resulting from the degeneration of a once open rill-system. Radiocarbon datings on correlated sandy sediments in the valleyheads permit to situate the rill phase before 2000 years BP. The last generation of microlandforms consists of belts of active forms that follow the rims of the plateaux and extend towards the centre, thus slowly consuming the fossil microrelief.

5. NOTES

1. Di-lungu, plural : malungu; a term from the Luba language. Here the term is used in a morphological sense (including form, substratum and vegetation) that is slightly different from the more commonly used botanical-geographical sense.

2. Following Malaisse (1975) the inverse process can be observed nowadays on the sand-covered plateaux; "...La steepe seche se caracterise principalement par l'abondance des geofrutex qui lui donne son aspect caracteristique... Des que le niveau de la nappe phreatique s'abaisse quelque peu, cette formation evolue en steppe seche arbustive ou arboree, formation vegetale differente, mais a composition floristique tres voisine..." (p.20)

3. Numerous observations in profile pits and borings prove that the typical intertropical tri-partite layer build-up: fine cover, stone line, substratum.

4. Wina, plural : mena; term from the Chokwe language.

5. Determination by Prof. Dr. F. Malaisse, National University of Zaire, Lubum-bashi.

6. Determination by Dr. G. Goffinet, State University Liege, Belgium.

7. A palynological survey was undertaken by Dr. E. Roche of the Royal Museum for Central Africa at Tervuren, Belgium.

8. GrN-7582 and GrN-7683 by Dr. W. Mook of Groningen, The Netherlands.

6. REFERENCES

Alexandre-Pyre, S. 1971. Le plateau des Biano (Katanga). Geologie et Geomor-phologie. Ac. Roy. Sc. O.M. Sc. Nat. Med. N.S., XV111, 3, 151 pp.

De Dapper, M. 1979. The microrelief of the sandcovered plateaux near Kolwezi (Shaba, Zaire). 1. The microrelief of the over-all dilungu. Geo-Eco-Trop, 3, 1-18.

De Dapper, M. 1981a. Geomorfologische studie van het plateaucomplex rond Kolwezi (Shaba-Zaire). Verh. Kon. Acad. Wet. Lett. Sch. K. Belgie, Kl. Wet., 43, 172, 203pp.

De Dapper, M. 1981b. The sandcovered plateaux near Kolwezi (Shaba, Zaire). 11. The microrelief of the crest dilungu. Geo-Eco-Trop., 5, 1-12.

De Dapper, M. 1985. Quaternary aridity in the tropics as evidenced from geomorphological research using conventional panchromatic aerial photographs (examples from Peninsular Malaysia and Zaire). Bull. Soc. belg. Geol - Bull. Belg. Ver. Geol., 94, 199-207.

De Dapper, M., De Moor, G., Corne, E., Thorree, E. 1987. Het mekrorelief op de zandige delen van het Midden-Kundelungu plateau (Shaba-Zaire). Natuurwet. Ts. (in preparation).

De PLoey, J. 1965. Position geomorphologique, genese et chronologie de certains depots superficiels au Congo occidental. Quaternaria, 7, 131-154.

Flint, R. and Bond, G. 1968. Pleistocene sand ridges and pans in Western-Rhodesia. Bulletin of the Geological Society of America, 79, 299-314.

Goudie, A. 1973. Duricrusts in Tropical and Subtropical Landscapes. Clarendon Press, Oxford, 174 pp.

Heine, K. 1982. The main stages of the late Quaternary evolution of the Kalahari region, southern Africa. Palaeoecology of Africa, 15, 53-76.

Lancaster, N. 1981. Palaeoenvironmental implications of fixed dune systems in southern Africa. Palaeogeography, Palaeoclimatology, Palaeoecology, 33, 327-346.

Mbenza, M. Rosche, E. Doutrelepont, H. 1984. Note sur les apprts de la palynologie et de l'etude des bois fossiles aux récherchés geomorphologiqués sur la vallée de la Lupembashi (Shaba-Zaire). Rev. Paleobiol., vol. spec., avril 184, 149-154.

Malaisse, F. 1975. Carte de la vegetation du bassin de la Luanza. In: Symoens, J.J. (ed.), Exploration Hydrobiologique du Bassin du Lac Bangweolo et du Luapula, XV111, 2, 1-41.

Rosche, E. 1979. Vegetation ançienne et actuellède l'Afrique centralé. African Economic History, 7, 30-37.

Rosche, E. and Mbenza, M. 1980. Exêmple d'evolution paleoclimatique au Pleistocene terminal et a l'Holocene au Shaba (Zaire). Mem. Mus. Nation. Hist. Nat., n. s., Ser. B, XXV11, 137-148.

Sterckx, J. 1974. Geographie et developpement. Analyse geographique du degre carre de Ruwe-Kolwezi (Zaire-Shaba). Cult. et Developm., V1, 3, 501-577.

Thomas, D. 1984. Ancient ergs of the former arid zones of Zimbabwe, Zambia and Angola. Transactions of the Institute of British Geographers, N.S., 9, 75-88.

Thomas, D. 1985. Evidence of aeolian processes in the Zimbabwean landscape. Transactions of the Zimbabwe Science Association, 62(8), 45-55.

Verboom, W., Brunt, M. 1970. An ecological survey of Western Province, Zambia, with special reference to the fodder resources. Vol. 1: The Environment. Land Resource Studies Division, Directorate of Overseas Survey, Tolworth, 8.

Verstappen, H. 1968. On the origin of longitudinal (seif) dunes. Zeitschrift für Geomorphologie N. F., 12, 200-220.

Verstappen, H. 1972. On the dune types, families and sequences in areas of undirectional winds. Gott. Geogr. Abh., 60, 341-353.

THE MOBILITY OF AEOLIAN DEPOSITS ON THE SOUTH AFRICAN WEST COAST

J.T. HARMSE
Department of Geography, Rand Afrikaans University

1. INTRODUCTION

The study focuses on the Sandveld (which literally means sandy field or area) on the west coast of the Republic of South Africa (Fig. 1). This area forms part of the extensive aeolian accumulations that cover the southwestern coastal lowlands of the African continent. These extend almost continuously from the Cape Flats in the south, over a distance of more than 2200 km to the town of Namibe in southern Angola, through Namibia.

The Sandveld is fringed in the north (at Latitude 28°24'S) by the perennial Orange River, which drains the larger part of the southern African interior, and in the south (at latitude 32°20'S) by a sandstone ridge ("Bobbejaansberg") of Silurian age, which is up to 200 m in height. The subaerial aeolian deposits as such range in width from 10km to 47km eastward from the coast. These deposits were formed during the Late Pliocene to Pleistocene Periods (Hendey, 1980; Harmse, 1987; Odendaal, 1983), whereas the well-defined coastal dune system that forms part thereof accumulated in the Holocene (Rogers, 1980). The depth of the Sandveld deposits varies from 9 to 67 m.

The climate of the area may be described as that of semi-desert under the present interglacial conditions. The vegetation is mainly Karoo scrub, ranging in surface cover from 0 - 80 % Average annual precipitation varies from approximately 250 mm in the south to less than 50 mm in the north (Pitman et al., 1981), with little east-west variation. The average annual temperature along the coast ranges from 14.1° C in the north to 19.3° C in the south (Nieman, 1981). To the interior, where the cooling effect of the Benguela upwelling current is minimized or frequently virtually absent, and the above mentioned temperatures are often doubled or, in the north, even trebled (Schulze, 1965). Evidence of the former aeolian action is, however, prevalent in the "hummocky" topography (King, 1963, p.78) that the area exhibits. This is due to thousands upon thousands of coppice dunes that are encountered throughout the Sandveld.

Geomorphological Studies in Southern Africa, G.F.Dardis & B.P.Moon (eds)
© *1988 Balkema, Rotterdam. ISBN 90 6191 831 6*

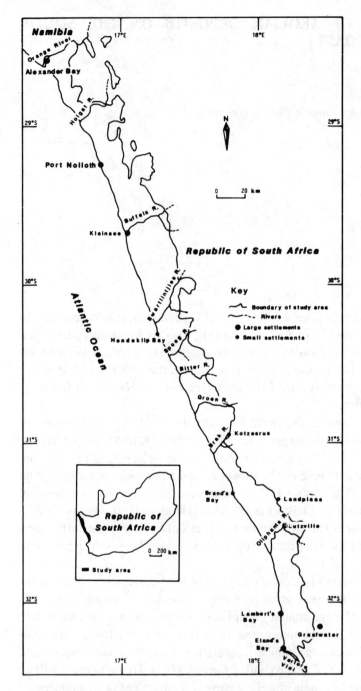

Fig. 1. The distribution of subaerial arenaceous deposits in the Sandveld.

Table 1. Some properties of Sandveld aeolianites

Number of sites	353
Number of mobile dunes	35
Percentage of mobile dunes	9.92
Number of immobile dunes	318
Percentage of immobile dunes	90.08
Mean grain size (phi)	1.997
Standard deviation grain size (phi)	0.349
Mean Sorting (phi)	0.621
Standard deviation sorting (phi)	0.192

2. TYPES OF DEPOSITS

The coppice dunes are accumulation of aeolian deposits without a sectional or plan shape similar to other known dune forms, and range in size from 50-70 m in length, 30-40 m in width, and 4-8 m in height. These are, however, not the only type of sub-aerial aeolian accumulation. In areas where aeolian activity can be observed, longitudinal dunes of up to 2 km in length, 50 m in width and 15 m in height are found, albeit in isolated localities. Parabolic and tranverse dunes of a much smaller size than the former are frequently encountered close to the coast. These are shaped in response to the dominating southerly and northerly winds respectively associated with the summer South Atlantic semi-permanent anticyclonic condition, and the winter cyclonic circulation pattern (Tyson, 1969).

3. SAMPLING AND ANALYSIS

A total of 353 sand samples were located on the basis of a stratified sytematic sampling procedure. This resulted in samples being collected at least 6 km apart. Care was exercised throughout to take each sample on a dune crest or, in the case of the

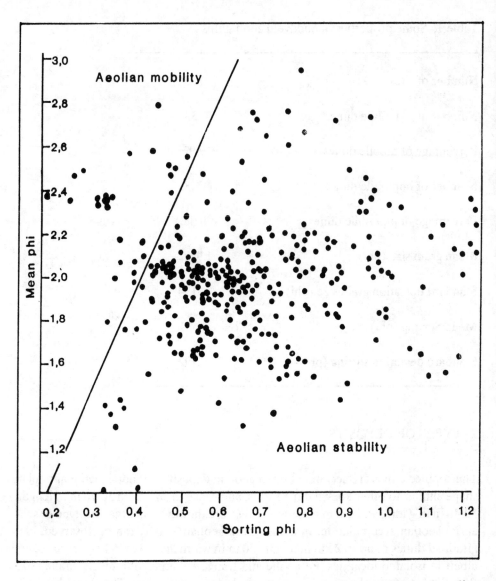

Fig. 2. Besler (1983) diagram to distinguish between aeolian mobility and aeolian immobility.

coppice dunes, at the dune apex (these coppice dunes do not display definite windward or leeward slopes).

The samples were all wet-sieved at 0.25 phi interval after the silt and clay fractions (average at 1.64 % by mass) had been removed by the settling tube method. The granulometric measures of mean grain size phi and sorting phi (Folk and Ward, 1957) were subsequently calculated for each sample individually, giving the results listed in Table 1.

The mean grain size phi versus sorting phi value for each sample was plotted on a

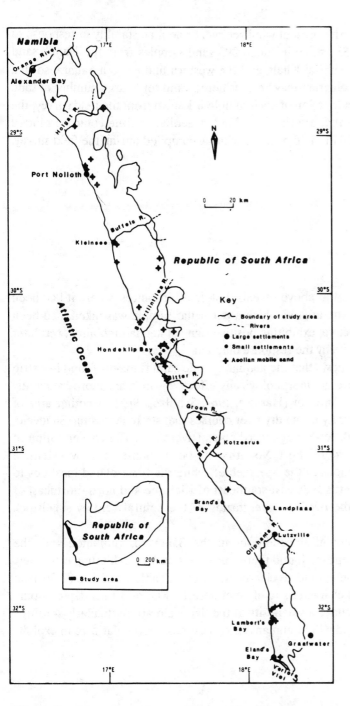

Fig. 3. Location of aeolian mobile sand bodies

Response diagram (Fig. 2) quite similar to that developed by Besler (1983). This
was done in order to distinguish between sand particles that exhibit aeolian

mobility in response to the present wind regime, or aeolian stability in response to the same. Besler's (1983) study included 393 sand samples from the Namib sand sea, the Kalahari, the Rub'-al-Khali, and the western and eastern Sahara, so that the resulting response diagram may be considered valid for a large number of sand bodies. Since the exact location of each sample is known from the field survey, the areal position of those that are thus classified as aeolian mobile could be plotted (Fig. 3) and compared with field notes previously compiled during the field survey.

4. CONCLUSIONS

The comparison referred to above reveals that, in all instances, where it had been noted in the field that the area in which a particular sample was taken, had been described as "aeolian active, exhibiting positive signs of aeolian mobility", was later found to be eolian mobile by the response diagram.

An analysis of Fig. 3 shows that the aeolian mobile sand is mainly found in a strip close to the coast, where geomorphologically effective winds are more frequently encountered than in the interior (Harmse, 1987; Schulze, 1965). Another area of occurrence is in the vicinity of the dry river courses that strike across the Sandveld, notably in the Groen-Bitter-Spoeg-Swartlintjies River zone. The concentration of aeolian mobile sand around Port Nolloth may be accounted for by extensive diamond mining operations in the Sperrgebiet, while the few isolated localities to the northeast of Lambert's Bay, east-southeast of Kleinsee and north-northeast of Port Nolloth may be the result of overgrazing of the natural veld by smallstock farmers.

The present climatological conditions in the Sandveld, combined with the arenaceous subaerial deposits found there, result in this area exhibiting rather few localities of aeolian mobile sand, as clearly indicated in Table 1 and Fig. 3. Such an occurrence is a typical characteristic of semi-desert. Figure 3 thus depicts some localities where the arenaceous deposits of the Sandveld are particularly sensitive to aeolian processes, and where humans should exercise particular care in exploiting the land.

5. SUMMARY

The present study focuses on the West coast of South Africa. This forms part of the aeolian accumulations that covers the southwestern coastal lowlands of the

African continent, extending almost continuously from Cape Town to Namibe in southern Angola. The aeolian deposits in the Sandveld range in width from 10 - 47 km eastward from the coast. Most of this is probably late-Pliocene to Pleistocene in age, while the coastal dunes themselves accumulated in the Holocene. In depth the aeolinites varies from 9 - 67 m. Evidence of former extensive aeolian action is prevalent in the form of coppice, longitudinal, parabolic, and transverse dunes, many of these corresponding in orientation to the present wind regime. Sand samples (353) were collected from dune crests or, in the cases of coppice dunes, on top thereof. These were all wet-sieved at 0.25 phi interval after the silt and clay fractions had been removed by tube setting. Folk and Ward's granulometric measures of mean grain size and sorting were calculated for all samples. These values were plotted on a response diagram quite similar to that developed by Besler (1983), to distinguish between sand and exhibit aeolian mobility in response to the present wind regime, or aeolian stability in response to the same. This revealed that only 9.92 % of Sandveld sand is mobile in response to the present wind regime, the locations thereof correspond with the field observations. The Sandveld thus exhibits typical semi-desert characteristics.

6. ACKNOWLEDGEMENTS

This research formed part of the authors Ph.D. thesis at the University of Stellenbosch, under the supervision of Prof. C.J. Swanvelder.

7. REFERENCES

Besler, H. 1983. The response diagram: distinction between aeolian mobility and stability of sands and eolian residuals by grain size parameters. Zeitschrift für Geomorphologie, Supplement Band 45, 287-301.

Folk, R.L. and Ward, W.V. 1957. Brazos River bar: A study in the significance of grain size parameters. Journal of Sedimentary Petrology, 27, 3-26.

Harmse, J.T. 1987. The Geomorphological History and Present Deformation of Sub-aerial Deposits along the West Coast of South Africa. (In Afrikaans). Unpublished Ph.D. thesis, Department of Geography, University of Stellenbosch, 295p.

Hendey, Q.B. 1982. Langebaanweg: A Record of Past Life. South African Museum, Cape Town.

King, L.C. 1963. South African Scenery, (3rd Ed.), Oliver and Boyd, London, 308p.

Nieman, W.A. 1981. Climate of the coastal lowlands of the Western Cape. In: Moll, E. (ed.), Proceedings of a Symposium on Coastal Lowlands of the Western Cape, March, 1981, University of the Western Cape, 11-15.

Odendaal, J.P. le G. 1983. The Origin and Development of the Coastal Plain of the Southwestern Cape: A Geomorphological Investigation. (In Afrikaans). Unpublished D.Phil. thesis, Department of Geography, University of Stellenbosch, 296p.

Rogers, J. 1980. First Report on the Cenozoic sediments between Cape Town and the Elands Bay, Open File Report No. 165, Geological Survey of South Africa, Pretoria, 100p.

Pitman, W.V., Middleton, B.J. and Midgley, D.C. 1981. Surface Water Resources of South Africa, vol.111, Part 1, Report No. 11/81, Hydrological Research Unit, University of the Witwatersrand, 142p.

Schulze, B.R. 1965. Climate of South Africa, Part 8: General review, Weather Bureau Publication 28, Government Printer, Pretoria, 330p.

Tyson, P.D. 1969. Atmospheric Circulation and Precipitation over South Africa, Occasional Paper No. 2, Department of Geography, University of the Witwatersrand, 44p.

THE GEOMORPHOLOGICAL ROLE OF VEGETATION IN THE DUNE SYSTEMS OF THE KALAHARI

DAVID S.G. THOMAS

Department of Geography, University of Sheffield

1. INTRODUCTION

Dunes are an important landscape component of the Kalahari sandveld. They have been documented by, amongst others, Hodson (1912) in part of the then Bechuanaland Protectorate (Botswana), by Lewis (1936) in the Cape Province of South Africa and by Flint and Bond (1968) in part of western Zimbabwe. The availability of aerial imagery covering an extensive area of the Kalahari, together with detailed field observations, enabled Grove (1969) to write his seminal paper on the landforms of much of Botswana. This included an account of linear, transverse and lunette dunes, and was subsequently followed by studies devoted to the distribution and significance of the dune systems in their own right (Lancaster, 1981a; Thomas, 1984).

An important characteristic of the Kalahari dunes is that they are commonly vegetated. One of the first attempts at desert dune classification (Hack, 1941) included vegetation as a major determinant of dune form and type. Nevertheless, vegetation has been relatively neglected in geomorphological studies of dune development. It is widely regarded as a sand trapping and stabilising agent (e.g. Mabbutt, 1977). In this way it contributes to the formation of parabolic dunes together with other more minor forms, especially coppice dunes, and stabilises other dune types which are then interpreted as 'fossilised' and given palaeoenvironmental significance (e.g. Thomas, 1984).

Recent investigations in Australia (Ash and Wasson, 1983; Wasson and Nanninga, 1986) and in Israel (Tsoar and Moller, 1986) have however begun to examine the geomorphological importance of dune vegetation in greater depth. The purpose of this paper is to provide a preliminary consideration of the implications of these studies for the geomorphology of the Kalahari dunes.

Geomorphological Studies in Southern Africa, G.F.Dardis & B.P.Moon (eds)
© *1988 Balkema, Rotterdam. ISBN 90 6191 831 6*

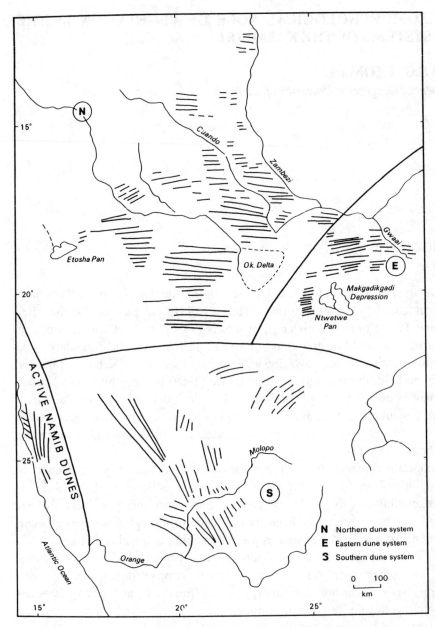

Fig. 1. Dune systems in southern Africa.

2. DUNE SYSTEMS OF THE KALAHARI

According to Fryberger and Goudie (1981), over 85 per cent of the Kalahari dunes are linear forms. These occur in three systems (Fig. 1, Thomas 1984) which together form an approximately semi-circular arc of dunes. Dune remnants have also been identified further to the north, in the Shaba region of Zaire (De Dapper, 1985).

146

The southern system, or Kalahari dune desert (Thomas, 1986a) occupies south-western Botswana and adjoining parts of South Africa and Namibia. Linear dunes in this area occur as parallel to sub-parallel, straight or slightly sinuous asymmetric ridges which may merge or bifurcate at 'Y' junctions (Thomas, 1986a). Ridges range in height from 5 to 20 m and are 20 m to 2 km apart, with spacing generally decreasing in a south to southeasterly direction.

The northern system is found in northern Botswana, Angola and Zambia; and the eastern system is centred upon western Zimbabwe. Dunes in both these systems are broader than in the Kalahari dune desert with spacing between 1 and 2.5 km. Ridges are up to 25 m high in the northern system (Lancaster, 1981a) and 35 m in the eastern system (Thomas, 1984, 1985). They are frequently much lower, however, and a height of 5 m is perhaps most representative. These broad sand ridges are widely believed to be degraded, principally under the effect of sheetwash (Flint and Bond, 1968).

The dunes of the Kalahari are almost without exception regarded as stabilised features. Recent research has examined the palaeoenvironmental significance of the Kalahari dunes (Lancaster, 1980, 1981a; Heine, 1982; Thomas, 1984); sedimentary properties (Van Rooyen and Verster, 1985; Thomas ,1985, 1986b; Lancaster, 1986; Thomas and Martin, 1987); dune patterns (Goudie, 1969; Thomas, 1986a) and dune form (Thomas, in press). In contrast, no direct evaluation has been made of the geomorphological significance of vegetation, despite plant cover being a notable component of the dune systems.

4. PALAEOENVIRONMENTAL STUDIES OF THE KALAHARI DUNE SYSTEMS

Following studies on the south side of the Sahara, (Grove, 1958; Grove and Warren, 1968), it has been suggested that the present limit of dune activity in southern Africa approximately corresponds with the 100 - 150 mm annual isohyte (Lancaster, 1981a). Consequently, modern dune mobility is believed to be confined to the Namib desert (e.g. Lancaster, 1980).

According to Meigs'(1953) classification of aridity which takes account of precipitation and evapotranspiration, the Kalahari dune desert occupies an arid climatic zone today, whilst both the northern and eastern dune systems fringe the northern limit of semi-aridity, and occur predominantly in sub-humid to humid climates. Rainfall in the vicinity of the Kalahari dunes increases in a north to north-easterly direction. In the southwestern Kalahari, the mean annual precipitation today is approximately 259-300 mm (Lancaster, 1980), whilst at Mongu in western Zambia, close to northern system dunes, mean precipitation is 972 mm/yr (Thomas, 1984). The Kalahari dunes have therefore been widely regarded as inactive and indicative of more extensive southern African aridity during one (Heine, 1982) or more (Lancaster, 1981a; Thomas, 1984) periods of reduced moisture availability in the Quaternary period.

Comparisons have also been made between dune alignments and the resultant direction of modern potential sand-moving winds. Consequently, it has been suggested that changes have occurred in the size and position of atmospheric circulation systems since the times of dune development (Lancaster, 1980, 1981a). Heine (1982) also proposed that circulation strengths were greater when dunes were active. Analysis of windspeed data from southern Africa, and comparison with values recorded in locations worldwide with modern active linear dunes (Fryberger, 1979), suggest that the present Kalahari wind environment is indeed one of low energy (Thomas, 1984).

5. VEGETATION AND LINEAR DUNE DEVELOPMENT

Linear dunes occur in a variety of forms which dune classification schemes tend to mask. There are perhaps two basic forms: *seif dunes*, which have sharp, often sinuous crests, and *linear* ridges, which are more subdued features, sometimes lacking slip faces, and often at least partially colonised by vegetation. The latter type dominate the dune systems of the Australian interior, and also the Kalahari.

Ash and Wasson (1983) examined sand movement on Australian linear dunes. They found that dunes in semi-arid areas possessed a vegetation cover which was sufficiently dense (often two storey) to control aeolian sand movement. In more arid areas, however, their conclusions suggest that earlier studies have over-estimated the ability of a partial vegetation cover to protect the ground surface and impede aeolian entrainment and sediment movement. In the absence of a continuous cover, individual shrubs and grass clumps streamline airflow in such a way that they create zones of airflow acceleration around the sides of the plant and deceleration in the downwind wake.

Although modern wind speeds are not strong enough to permit substantial aeolian sediment transport throughout the central Australian dunefields, Ash and Wasson (1983) observed that significant sandflow occurred on dune crests with up to 35 per cent of the ground surface colonised by plants. Sand movement is further facilitated on the partially vegetated crests because wind velocities increase significantly up dune flanks (Tsoar, 1978, 1983; Watson, 1987) due to boundary layer streamline compression. A subsequent investigation (Wasson and Nanninga, 1986) has shown that sand transport rates calculated for a partially vegetated surface using a modified Bagnold equation agree well with those measured in the field (Fig. 2).

Wind speeds, not the vegetation cover, limit present day sand movement on the linear dunes of the Australian arid zone (Ash and Wasson 1983). Furthermore, their conclusions suggest that there is not an abrupt precipitation threshold be-

148

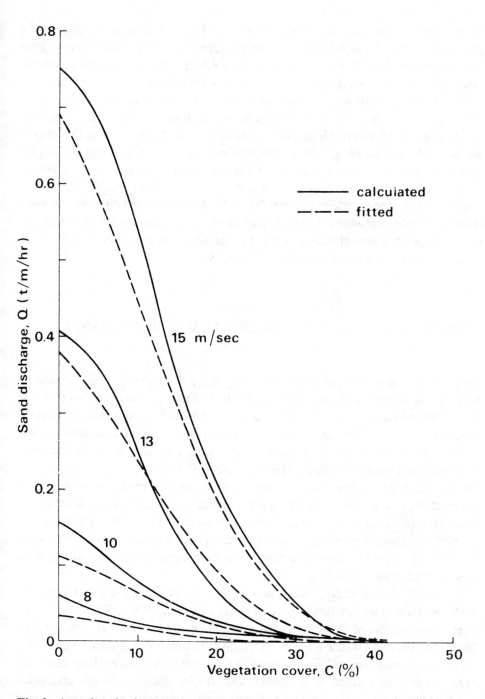

Fig. 2. Actual and calculated rates of sand transport on partially vegetated surfaces, at a range of wind velocities. From Wasson and Nanninga (1986). (Reprinted by permission of John Wiley & Sons Ltd).

tween dune activity and dune stabilisation, as invoked in many studies of the palaeoenvironmental significance of vegetated desert dunes. Rather, where vegetation covers less than 35 per cent of the dune crestal surface, it indicates an inverse relationship between sand movement and vegetation density.

Following a study of linear dunes in the Negev desert, Tsoar and Moller (1986) proposed that vegetation is a normal component of dunes in areas where precipitation is as low as 50 mm/yr. Linear dunes may be especially significant in this respect because they are *extending*, rather than *mobile* dune forms, affording good opportunity for plant colonisation. However, if the vegetation cover is destroyed, aeolian activity increases and dunes may change in form (Tsoar and Moller, 1986) from rounded linear ridges, orientated parallel to the dominant wind direction, to more dynamic braided linear of seif dunes. The latter occur under the effect of bimodal wind regimes and are orientated parallel to the resultant direction of sand movement, not the dominant wind direction (Tsoar, 1983).

6. DUNEFIELD VEGETATION IN CENTRAL SOUTHERN AFRICA

Vegetation density and type differ according to the dune system under consideration, due to variations in moisture availability. In line with other studies and with respect to the topography of linear dunefields, it is most appropriate to examine vegetation in terms of the contrasts between dune ridges and the interdune areas or straats. These dune/interdune contrasts are important for the identification of dune systems from aerial imagery (Fig. 3, Thomas, 1986c).

In the northern and eastern dune systems, dune ridges support the densest plant cover, though it is erroneous to regard the ridges as completely covered with vegetation (cf. Lancaster, 1981a). North and west of the Okavango delta, and probably in other parts of the northern system, dunes support open savanna communities dominated by *Burkea africana* and *Pterocarpus angolensis*, with grasses present in interdune areas (Grove 1969).

In western Zimbabwe, dune crests also have a relatively dense woodland cover in which *Baikiaea plurijuga* (teak) predominates (Thomas, 1984). The mature teak trees attain heights of up to 20m (Boughey, 1963; Flint and Bond, 1968). Ridge flanks, and locations where fire has affected vegetation communities, often have a mixed scrub savanna woodland characterised by *Terminalea sericea*. Interdunes are again grassed, dominated by *Erogrostis* and *Aristida* species, though occasional woody species are present. An under-storey of grass is also present in some areas of wooded ridges.

Vegetation is less well developed in the Kalahari dune desert, in line with the reduced moisture availability in this arid area. Ridge crests are often bare (e.g. Grove, 1969, Van Rooyen and Verster, 1983) supporting some clumped grasses. Interdune areas are better vegetated with what has been described as an open to

Fig. 3. Landstat image (August 1973) of part of the northern dune system. Dune ridges trend east - west, and are distinguished by the zonation of vegetation between dune ridges and interdune areas. Burn scars are also evident in the centre of the image. The Zambezi River is also indicated in the image, trending diagonally from centre top to centre right.

sparse low *Acacia* savanna woodland (Huntley, 1982), where individual trees are about 20 m apart (Grove, 1969). Appreciable regional differences exist, however, in interdune tree densities, with greater numbers in the northern part of the system (Van der Meulen and Van Gils, 1983) and in the vicinity of dry river valleys (Buckley et al., 1987).

Fig. 4. Vegetated dune ridges, western Zimbabwe. Dense secondary climax Terminalia scrub woodland dominates the ridges, with a grass under-storey, whilst the interdune straat is grassed.

7. GEOMORPHOLOGICAL IMPLICATIONS OF DUNE VEGETATION

7.1. Northern and eastern dune systems

Despite the extent of the dune ridge tree cover in the northern and eastern dune systems, much of the ground surface is devoid of vegetation cover. Flint and Bond (1968) estimated that in western Zimbabwe 50 % of the ridge surface were bare, well within the limits identified by Ash and Wasson (1983) as permitting aeolian activity to operate on linear dunes.

There are, however many lines of evidence which indicate that these dune ridges, occurring beyond the southern African arid zone, are indeed inactive in terms of aeolian processes (e.g. Thomas, 1985). Most significant are the degree of maturity of the tree cover (which also provides an extensive litter cover during the dry season), the lack of internal aeolian bedding structures, and sedimentary characteristics. The last of these include a high silt and clay content (not in the form of deflatable pellets: cf. Wasson, 1983), which is much greater than that present in active aeolian sands. This may result from post depositional pedogenic weathering of the sands (Thomas, 1985). The systematic grain size distribution variations which occur across the profile of active dunes (e.g. Lancaster, 1981b) are also absent.

7.2. Kalahari dune desert

The linear ridges in the Kalahari dune desert are in many respects analogous with those investigated in Australia and Israel by Ash and Wasson (1983) and Tsoar and Moller (1986). There is evidence of significant modern sand movement on ridge crests (Fig. 5a). The criteria used by Ash and Wasson (1983) (i.e. bare rippled sand between mounds of sand within shrubs and grasses) are clearly satisfied on most, perhaps all, dune ridges. In addition, ridges also possess small slip faces, which are a further indication of aeolian activity. Sedimentary characteristics (Lancaster, 1986) are also comparable with those of active linear dunes.

Grove (1969) also noted that aeolian sand movement occurred on ridge crests in the southwestern Kalahari, but subsequent researchers have regarded these dunes essentially as palaeoforms inherited from an episode of greater aridity in the past. However, if Tsoar and Moller's (1986) assertion that partially vegetated linear ridges are a normal *active* component of less arid sand seas is correct, the palaeoenvironmental interpretation of these Kalahari ridges is inapplicable.

Tsoar and Moller (1986) noted that overgrazing on linear ridges could result in a change in duneform to seif or linear braided dunes. In parts of the Kalahari dune desert overgrazing has occurred, significantly reducing the already limited vegetation cover on dune crests and also in interdune straats. Enhanced aeolian activity has resulted in the extension of the active crestal zone of dune ridges, a corresponding enlargement of slip faces (Fig. 5b), and the construction of small coppice dunes in interdunes (Fig. 5c). In other areas, the linear dunes display the sinuous, active crests of seif dunes (Fig. 5d).

It is not yet apparent whether these dunes represent a transformation of dune type through the effects of overgrazing, as identified by Tsoar and Moller (1986), or whether they are more 'natural' forms. Unlike wind regimes in the Negev, wind patterns are relatively simple in much of the Kalahari dune desert, so that the dominant direction and the resultant sandflow direction are very closely aligned (Lancaster, 1981a). Consequently, a transformation in dune type from linear ridge to seif dune is unlikely to be accompanied by the change in dune orientation which

a

b

c

154

d

Fig. 5. (a). Typical sparsely vegetated dune ridge crest, Kalahari Dune Desert. Note evidence of aeolian activity (see text). (b). Dune ridge and interdune devegetated through overgrazing, displaying enhanced aeolian activity, Kalahari Dune Desert. (c). Small coppice dune developed in interdune straat where overgrazing has destroyed much of the vegetation cover. (d). Linear ridges in the Kalahari Dune Desert which display sinuous, seif-like crests.

Tsoar and Moller (1986) recognised.

8. CONCLUSIONS

Following recent studies in Australia and Israel, the geomorphological role of vegetation in the Kalahari linear dune systems has been assessed. The following conclusions can be drawn.

(1). There is sufficient evidence to suggest that the northern and eastern dune systems are relict aeolian landforms, indicative of aridity and consequential aeolian processes having greater effect on one or more occassions in the Quaternary period.

(2). Vegetation densities are sufficiently low on the crests of dunes in the Kalahari dune desert to permit significant aeolian activity to occur. That it does is indicated by bare rippled surfaces, mounds of sand trapped within vegetation clumps and slip face development.

(3). Tsoar and Moller (1986) suggest that linear ridges, partially vegetated and displaying crestal sand movement, are a normal, active type of dune in desert environments. If this is the case, the dune in the Kalahari dune desert are not palaeoforms.

(4). Some linear dunes display greater activity than described in point 2. In some (but not necessarily all) cases, this can be attributed to the effects of overgrazing.

9. SUMMARY

In central southern Africa, extensive dune systems occur between latitudes 28° and 12° South. Linear dunes predominate, occurring in the three systems. The dunes commonly possess a vegetation cover, the presence of which is widely interpreted as indicative of present day dune stability and more extensive aridity in the subcontinent at various times in the Quaternary period. Drawing on recent research in Australia and Israel, which suggests that a vegetation cover of up to 35 % does not necessarily inhibit aeolian sand transport on dune crests and that vegetation may be a normal component of some types of active linear dunes, the geomorphological role of vegetation on the Kalahari linear dunes is re-evaluated.

10. ACKNOWLEDGEMENTS

Research in the Kalahari over the period 1982 - 1987 has been funded by the Natural Environment Research Council (U.K.), a Strakosch Fellowship and The Royal Society. Fieldwork has been assisted by many people, all of whom are thanked, especially Dr Paul Shaw (University of Botswana) and Elizabeth Thomas.

11. REFERENCES

Ash, J.E. and Wasson, R.J. 1983. Vegetation and sand mobility in the Australian desert dunefield'. Zeitschrift für Geomorphologie Supplementband 45, 7-25.

Boughey, A.S. 1963. Interaction between animals, vegetation and fire in southern Rhodesia. Ohio Journal of Science, 63, 193-209.

Buckley, R., Gubb, A. and Wasson, R. 1987. Parallel dunefield ecosystems: predicted soil nitrogen gradient tested. Journal of Arid Environments, 13, 105-110.

De Dapper, M. 1985. Quaternary aridity in the tropics as evidenced from geomorphological research using conventional panochromatic aerial photographs (examples from Peninsular Malaysia and Zaire). Bulletin de la Societie Belge de Geologie, 94, 199-207.

Flint, R.F. and Bond, G. 1968. Pleistocene sand ridges and pans in western Rhodesia. Bulletin of the Geological Society of America, 79, 299-314.

Fryberger, S. 1979. Dune forms and wind regime. In: McKee, E.D. (ed.), A Study of Global Sand Seas. U.S. Geological Survey Professional Paper 1052, 137-169.

Fryberger, S.G. and Goudie, A.S. 1981. Arid geomorphology. Progress in Physical Geography, 5, 420-428.

Goudie, A.S. 1969. Statistical laws and dune ridges in southern Africa. Geographical Journal, 135, 404-406.

Grove, A.T. 1958. The ancient erg of Hausasaland and similar formations on the southern side of the Sahara. Geographical Journal, 124, 526-533.

Grove, A.T. 1969. Landforms and climatic change in the Kalahari and Ngamiland. Geographical Journal, 135, 191-212.

Grove, A.T. and Warren, A. 1968. Quaternary landforms and climate on the south side of the Sahara. Geographical Journal, 134, 194-208.

Hack, J.T. 1941. Dunes of the western Navajo Country. Geographical Review, 31, 240-263.

Heine, K. 1982. The main stages of the late Quaternary evolution of the Kalahari region, southern Africa. Palaeoecology of Africa, 15, 53-76.

Hodson, Lieut. A.W. 1912. Trekking the Great Thirst. Travel and sport in the Kalahari Desert. T. Fisher Unwin, London, 359pp.

Huntley, B.J. 1982. Southern African savannas. In: Huntley, B.J. and Walker, B.H. (eds.), Ecology of Tropical Savannas, Springer, New York, 101-119.

Lancaster, N. 1980. Dune systems and palaeoenvironments in southern Africa. Palaeoentology Africana, 23, 185-189.

Lancaster, N. 1981a. Palaeoenvironmental implications of fixed dune systems in southern Africa. Palaeogeography, Palaeoclimatology, Palaeoecology, 33, 327-346.

Lancaster, N. 1981b. Grain size characteristics of Namib Desert linear sand dunes. Sedimentology, 28, 115-122.

Lancaster, N. 1986. Grain size characteristics of linear dunes in the southwestern Kalahari. Journal of Sedimentary Petrology, 56, 395-400.

Lewis, A.D. 1936. Sand dunes of the Kalahari within the borders of the Union. South African Geographical Journal, 19, 22-32.

Mabbutt, J.A. 1977. Desert Landforms. A.N.U. Press, Canberra: 340pp.

Meigs, P. 1953. World distribution of arid and semi-arid homoclimates. In: Arid Zone Hydrology. UNESCO Arid Zone Research Series 1, 203-209.

Thomas, D.S.G. 1984. Ancient ergs of the former arid zones of Zimbabwe, Zambia and Angola. Transactions of the Institute of British Geographers, New Series, 9, 75-88.

Thomas, D.S.G. 1985. Evidence of aeolian processes in the Zimbabwean Landscape. Zimbabwe Scientific Association, Transactions, 62, 45-55.

Thomas, D.S.G. 1986a. Dune pattern statistics applied to the Kalahari Dune Desert, southern Africa. Zeitschrift für Geomorphologie NF, 30, 231-242.

Thomas, D.S.G. 1986b. The response diagram and ancient desert sands. Zeitschrift für Geomorphologie NF, 30, 363-369.

Thomas, D.S.G. 1986c. Ancient deserts revealed. Geographical Magazine, 58, 11-15.

Thomas, D.S.G. In press. Analysis of linear dune sediment-form relationships in the Kalahari Dune Desert. Earth Surface Processes and Landforms.

Thomas, D.S.G. and Martin, H.E. 1987. Grain-size characteristics of linear dunes in the southwestern Kalahari - discussion. Journal of Sedimentary Petrology, 57, 572-573.

Tsoar, H. 1978. The dynamics of longitudinal dunes. Final technical report DA-ERO 76-G-072 European Research Office, U.S. Army, London, 171pp.

Tsoar, H. 1983. Dynamic processes acting on a longitudinal (seif) dune. Sedimentology, 30, 566-578.

Tsoar, H. and Moller, J.T., 1986. The role of vegetation in the formation of linear sand dunes. In: Nickling, W.G. (ed.), Aeolian Geomorphology, Allen and Unwin, Boston, 75-95.

Van der Meulen, F. and Van Gils, H.A.M.J. 1983. Savannas of southern Africa: attributes and use of some types along the Tropic of Capricorn. Bothalia, 14, 675-681.

Van Rooyen, T.H. and Verster, E. 1983. Granulometric properties of the roaring sands in the south-eastern Kalahari. Journal of Arid Environments, 6, 215-222.

Wasson, R.J. 1983. The Cainozoic history of the Strezelecfi and Simpson dunefields (Australia), and the origin of the desert dunes. Zeitschrift für Geomorphlogie Supplementband, 45, 85-115.

Wasson, R.J. and Nanninga, P.M. 1986. Estimating wind transport of sand on vegetated surfaces. Earth Surface Processes and Landforms, 11, 505-514.

Watson, A. 1987. Variations in wind velocity and sand transport on the windward flanks of desert sand dunes. Sedimentology, 34, 511-516.

DIATOMS AND OTHER FOSSIL REMAINS IN CALCAREOUS LACUSTRINE SEDIMENTS OF THE NORTHERN NAMIB SAND SEA, SOUTH WEST AFRICA/NAMIBIA

JAMES T. TELLER
Department of Geological Sciences, University of Manitoba
M. RYBAK, I. RYBAK
Aquatic Research and Ecological Consultants, Mississauga
N. LANCASTER
Department of Geology, Arizona State University
N.W. RUTTER
Department of Geology, University of Alberta
JOHN D. WARD
Geological Survey of SWA, Windhoek, Namibia

1. INTRODUCTION

The Namib Sand Sea lies along the coast of South West Africa/Namibia between about Walvis Bay and Luderitz, and extends inland for about 100 km (Fig. 1). Dunes of the sand sea are dominantly linear, elongated in a north-south direction and typically rise more than 100 m above the broad interdune corridors. Lancaster (1983; 1988) describes these dunes in detail and the transverse, barchanoid, and star dunes that are present in some areas.

The Kuiseb River forms the northern boundary of the sand sea (Fig. 1), and appears to have established its role in limiting the northward migration of the dunes in early Pleistocene time (cf. Ward, 1982). Other ephemeral rivers such as the Tsondab, Tsauchab, and Koichab flow westward toward the Atlantic Ocean from the interior plateau and Great Escarpment, but terminate in the dunes.

The climate of the area is warm to hot and extremely arid, with rainfall typically no more than 50 mm/yr, and less than half of that in the western part of the sand sea (Lancaster et al., 1984). Fog precipitation, mainly within 50 km of the coast, adds significantly to the total.

Thin beds of lacustrine carbonate are locally exposed between linear dunes of the sand sea. Ward (1984) has mapped the occurrence of these beds, distinguishing between those of probable Tertiary-age, which are weathered out of the Tsondab Sandstone, and those which are related to the Quaternary-age Sossus Sand Formation of the present Sand Sea (Fig. 1). Ward (1984) names the younger carbonate the Khommabes Carbonate Member (of the Sossus Sand Formation) after the type locality at Khommabes, which lies just south of the Kuiseb River near Gobabeb

Geomorphological Studies in Southern Africa, G.F.Dardis & B.P.Moon (eds)
© 1988 Balkema, Rotterdam. ISBN 90 6191 831 6

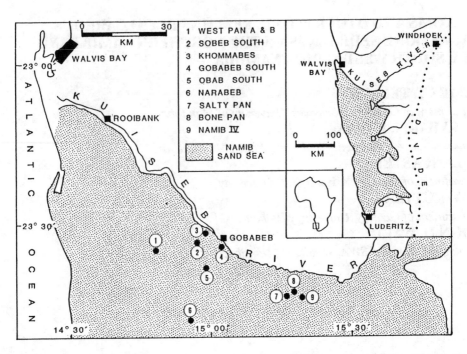

Fig. 1. Locations of Quaternary lacustrine pan deposits in the northern Namib Sand Sea in this study. Inset shows the entire Sand Sea (stippled) and the divide that directs runoff westwards.

(Fig. 1). The calcareous pan sediments at this locality and at several others have been studied by Ward (1984), Teller and Lancaster (1986a,b; 1987), Seely and Sandelowsky (1974), and Selby et al. (1979), with radiocarbon dates and brief description from several pans reported in Vogel and Visser (1981). A major study of both Quaternary and Tertiary calcareous lake deposits is underway by J.T. Teller, J.D. Ward, N. Lancaster and N.W. Rutter; only preliminary results have been published to date (Ward, 1986; Teller and Lancaster, 1986c; Ward et al., 1987).

The Quatenary pan deposits consist of less than 1 m of interbedded calcareous mudstones, silty to sandy limestones, and calcareous silts and sands. The carbonates in these pans are mainly calcite and magnesian calcite, and have been radiocarbon dated at 20000 to 40000 years B.P., except for one date of 11130 years B.P. at Salty Pan (Table 1). Thus, their age falls in the range of most calcareous deposits in the central Namib Desert (cf. Vogel and Visser, 1981). These lacustrine beds are locally exposed on the floor of the interdune depressions in the northern part of the Namib Sand Sea, and commonly form an eroded, slightly elevated resistant cap over less calcareous sands, especially where case-hardening of the upper few millimetres has occurred. In many places, Tsondab Sandstone or Salem granite underlies the calcareous sequence at shallow depths, or crops out nearby. In all cases, the real extent of the pan deposits is small, typically comprising a few thousand square

160

Table 1. Radiocarbon dates and δ^{13}C values from pans in this study. All dates on carbonate.

Site/^{14}Cdate/lab no.	δ^{13}C (o/oo)	Material
Gobabeb South (Vogel & Visser, 1981, p. 76)		
21,300 ± 260 (Pta-2651)	+1.3	reed cast
21,500 ± 260 (Pta-2652)	+1.3	reed cast
Khommabes (Vogel & Visser, 1981, p. 76-77)		
20,900 ± 230 (Pta-1091)	-5.2	root cast
21,500 ± 190 (Pta-2604)	-3.6	termite nest
22,400 ± 210 (Pta-2584)	-2.8	worm channel
27,400 ± 310 (Pta-2590)	-4.5	reed cast
28,500 ± 370 (Pta-2591)	-4.4	reed cast
31,600 ± 430 (Pta-2589)	-8.6	reed cast
31,900 ± 460 (Pta-2588)	-8.5	reed cast
Narabeb (Teller & Lancaster, 1986b; Vogel & Visser, 1981, p. 73)		
20,320 ± 300 (Beta-9115)	+6.6	mudstone
22,330 ± 600 (Beta-9116)	+6.4	mudstone
22,500 ± 280 (Pta-3704)	+0.7	mudstone
26,400 ± 340 (Pta-3759)	-0.5	calc silt
28,500 ± 500 (Pta-1197)	-3.2	root cast
39,800 ± 1700 (Pta-3770)	-1.1	mudstone
Salty Pan		
11,130 ± 125 (BGS-1129)	-4.6	calcite crust
West Pan A		
27,500 ± 1000 (BGS-1128)	+0.4	limestone

metres of eroded outcrop.

The main purpose of this paper is to describe the diatoms of two pans and to discuss their paleoecological significance. In addition, other organic remains in the sediments of other Quaternary-age pans of the northern Namib Sand Sea are described and used to help with interpretation of the paleoenvironmental setting.

2. ORGANIC REMAINS IN THE PAN SEDIMENTS

Fossil organic material occurs in all Quaternary pans in the northern Namib Sand Sea discussed in this paper (Table 2). The calcified stems of the reed *Phragmites* sp. occur in nine of the 10 pans, gastropods in four pans, and diatoms in five pans. In addition, Early to Late Stone Age artefacts are known from four of these locations.

Table 2. Organisms and archaeological remains identified from pans in this study.

Site	Diatoms	Gastropods	Phragmites	Artefacts
Bone Pan			X	X
Gobabeb South	X	X^2	X	
Khommabes	X		X	X^3
Namib IV	X		X	X^4
Narabeb	X^1			X^5
Obab South			X	
Salty Pan			X	
Sobeb South		X	X	
West Pan A		X	X	
West Pan B	X	X	X	

----

[1] Teller and Lancaster (1986b)

[2] Ward (1984, p. 268)

[3] Ward (1984, p. 271)

[4] Shackley (1980)

[5] Seely and Sandelowsky (1974), Shackley (1985)

A variety of other organic remains have been reported from several of the pans. At the Namib IV site, bones and tooth fragments of *Elephas recki* have been found in calcareous sands (Shackley, 1980). Calcified root casts of the !Nara plant (*Acanthosicyos horrida*) occur at Khommabes (Vogel and Visser, 1981, p.76; Ward et al., 1983, p.180) and possibly at Narabeb (Vogel and Visser, 1981, p. 73), and "pedotubles" are known from deposits in all ten pans. Vogel and Visser (1981, p. 77) describe a "calcified termite nest" at Khommabes, and Seely and Mitchell (1986) suggest that the pedotubles at Narabeb are the burrows of termites.

In most cases, the *Phragmites* stems were found associated with the highly calcareous, uppermost lacustrine unit in the pans. Commonly, such as at Khommabes, they occur as loose calcified stem-cast fragments (without the original vegetal material) on a case-hardened, nearly pure calcite unit that forms a thin resistant cap over the older calcareous part of the sequence (Ward, 1984; Ward et al., 1983, p.180; Teller and Lancaster, 1986a; Vogel and Visser, 1981, p.76-77). Occasionally, root nodes or broken stems are found cemented within this limestone unit.

Gastropods have been found in the sediments of four pans (Table 2). Identification of the snails in a buried calcareous silty sand unit at West Pan B by C. Appleton (University of Natal) shows fresh to brackish water types, *Biomphalaria pfeifferi, Melanoides turbercula, Bulinus* sp., and *Tomichia* (cf. *T. ventricosa* sp. is present in the sandy limestone of the surface unit at West Pan A. Ward (1984, p.268) reports gastropods from the Gobabeb South site that were tentatively identified as *Succinia* sp. The gastropods collected from a carbonate-cemented (case-hardened) sand, 17 cm below the *Phragmites*-bearing case-hardened surface crust (in sand) at Sobeb South were not identified.

The following section discusses the 34 species of diatoms found in sediments at West Pan B, Namib IV, Gobabeb South, and Khommabes.

3. DIATOMS

3.1 Sampling and analysis

A total of 89 samples from 17 sites in the northern Namib Sand Sea were investigated for diatoms. Of these, 61 samples are from the ten pans interpreted to be Quaternary in age; the remainder probably are Tertairy-age beds. With only one exception, all diatom remains come from Quaternary lacustrine deposits. However, diatoms were found in only four of the 10 Quaternary-age pans and in only fourteen of the 61 samples from these pans (Table 3). In addition Teller and Lancaster (1986b) reported diatoms in one unit at a fifth pan, Narabeb, although no frustules could be identified in a second split examined by M. Rybak. In seven of the 14 diatom-bearing samples (West Pan B [2 samples], Gobabeb South [1 sample], Namib IV [3 samples], and Khommabes [1 sample]) the abundance was too low to allow a quantitative analysis. Partial dissolution of frustules was common. Only in two beds at Namib IV and three beds at West Pan B was preservation and abun-

Table 3. Percentages of individual diatom species identified sediments in four pans in the northern Namib Sand Sea. Numbers heading columns refer to sample numbers in stratigraphic sections of Figs. 3 and 4. Presence of Chrysophycean cysts and Porifera spicules indicated by +, as are diatoms in samples with low totals. (G = Gobabeb South, K = Khommabes).

SPECIES	WEST PAN B							NAMIB IV				
	1	2	3	4	5	6	7	c	b	a	G	K
1. Achnanthes sp.					0.1	+		+	2.4	0.8		
2. Amphora mexicana var major (Cl.)				0.7	0.1	1.6				0.1		
3. Amphora sp. (fragments)				1.0	0.7	1.1						
4. Anomoeoneis sphaerophora (Kutz.) Pfitz.			2.0	1.0								+
5. Campylodiscus clypeus Ehr.	+	99.7	96.0	9.8	0.7	5.5	+	+	94.6	87.4		
6. Campylodiscus sp. (fragments)												
7. Cyclotella meneghiniana Kutz.				1.4	4.0	2.7			0.7	2.1		
8. Cyclotella sp.		0.1			0.5				0.2	1.8		
9. Cymbella cistula (Hemp.) Grun.				2.1					0.1	0.2		
10. Cymbella sp. (fragments)										0.1		
11. Diploneis smithii (Breb.) Cl.					0.6				0.1	0.4		
12. D. subovalis Cl.		0.1										
13. Diploneis sp. (fragments)					0.1							
14. Epithemia sorex Kutz.				0.7	0.1					0.3		
15. E. zebra (Ehr.) Kutz.				3.2	1.6	4.4						
16. E. zebra var saxonica (Kutz.) Grun.				1.4	3.4	4.4			0.5	0.8		
17. E. turgida var granulata (Ehr.) Grun.				12.2	9.0	1.6				0.3		
18. Epithemia sp. (fragments)					0.8	15.4				2.0		
19. Fragilaria brevistriata Grun.												
20. F. brevistriata var linearis May.				4.2	1.8	2.7			0.7			
21. F. pinnata Ehr.	+											
22. Fragilaria sp. (fragment)												
23. Mastogloia sp. (fragment)										0.2		
24. Navicula oblonga Kutz.		0.1	2.0	0.4	0.5	6.0		+		0.4		
25. Navicula sp. A (fragments)				0.3					0.2	1.1		
26. Navicula sp. B					0.5							
27. Neidium sp. (fragments)					0.5							
28. Nitzschia amphibia Grun.									0.2			
29. N. denticula Grun.					1.2							
30. N. frustulum (Kutz.) Grun.				1.0	2.2	1.1				1.1		
31. N. hybridaeformis Hust.										0.6		
32. Nitzschia sp. (fragments)				4.5	0.6	2.2			0.1			
33. Pinnularia sp. (fragments)				0.7		+			0.2			
34. Synedra ulna (Nitzsch.) Ehr.	+			55.2	71.5	51.3			+	0.3		
Chrysophycean cysts		+	+	+	+	+	+	+		+		
Porifera spicules		+	+	+	+	+	+	+		+		
Total number of diatoms in one slide	4	853	100	285	825	182	3	12	1057	940	3	1

dance adequate to allow a paleoenvironmental reconstruction.

Samples were disaggregated by heating and cooling, and then decanted several times to remove fine particulates from the diatom frustules. The diatom suspension was settled on cover slips according to Battarbee's (1973) method and mounted on slides with Hyrax. Analyses were made using a Zeiss RA microscope, at a magnification of 1000 X. Because of the low abundance, all identifiable frustules on each slide were counted (Table 3). The same cleaned suspensions were used for SEM examination. An Hitachi S-530 SEM was used, with an accelerating voltage of 10 kV for photography. Diatom identification and ecological requirements of taxa were based on Hustedt (1930, 1957), Huber-Pestalozzi (1942), Cleve-Euler (1951-1955), Patrick and Reimer (1966, 1975, Cholnoky (1968), Gasse (1978, 1980, 1986a, 1986b), Gerloff and Cholnoky (1979), Hakansson (1979), Foged (1979), Lange-Bertalot (1980), Lange-Bertalot and Rumrich (1980), Beaver (1981), Germain (1981), and Gasse et al. (1983, 1988).

3.2. Diatom flora

The fossil diatom flora comprises 34 taxa, which represent 15 genera. Some of the observed diatoms have been assigned to particular taxa only tentatively, because of very poor preservation or great morphologic variation of their valves (e.g. *Nitzschia* sp.).

Campylodiscus clypeus (Fig. 2a) was the dominant form in the Namib IV samples and in 3 of the West Pan B samples, although its abundance ranged between 0.7 and 99.7 % (Table 3). The other common form was *Synedra ulna,* which ranged between 51 and 72 % in 3 samples at West Pan B. In the same three samples from West Pan B, the subdominant forms were *Epithemia zebra* and var. *saxonica.* *Nitzschia* was represented by *N. denticula, N. frustulum, N. hybridaeformis* and *N.(cf) amphibia.* The remaining diatom taxa were found in very small numbers. Figure 2 shows several of the species identified. Table 2 also shows the occurrence of chrysophycean cysts and Porifera spicules in the samples from West Pan A and Namib IV.

In a study by Teller and Lancaster (1986b) at Narabeb, Tabellaria fenestrata was the most abundant species. Several other planktonic and freshwater species were also identified at that site: *Navicula* sp., *Cyclotella kutzingiana, Synedra delicatissima* (var. *angustissima), Asterionella formosa, Nitzschia palea, Melosira italica,* and *Fragilaria vaucheriae* (var. *vaucheriae).*

3.3. Paleoecology

Because of the low abundance of diatoms in all samples except those at Namib IV and West Pan B, paleoenvironmental interpretations of the water bodies in which

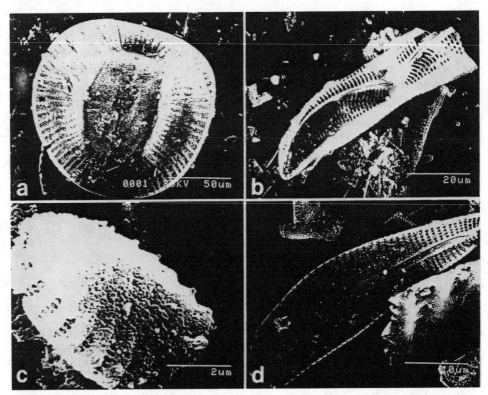

Fig. 2. Four species of diatoms identified in this study: (a) *Camplodiscus clypeus*, (b) *Amphora mexicana* var. major, (c) *Fragilaria pinnata*, (d) *Cymbella cistula.*

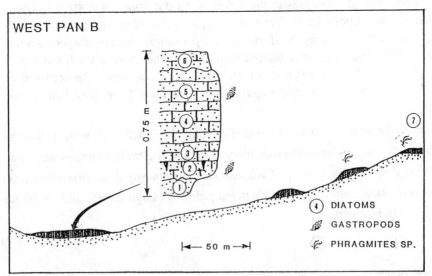

Fig. 3. Stratigraphy of West Pan B site. The calcareous lacustrine beds (black) overlie sand, but have been extensively eroded. Numbers in the stratigraphic section refer to samples that contain diatoms.

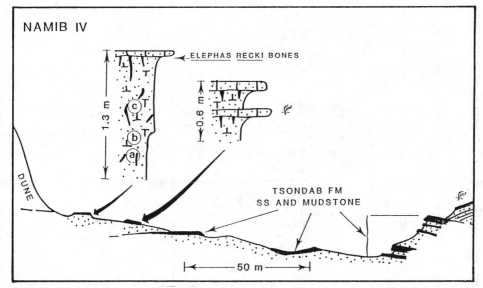

Fig. 4. Stratigraphy at the Namiv IV site. The lacustrine beds of both Tertiary (Tsonab Fm) and Quaternary age are shown in black, and both have been extensively eroded. Letters in the burrowed calcareous sand below where *Elephus reck*i was discovered (Shackley, 1980), indicate the only samples in which diatoms were identified. *Phragmites* sp. (reed symbols) are present at two sites overlying the Tsonab Formation.

the diatoms lived are limited to those two pans. At other sites, reconstruction of the aquatic conditions must be based on mineralogy or other organisms, and may then be supplemented with the sparse diatoms record.

The stratigraphic sections at West Pan B and Namib IV are shown in Figs. 3 and 4, and the beds in which diatoms were found are indicated. Table 4 shows a complication of the ecological requirements for most of the species identified in the Namib pans.

At West Pan B, diatoms were found in all calcareous sediments (#1 to 6, Fig. 3; Table 3). In the lower 20 cm of the calcareous section, which overlies a non-calcareous fine grained sand (#1, Fig. 3), silty to sandy limestone and limey sandstone contain both gastropods and diatoms (#2 and 3 of Fig. 3). The diatom *Campylodiscus clypeus* (Fig. 2a) comprises 96-99.7% of this assemblage, and is interpreted to indicate that water levels were relatively low and brackish, with a pH 8.5 and temperatures of 21-35°C (see Table 4). Brackish conditions are also indicated by the associated gastropods, Tomichia (cf. T. ventricosa), Melanoides turbeculata, Bulinus sp., and Biomphalaria pfefferi. The latter suggests a more-or-less permanent water bodies in which temperatures did not rise much above 30oC and where there is marginal vegetation present (C. Appleton, University of Natal, personal communication).

167

Table 4. Ecological requirements of selected diatoms in this study (after Beaver, 1981; Gasse, 1980, 1986; Gasse et al., 1987; Hustedt, 1957).

SPECIES	Hustedt's halobian system	SALINITY o/oo (fresh 0.5 – oligo-saline 5 – meso-saline 30 – poly-saline 40 – eu-saline 70 – meta-saline)	pH	habitats	nutrient	temp. °C	conductivity µS cm⁻¹	alkalinity
Amphora mexicana var. major	M, P	xxxxXXXXXXxxxx	>8	e, l	-	-	-	-
Anomoeoneis sphaerophora	Hf, M	xXXXXXXXXXXXXXXXXXX	>8.5	l	M	E	-	-
Campylodiscus clypeus	M	xxxxXXXXX	8.5	l	-	21- >35	3000- <10,000	2- >100
Cyclotella meneghiniana	Hf, M	xxxXXXXXXXXXXXxxxxx	8-9	l, p	E, M	E	1000- < 3,000	2- < 50
Cymbella cistula	I	XXXXXxxxxx	7	e	E, M	-	< 300- >10,000	-
Diploneis smithii	M, P	XXXXX	-	l, a	-	E	-	-
D. subovalis	-		7.3-8.0	-	-	10- >35	1000- <10,000	10- < 50
Epithemia sorex	-		6.0-9.5	l, e	E	10- >27	< 300- >10,000	<2- >100
E. zebra	-		7.0-9.5	l, e	-	21- <27	< 300- 3,000	<2- < 50
E. zebra var. saxonica	-		7.0-9.5	l, e	-	21- <27	< 300- 3,000	<2- < 50
E. turgida var. granulata	I	XXXXXXXXXXxxx	-	l, e	M	E	-	-
Fragilaria brevistriata	I		-	l, e	-	-	-	-
F. pinnata	-		-	l, p	E, M	10- <35	-	-
Navicula oblonga	Hf	xXXXXXXxxx	8.3-8.7	l	E	21- <27	-	-
Nitzschia amphibia	I	xxxXXXXXXXXXXXXxxxx	8.5	l	E	21- <27	< 300- 3,000	-
N. denticula	I	xxxXXXXXXXXXxxx	8.2-8.5	l	E	21- <27	-	-
N. frustulum	E, M	xxxXXXXXXXXXXXXxxxxx	>8.5	l, p	E	27- <35	3000- <10,000	-
N. hybridaeformis	-		7	l, p	-	E	-	-
Synedra ulna	I	XXXXXXXXXxxxx	7-8	l, e, p	E	10- <27	< 300- 1,000	<2- < 10

Hustedt's halobian System

P - polyhalobous
M - mesohalobous
E - euryhaline mesohalobous
Hf - halophilous
I - indifferent

Habitats

e - epiphytic
l - littoral
p - planktonic
a - aerophile

Nutrient

E - eutrophic
M - mesotrophic
0 - oligotrophic

Temperature

E - eurythermic

In the sandy to silty limestone overlying this at West Pan B, the diatom assemblage changes drammatically and *Synedra ulna* becomes dominant (samples #4 and 5 of Fig. 3; Table 3). The presence of *S. ulna* and the subdominant species of *Epithemia*, spp., plus several freshwater and epiphytic forms (Table 3), suggests that the salinities and pH may have been less and water depths greater than during deposition of the underlying sediments. Table 4 indicates other paleoenvironmental conditions likely during this lake stage. Capping this section is a case-hardened calcareous sand (#6 of Fig. 3) that is also dominated by *Synedra ulna* and *Epithemia* spp. (Table 3). Calcified stems of the reed *Phragmites* are present in this case-hardened and along the higher marginal zones of West Pan B (Fig. 3), suggesting fresh to brackish conditions and water depths of at least 6 m above the floor of the pan. Alternatively this thin *Phragmites*-bearing calcareous unit may be related to groundwater seepage.

Diatoms at Namib IV were found only in the fine-grained sand that is low (less than 3 %) in calcium carbonate (a and b in section, Fig. 4); the overlying metre of more highly calcareous fine sand (greater than 25 % carbonate) contains no diatoms, except in sample c (Fig. 4), where 12 frustules were identified (Table 3). Bones of the extinct elephant, *Elephas recki*, were collected from the upper part of this section by Shackley (1980).

Campylodiscus clypeus comprises more than 87 % of the assemblage at Namib IV (table 3). *C. clypeus* and associated species indicate that the water was brackish, with less than 30$^{o/oo}$ salinity. Table 4 indicates the ecological conditions associated with all diatoms at this site. The presence of calcified *Phragmites* stems higher on the pan margin (Fig. 4) suggests that water may have reached depths of at least 2 m in this pan, although groundwater seepage could have supplied enough water to sustain these reeds and allow the precipitation of a thin calcareous crust over the sand there.

In contrast to the dominantly benthic species that are newly reported in this paper, the diatoms identified at Narabeb are largely planktonic, as well as being freshwater types (Teller and Lancaster, 1986b). This, along with the distinctly different nature of the sediments at Narabeb, suggests that environmental conditions may not have been the same there; possibly water depths were greater.

4. PALEOCLIMATIC AND PALEOHYDROLOGIC IMPLICATIONS OF BIOTA

The presence of diatoms and aquatic gastropods in highly calcareous sediments at a number of interdune locations in the Namib Sand Sea that are continuously dry today indicates that regional and/or local conditions were at one time substantially wetter. The association of elephant bones, artefacts, and calcified root casts, termite burrows, and Phragmites at these sites support this conclusion. It must be remembered, however, that even today, river valleys such as the Kuiseb, and playas such as Tsondab Vlei and Sossus Vlei, are anomalously wet in contrast to the surrounding hyperarid Namib Desert. The fact that such environments may serve as "oases" for fauna and flora in an otherwise unsuitable climatic environment has been pointed out in Vogel et al. (1981) and Ward et al.(1983).

During the Pleistocene, as today, rainfall in the wetter uplands to the east provided sporadic incursions of water to the Sand Sea, either by runoff or by groundwater recharge. As noted by Teller and Lancaster (1986c) and Ward et al.(1987), water reaching the interdune areas where these pan deposits are found may have been the result of; (1) former rivers extending their termini farther westward into the Sand Sea (e.g. Narabeb {Seely and Sandelowsky, 1974; Teller and Lancaster, 1986b}), (2) lateral extensions of river floodwaters into interdune corridors (e.g. Khommabes [Teller and Lancaster, 1986a]), or (3) groundwater seepage, either through aquifers such as the Tertiary Tsondab Sandstone or through the dunes of the Sand Sea. For example, some sites adjacent to ancient courses of the Kuiseb and Tsondab rivers would have been influenced by increased water levels in those valleys, and groundwater seepage may have been caused ponding in closed depressions. Local runoff and groundwater recharge may have played a role in contributing water to all pan sites. It may be significant that pan deposits at Namib IV, West Pan A and B, Salty Pan, Sobeb South, Obab South, and Narabeb directly overlie the Tsondab Sandstone, which could have served as an aquifer supplying water to those sites, although Namib IV and Salty Pan lie at elevations well above the Kuiseb or Tsondab valleys. Gobabeb South and Khommabes overlie relatively impermeable granite, which would have limited downward percolation of groundwater and have directed its lateral migration through overlying beds into those depressions.

Therefore, an increase in precipitation in the headwaters of the Kuiseb and Tsondab rivers would have increased in those valleys, possibly resulting in back-flooding into inter-dune corridors and, in the case of the Tsondab River, causing water to penetrate into the Sand Sea far beyond the present end point of the river. Influent water from the Kuiseb and Tsondab valleys probably also raised groundwater levels adjacent to these rivers, which may have led to ponding in some interdune depressions. The clustering of radiocarbon dates between 10000 and 40000 years B.P. from these pans, as well as from other calcareous deposits such as tufa, calcrete, and carbonate cement (see Vogel and Visser, 1981; Vogel, 1987), strongly suggests that rainfall increased across the Namib at this time (cf. Deacon and Lancaster, 1988). In the absence of younger radiocarbon dates from the pans in this study, plus the fact that ponding is unknown at those sites in historic time, it seems likely that lakes formed in these depressions during the Holocene.

5. CONCLUSIONS

Thin beds of sandy limestone and calcareous mudstones, silts, and sands at ten interdune sites in the northern Namib Sand Sea contain fossil organic material. The carbonates in some of these lacustrine deposits have been dated, and fall in the range between 10000 and 40000 years B.P., as do most dates in the Namib Desert.

The calcified stems of *Phragmites* sp. were found in nine of the 10 pans, gastropods in 4 pans, and diatoms in 5 pans. The ecological requirements of all identified species indicate that these depressions once contained fresh to brackish water, and in the presence of calcified root casts and other organic remains support the conclusion that conditions were, at least periodically, wetter at these sites during the late Pleistocene.

Because of their present isolation, it seems unlikely that water could have reached these interdune pan locations under the present climatic and hydrological conditions. It is possible that rivers draining the highlands to the east may have penetrated farther into the Sand Sea during the Pleistocene, either because their termini extended farther west or as a result of lateral flooding into inter-dune corridors. It is also possible that groundwaters may have been discharged into interdune depressions during this time, either from the permeable Tertiary Tsondab Sandstone, which underlies most of the pans or by seepage through the dunes confining the pans.

6. SUMMARY

A number of small interdune pans in the northern Namib Sane Sea contain fossil remains. Of the ten Quaternary-age pans studied, 5 contained diatoms, 9 contained calcified *Phragmites'* stems, and 4 contained gastropods. The sediments of pans where fossils were found are composed of thin bedded to laminated calcareous silts and sands, calcareous mudstones, and or silty to sandy carbonate. In most cases these lacustrine beds are less than a metre thick.

Although the fossil diatom flora of this study comprises 34 taxa, which represent 15 genera, diatoms were present in only 14 of the 61 samples, and there was partial dissolution of many frustules. *Campylodiscus clypeus* or *Synedra ulna* are dominant in all samples. A previous study of one pan identified 7 additional species, with *Tabellaria fenestrata* being dominant. At only two sites, West Pan B and Namib IV, was the abundance of diatoms adequate to allow palaeoenvironmental reconstruction. Water levels there appear to have fluctuated, with the salinity varying between about 300 and 10000 microsiemens (fresh to brackish conditions); pH ranged from 7 to 8.5. Eutrophic conditions with aquatic macrophytes occurred during part of the history of West Pan B.

Similarly the gastropods *Melanoides turberculata, Succinea* sp., *Bulinus* sp., and *Tomicha* sp. at West Pan B suggest a fresh to slightly saline (brackish) environment, as do *Phragmites* sp. that were found in most pans. The gastropod *Biomphalaria pfeifferi* at West Pan B suggests a more-or-less permanent water body that was relatively fresh (300-400 microsiemens), and which had marginal vegetation.

Sixteen radiocarbon dates on carbonate from four of the 10 pans in this study fall between 20000 and 40000 years B.P., which is almost the common age range of other carbonates dated in the region. One date from Salty Pan falls in the other common age range for carbonate sediments of the region, 10000-15000 years B.P. Thus, two relatively wet phases are indicated for the pans of the northern Namib Sand Sea. Some interdune depressions were affected by this late Pleistocene increase in precipitation, as the Tsondab and Kuiseb rivers expanded their flooding farther into the Sand Sea, as groundwater levels rose, and as local runoff increased. Although lake levels fluctuated during these periods, water remained ponded long enough to establish a fresh to brackish water community of organisms.

7. ACKNOWLEDGEMENTS

Out thanks to C. Appleton, University of Natal, for identification and interpretation of the gastropods. John Vogel supplies several radiocarbon dates as well as useful discussions on the pans. Mary Seely and the Desert Ecological Research Unit at Gobabeb and the Geological Survey of SWA/Namibia provided field support for this project.

8. REFERENCES

Battarbee, R.W. 1973. A new method for estimating absolute microfossil numbers with special reference to diatoms. Limnology and Oceanography, 18, 647-653.

Beaver, J. 1981. Aparent ecological characteristics of some common freshwater diatoms. Ontario Ministry of Environment, Rexdele, Ontario, 517 pp.

Cholnoky, F.J. 1968. The ecology of diatoms in inland water. J. Cramer, Lehre, 699 pp.

Cleve-Euler, A. 1951-1955. Die Diatommen von Schweden und Finnland: Kungl. Svenska Vetensk. Handl. Sev. 4, 2(1), 3-163; 4(1), 3-158; 4(5), 3-255; 5(4), 3-231; 3(3), 3-153.

Deacon, J. and Lancaster, N. 1988. Late Quaternary Palaeoenvironments of Southern Africa. Clarendon Press, Oxford, 209 pp.

Foged, N. 1979. Diatoms in New Zealand, the Northern Island: Bibliotheca Phycol., 45, 1-255.

Gasse, F. 1978. Les diatomees holocenes d'une tourbiere (4040m) d'unemontagne ethiopienne. le Mont Badda. Rev. Algol., N.S., 13, 105-149.

Gasse, F. 1980. Les diatomees lacustres plio-pleistocenes du Gaded (Ethiopie). Systematique, paleoecologie, biostratigraphie: Revue Algologique, Serie 3, 250 pp.

Gasse, F. 1986a. East African diatoms and water pH. In: Smol, J.P., Battarbee, R.W., Davis, R.B. and Merilainen, J. (eds.), Diatom and Lake Acidity, Dr. W. Junk Publishers, Dorderecht, 149-168.

Gasse, F. 1986b. East African Diatoms. Taxonomy, Ecological Distribution: J. Cramer, Berlin, Germany, 291 pp.

Gasse, F., Fontes, J.Ch., Plaziat, J.C., Carbonet, P., Kaczmaeska, I., DeDeckker, P., Soulile-Marsche, I., Callot, Y.and Dupeuble, P. 1988. Biological remains, geochemistry, and stable isotopes for the reconstruction of environmental and hydrological changes in the Holocene lakes from North Sahara. Palaeogeography, Palaeoclimatology, Palaeoecology, (in press).

Gasse, F., Talling, J.F. and Kilham, P. 1983. Diatom assemblages in East Africa: classification, distribution and ecology: Rev. Hydrobiol. trop., 16, 3-34.

Gerloff, J. and Cholnoky, B. 1979. Diatomaceae II. Verlag Von J. Cramer, Girschberg, Germany, 85 pp.

Germain, H. 1981. Flore des Diatomees. Societe Nouvelle des Editions, Boubee, Paris: 444 pp.

Hakansson, H. 1979. Examination of diatom type of material of C.A, Agardh. Nova Hedwigia, 64, 163-168.

Huber-Pestalozzi, G. 1942. Das Phytoplankton des Süsswassers: Binnengewässer 16. 549 pp.

Hustedt, F. 1930. Die Süsswassers flora Mitteleuropas: Heft 10, p. 1-466. Bacillariophyta (diatomeae). Jena. Mitteleurpas.

Hustedt, F 1957. Die Diatomeen flora des Fluss-systems der Weser im Gebiet der Hansestadt Bremen: Ab. Naturw. Ver. Bremen 34, 181-440.

Lancaster, J., Lancaster, N. and Seely, M.K. 1984. Climate of the central Namib Desert. Madoqua, 14, 5-61.

Lancaster, N. 1983. Linear dunes of the Namib Sand Sea. Zeitschrift für Geomorphologie N.F., Suppl., 45, 27-49.

Lancaster, N. 1988. The Namib Sand Sea; Dune forms, Processes, and Sediments: A.A. Balkema, Rotterdam, 170 pp.

Lange-Bertalot, H. 1980. New species, combinations and synonyms in the genus Nitzschia. Bacillaria, 3, 41-77.

Lange-Bertalot, H. and Rumrich, U. 1980. The taxomic identity of some ecological important small Naviculae: 6th Diatom-Symposium, Nova Hedwigia, 135-153.

Patrick, R. and Reimer, C. 1966. The Diatoms of the United States, Exclusive Alaska and Hawii, Part 1: Acad. Nat. Science Philadelphia Monograph 13, 1-688.

Patrick, R. and Reimer, C. 1975. The Diatoms of the United States Exclusive Alaska and Hawaii, Part II: Acad. Nat. Science Philadelphia Monograph. 13, 1-213.

Seely, M.K. and Mitchell, D. 1986. Termite casts in Tsondab Sandstone? Palaeoecology of Africa, 17, 109-112.

Seely, M.K. and Sandelowsky, B.H. 1974. Dating the regression of a river's end point: South African Archaeological Bulletin, Goodwin Series, 2, 61-64.

Selby, M.J., Hendey, C.H. and Seely, M.K. 1979. A late Quaternary lake in the central Namib Desert, southern Africa, and some implications. Palaeogeography, Palaeoclimatology, Palaeoecology, 26, 37-41.

Shackley, M. 1980. An Acheulean industry with Elephas recki fauna from Namib IV, South West Africa (Namibia): Nature, 284, 340-341.

Shackley, M. 1985. Palaeolithic archaeology of the central Namib Desert: Cimbebasia B. Memoir, 6, 84 pp.

Teller, J.T. and Lancaster, N. 1986a. History of sediments at Khommabes, central Namib Desert. Madoqua, 14, 409-420.

Teller, J.T. and Lancaster, N. 1986b. Lacustrine sediments at Narabeb in the central Namib Desert, Namibia. Palaeogeography, Palaeoclimatology, Palaeoecology, 56, 177-195.

Teller, J.T. and Lancaster, N. 1986c. Interdune lacustrine deposits in the Namib Sand Sea, Namibia. 12th International Sedimentologists Congress. Canberra, Abstracts, p. 298.

Teller, J.T. and Lancaster, N. 1987. Description of late Cenozoic sediments at Narabeb, central Namib Desert. Madoqua, 15 (in press).

Vogel, J.C. 1987. Chronological Framework for Palaeoclimatic Events in the Namib. National Physical Research Laboratory Research CFIS 145. 20 pp.

Vogel, J.C. and Visser, E. 1981. Pretoria radiocarbon dates II: Radicarbon, 23, 43-80.

Ward, J.D. 1982. Aspects of a suite of Quaternary conglomerate sediments in the Kuiseb Valley, Namibia. Palaeoecology of Africa, 15, 211-216.

Ward, J.D. 1984. Aspects of the Cenozoic geology in the Kuiseb Valley, central Namib Desert: Unpublished Ph.D. Thesis, University of Natal, Pietermaritzburg, 310 pp.

Ward, J.D. 1986. Aeolian, fluvial and playa facies of the Tertiary Tsondab Sandstone Formation in the central Namib Desert, South West Africa/Namibia. 12th International Sedimentologists Congress, Canberra, Abstracts, 320.

Ward, J.D., Seely, M.K. and Lancaster, N. 1983. On the antiquity of the Namib. South African Journal of Science, 79, 175-183.

Ward, J.D., Teller, J.T., Rutter, N.W. and Lancaster, N. 1987. Quaternary lacustrine deposits in the central Namib Desert: XII INQUA Congress. Ottawa, Programme and Abstracts, p. 284.

EQUILIBRIUM CONCEPTS, VEGETATION CHANGE AND SOIL EROSION IN SEMI-ARID AREAS: SOME CONSIDERATIONS FOR THE KAROO

KATE M. ROWNTREE
Department of Geography, Rhodes University

1. INTRODUCTION

To many ecologists and pasture scientists in South Africa, Karoo, land degradation and desertification are synonymous terms. The Karoo is the name given to the region of semi-desert which supports a characteristic vegetation cover of dwarf shrubs, grasses and succulent scrub (Meadows, 1985) covering an area of approximately 427015 sq. km in the Cape Province of South Africa. Despite the low mean annual rainfall of between 500 to 100 mm (Venter et al. 1986) the Karoo has long been exploited for its grazing potential: firstly by the Khoi San who utilised the extensive game herds, secondly by Khoikhoi pasturalists and lastly by settled European farmers. It is widely believed that as a result of over exploitation of this semi-arid ecosystem the vegetation has undergone an irreversible change towards increasing aridity coupled with a greatly increased potential for soil erosion (Roux and Vorster, 1983).

The ubiquitous nature of soil erosion in the Karoo is self evident (Roux and Opperman, 1986) and has been recognised at least since 1923 (Anon, 1923). The Vlekpoort (1941) and Lake Arthur (1946) conservation areas, both in the eastern Karoo, were amongst the first to be proclaimed in the country (Ross, 1947). Sheet, rill and gully erosion are especially prevalent in the eastern parts of the Karoo (as for example near De Aar and Middelburg) due to the occurrence of duplex soils with poor surface infiltration and dispersed subsoils (Ellis and Lambrechts, 1986). Erosion is seen to be both a consequence and a cause of the veld deterioration (Acocks, 1975) which has been particularly pronounced since the beginning of European settlement in the 1700's. The main thrust of ecological research within the Karoo biome is therefore towards reversing this trend wherever possible, or at least towards preventing   further deterioration.

Although poor veld management has almost certainly contributed to soil degradation in the Karoo, the problem must also be seen in the context of natural environmental conditions and long term geomorphological process. The worldwide susceptibility of semi-arid climates to erosion, due to the combined effects of erosive rainfall and a low biomass, is well established (Langbein and Schumm, 1958; Thornes, 1976; Wilson, 1973). The widespread occurrence of surface water erosion in the Karoo can be explained primarily in terms of the aridity of the climate and the instability of the vegetation cover; within these environmental limits grazing systems which further reduce and alter the vegetation become critical in accelerating erosion rates.

Erosion may take a number of forms, including sheet wash, rilling and gullying. Sheet wash and rilling represent the means by which sediment is transported across the slope by overland flow and are important slope-forming processes in semi-arid areas (King, 1963; Carson and Kirkby, 1972). In contrast gullies are part of a drainage network and an actively eroding gully represents an unstable landform which is part of a drainage network transformation (Schumm et al., 1984). Whereas sheet wash and rilling may be compatible with a stable slope system, gully erosion must represent instability at least in the short term. It is therefore thought appropriate to consider the relationship between instability in soil-slope systems as represented by gully erosion and instability in the plant community as induced by poor veld management. Gully erosion and its relation to vegetation cover therefore forms the focus of this paper.

A comprehensive review of gully phenomena is represented by Schumm et al. (1984). They consider gullies to be part of a broader group of incised channels which include both entrenched streams and gullies themselves. An entrenched stream is a term given to the deep incision and widening of a pre-existing channel situated in the gully floor, whereas a gully is an extension of the drainage network into an area where no channel previously existed and therefore represents an increase in drainage density, either as an extension of the main channel or as lateral gullies into the valley side slopes. The colloquial term 'donga' would appear to include both types of incised channel.

Both entrenched channels and gullies are related to rejuvenation of the stream network and are closely interlinked. Entrenchment of the stream channel may lead to gullying of the valley head or the valley side slopes, whereas the increased water and sediment discharge from the stream gullies in turn affect the downstream channel. Rejuvenation of the network inevitably leads to continued erosion until a new equilibrium form is achieved. The evolution of the gully form has been studied by a number of researchers (Ireland et al., 1939; Leopold and Miller, 1956; Brice, 1966; Heede, 1974; Schumm et al., 1984). Common to most discussions is a general evolutionary sequence which is initiated by incision followed by upslope migration of the resulting headcut to form a deep, narrow channel. Stability is achieved as the gully floor widens, the slopes flatten to a stable angle and the gully is recolonised by vegetation. The morphology of the gully system depends also on the host materials (Patton and Schumm, 1981; Ireland et al., 1939) and the dominant erosion process (Bradford and Piest, 1980); entrenched gullies are dominated by mass failure processes whereas v-shaped gullies are dominated by surface water erosion.

The main body of this paper is divided into three parts. The first presents a review of factors which are associated with rejuvenation processes and considers the available evidence for the Karoo. The second examines the relationship between vegetation cover and erosion processes, while the third section explores alternative equilibrium models relevant to semi-arid erosional systems.

2. REJUVENATION OF THE DRAINAGE NETWORK

Rejuvenation of a drainage network is the result of some geomorphological threshold being exceeded so that rapid erosion is initiated. Schumm et al. (1984 p. 14) defined a geomorphic threshold as "....a threshold of landform stability that is exceeded either by intrinsic change of the landform itself.... or by a change of an external variable....". Intrinsic changes include changes in shear strength due to weathering (Carson and Kirkby, 1972), or changes in form as for example when the slope is steepened due to aggradation (Patton and Schumm, 1981). External factors include those related to climatic changes or land use changes which affect slope hydrology and/or surface resistance to erosion.

Gullies are part of an expanding drainage network and therefore cannot be studied in isolation from the catchment area within which they occur. The occurrence of gullies must therefore be explained firstly in terms of catchment characteristics which control inputs of water and sediment to the valley floor and valley floor characteristics (or those of the valley side drainage line) which determine resistance to erosion (Graf, 1979). Catchment variables found to be important in explaining the distribution of gullies include area (Graf, 1979; Patton and Schumm, 1975; Stocking, 1980), and vegetation cover (Graf, 1979; Melton, 1965), while the most critical valley floor variables appear to be vegetation cover (Graf, 1979; Melton, 1965) and slope gradient (Patton and Schumm, 1975). As expected, vegetation cover is thus a key variable which determines both surface runoff volumes and valley floor resistance and factors which negatively effect vegetation cover are likely to promote rejuvenation of the drainage network.

Much of the main research on gully erosion has come from the United States where poor farm management and overgrazing in the early part of this century were widely blamed for the widespread arroya erosion throughout the arid southwest. More recent studies have challenged this assumption, and though it is recognised that pasture deterioration is one of the causes of gully initiation (Graf, 1979; Heede, 1974; Melton, 1965), other factors have also been found to be important. These include changes in the seasonal distribution of intense storms (Graf, 1977) and intrinsic changes due to natural cycles of cut and fill in a semi-arid climate (Denny, 1967; Melton, 1965; Patton and Schumm, 1975 and 1981; Womak and Schumm, 1977). The importance of applying multiple hypotheses to explain gully erosion is stressed by Cooke and Reeves (1976) in the conclusion to a comprehensive study of arroya erosion in the south-western United States.

Changes in the vegetation cover due to mismanagement have also been blamed for erosion in the Karoo. Changes described in the literature (Acocks, 1975; Roux and Vorster, 1983) include; (1) a general loss in canopy cover, (2) a change in species composition, with both a decrease in number of species and in increase in the percentage of unpalatable species, (3) a change in the structure of the vegetation with an increase in shrubs relative to grass, and (4) an eastward spread of karroid vegetation into the wetter eastern Cape, Ciskei and the Orange Free State.

Evidence for vegetation change comes largely from historical documents and must be viewed with caution. Karoo vegetation is highly dynamic and responds dramatically to rainfall (King, 1963; Cowling, 1986) so that reports made in high rainfall years would indicate a considerably better vegetation cover than that observed in dry years. Our knowledge of the former vegetation cover is therefore somewhat sketchy and there is a belief by some researchers that the present day vegetation communities are in fact less degraded than is often presumed. For example, ongoing field surveys by Palmer (pers. comm.) indicate that the floristics and structure of plant communities are determined more by intrinsic environmental conditions than by grazing intensities.

In view of the general uncertainty as to the causes of the gully initiation, to explain gully erosion in the Karoo in terms of veld deterioration alone may be too simplistic and alternative hypotheses should be investigated. In particular the possible relevance of intrinsic thresholds should be borne in mind. Elsewhere in southern Africa researchers have pointed to non-anthropogenic causes of gullies which are seen to be part of the natural drainage system in certain environments (Stocking, 1980; Broderick, 1986). None the less, the vegetation cover is obviously an important part of the erosional system and its relationship to runoff and erosion processes merits further attention.

3. VEGETATION AND SOIL EROSION

Vegetation and soil are intimately related to one another within the ecosystem, vegetation cover being dependent on a supply of minerals and water from the soil, whereas soil development is in large part a function of vegetation cover. The two components, soil and vegetation, must therefore be considered as closely interrelated parts of one system so that the stability of one is reflected in the stability of the other. It is not surprising, therefore, that vegetation cover is such a key variable in controlling erosion rates (Kirkby, 1980) and that erosion so often results as a consequence of cover changes. What is more surprising is the paucity of quantitative studies of the effect of vegetation on erosion (Thornes, 1985).

The qualitative effects of vegetation are well recognised: protection of the soil from raindrop impact, increases in infiltration rates both through improvements in soil structure and utilisation of soil moisture, and increased resistance of the soil surface to erosion (Thornes, 1985). The relative magnitude of these effects and

178

their relationship to different vegetation structures is not so well understood. Of the results that are available there appears to be a general agreement that a ground cover of between 20 to 30 per cent is critical for erosion by sheet wash (Dunne, 1977; Elwell and Stocking, 1976; Snyman, 1986). The importance of the structure and composition of the vegetation in controlling erosion rates has been stressed by a number of authors (Roux, 1981; Thornes, 1985). Roux, comparing erosion losses from slopes near Grootfontein, substantiated for the Karoo the widely held belief that a grass cover is more effective against sheet erosion than are shrubs. Crossby et al. (1981) provide estimates of the C factor for the Universal Soil Loss Equation (USLE) which can be applied to a wide range of South African cover types including bush and grassland. The USLE, however, is designed to estimate long term soil loss rates by sheet and rill erosion; it is not directly applicable to gully erosion.

Selby and Hosking (1973), point to the importance of separating the effects of vegetation on runoff generation and surface resistance when considering gully erosion. Like Roux (1981), they found that, compared to scrub, grassland provided greater resistance to simulated gully erosion of pumice soils in New Zealand, but it also produced a greater runoff volume and was therefore more intensely gullied. Also of interest are results obtained by Thompson (1935) who showed that grazing of erosion plots near Pretoria markedly increased surface runoff relative to that from intact veld, whereas the increase in erosion was only slight. The potential for gully erosion is therefore increased due to the high runoff volumes and low sediment load contributed by the grazed areas. These results emphasise the importance of considering gully erosion as one component of the total catchment system, erosion at the site being a function of upslope contributions of runoff and sediment. The distribution of the vegetation cover within the catchment area is also highly significant.

The relationship between the spatial distribution of vegetation cover and hydrogeomorphological processes is stressed by Zimmerman and Thom (1980). They illustrate how the pattern of vegetation is related to landform units through controls on soil stability, moisture availability and nutrient fluxes. In semi-arid areas such as the Karoo the primary factor limiting species composition, vegetation biomass and canopy structure is obviously moisture availability, itself a function primarily of precipitation less evapo-transpiration, and secondarily of topographic and pedological controls on soil moisture and runoff. In the Karoo the dwarf and low succulent shrubs dominate the winter rainfall regions of the south-west and west, whereas grasses become more important in the summer rainfall regions of the east (Cowling, 1986). Locally soil type, topography and aspect are important factors controlling moisture availability and therefore vegetation characteristics (Acocks, 1975). Meadows (1985) describes a general correlation between plant communities and topography in the Karoo with a mosaic of dwarf shrub and grass communities on the pediments or 'flagtes', larger shrub and small bush communities on the steep talus slopes and woodland or gallery forest along the main drainage lines where groundwater is available. The extent of gullying therefore should be correlated to the broad precipitation-vegetation zones, whereas the location of individual gullies and other erosional phenomena should be related to the topographic controls on moisture distribution, vegetation cover and runoff.

The vegetation-soil system cannot be considered in isolation from the herbivores which graze the vegetation and strongly influence its characteristics. Much effort has been put into devising veld management systems for the Karoo which optimise the returns from livestock without causing degradation of the vegetation and soil (Hugo, 1968). Grazing affects the density, structure and composition of vegetation. Selective grazing tends to encourage shrub growth at the expense of the herbaceous and grass species (Roux and Vorster 1983), thus reducing the soils resistance to sheet and rill erosion. Short period intensive grazing is thought to be better at maintaining species diversity and hence the resilience of the soil and vegetation against degradation (McCabe, 1987).

4. EVOLUTION OF GULLY SYSTEMS AND EQUILIBRIUM CONCEPTS

Research in the U.S.A. has shown that gully erosion and channel incisement either may be a response to a change in catchment conditions which cause the drainage systems to move towards a new equilibrium state, or may represent the long term dynamic equilibrium which is characteristic of semi-arid areas, as modelled by Schumm and Hadley's semi-arid cycle of erosion (Schumm and Hadley, 1957). Implicit in this cycle is the notion that contemporaneous erosion and deposition are important features of stability in semi-arid areas, but the locus of each shifts through time. Hence stability and eventual elimination of the gully in its downstream reaches depends on the deposition of sediment eroded from upslope sections by surface wash and rilling; without this erosion the system remains unstable and further incision may ensue. Eventually, oversteepening of the lower gully reach as a result of aggradation leads to rejuvenation and incision begins anew. The system therefore can be said to exhibit a form of equilibrium in which shifts in erosion and deposition are the result of intrinsic geomorphological changes rather than external in energy inputs.

The evolution of a particular gully system may be described in the context of one or other of the above equilibrium conditions or a combination of both. For example in disturbed areas there is initially a long term shift in dynamic equilibrium due to external factors such as poor veld management which induce a semi-arid type of condition. If the disturbed environmental conditions are maintained, however, the new dynamic equilibrium towards which the system moves may be that of the semi-arid cycle of erosion.

It has been shown above that, according to Hadley and Schumm (1957), the long term equilibrium state for semi-arid areas may be in the form of a continuous cycle of gully entrenchment and infilling which is dependent on the maintenance of an upslope sediment. It therefore follows that the stable vegetation in semi-arid areas must include at least some communities which allow continued erosion while other communities will be associated with sites of deposition. Preliminary field observation near Middelburg in the Karoo have shown that the vegetation of flats beneath heavily grazed slopes with a high sediment yield is more vigorous than that on flats below protected slopes. Sediment transport and deposition thus appear to be a positive factor in vegetation growth. The redistribution of finer sediments around vegetation clumps may represent an important source of nutrients in an environment where chemical weathering is believed to be slow (Dunne, 1978). The recolonisation of gullies by vegetation is also fundamental to any consideration of gully stabilisation and yet, as noted by Thornes (1985), there is a surprising lack of detailed research on this topic. Recolonisation be vegetation requires adequate soil moisture and a stable substrate, both conditions being negated by erosion. The idea of competition between erosion and plant colonisation is modelled by Thornes who points to the need for field research to substantiate his theoretical concepts.

It is appropriate at this point to return to a consideration of equilibrium conditions for the soil-vegetation system of the Karoo. Vegetation may infact be inherently unstable in the sense that it changes in the short term in response to rainfall, transient grazing pressures, fire and so on, but may exhibit 'resilience' in that the community persists through time (Ferrar, 1983; Holling, 1973). If this is the case, then soil erosion may occur in response to short term changes in vegetation cover, yet in the long term the overall balance of erosion and deposition must be such as to maintain the structure of the plant community. If this were not so, the community would show a long term transgression towards a degraded form in response to cumulative erosion. Hence if the vegetation of the Karoo is in dynamic equilibrium with a highly variable precipitation input, then any observed soil erosion must be part of that equilibrium and the erosion of the soil, its transport and deposition are important component of ecosystem dynamics. The all pervasive nature of erosion in semi-arid areas (Langbein and Schumm, 1958; Thornes, 1976; Wilson, 1973) supports the conclusion that soil erosion and redistribution of sediment must be intrinsic processes of semi-arid ecsytems to which the vegetation is adapted. Recognition and elucidation of these dynamic geomorphological factors would seem to be an essential component of any ecological studies within semi-arid areas such as the Karoo.

5. CONCLUSIONS

The discussion in this paper has underlined the importance of studying soil erosion in semi-arid areas in relation to the stability of the vegetation. Veld deterioration and accelerated soil erosion are undoubtably prevalent within many plant communities in the Karoo and adjacent areas, but before definitive conclusions can be made it is necessary to come to a better understanding of both the sediment related processes and the dynamics of vegetation change in the Karoo as well as the inter-relationships between the two. As Meadows (1985:101) points out "...remarkably little is known about the structure and function of these communities..." and until a more thorough understanding of one of the key factors controlling erosion (vegetation cover) is obtained it is perhaps dangerous to make definitive statements about erosion itself.

6. SUMMARY

This paper examines the relationship between vegetation cover and hydrogeomorphic processes with particular reference to soil erosion in the semi-arid Karoo of South Africa. The discussion focuses on the processes of gully erosion because it is seen as an expression of geomorphological instability which has direct implications for ecosystem equilibrium. The paper is divided into three parts. The first reviews factors associated with the rejuvenation of drainage systems responsible for the initiation of gully erosion and examines the evidence for the Karoo itself. The second outlines established relationships between vegetation cover and sediment processes within the context of gully erosion. Finally the paper brings together concepts of geomorphological stability and ecosystem stability in order to explore the interrelationships between them. The paper concludes that in semi-arid areas such as the Karoo, erosion may not necessarily represent degradation because cycles of erosion and deposition must be part of the dynamic equilibrium of semi-arid ecosystems. The need for further research into the relationship between plant community dynamics and sediment processes is emphasised.

7. REFERENCES

Acocks, J.P.H. 1975. Veld types of South Africa. Memoir of the Botanical Survey of South Africa, 40, Department of Agriculture Technical Services, Pretoria.

Anon 1923. Final Report of the Drought Investigation Commission. Cape times Ltd. Govt. Printers.

Bradford, J.M. and Piest, R.F. 1980. Erosional development in valley bottom gullies in the upper midwestern United States. In: D.R. Coates and J.D. Vitak (eds.), Thresholds in Geomorphology, Allen and Unwin, 75-101.

Broderick, D. 1986. A history of land degradation in the Tugela Basin, Natal. Sizwe, University of Natal, Durban.

Brice, J.C. 1966. Erosion and deposition in the loess-mantled Great Plains, Medicine Creek Drainage Basin, Nebraska. U.S. Geological Survey Professional Paper, 352H, 259-339.

Carson, M.A. and Kirby, M.J. 1972. Hillslope Form and Process. Cambridge University Press.

Crosby, C.T., Smithen, A.A. and McPhee, P.J. 1981. Role of soil loss equations in estimating sediment production. In: H. Maaren (ed.), Workshop on the Effect of Rural and Land Use and Catchment Management on Water Wesources. TR 113, Department of Water Affairs, Forestry and Environmental Conservation, Pretoria, 183-213.

Cooke, R. and Reeves, R. 1976. Arroyos and Environmental Change in the American South-west. Clarendon Press, Oxford.

Cowling, R.M. 1986. A description of the Karoo Biome Project. South African Scientific Programmes Report No. 122. FRD, CSIR, Pretoria.

Denny, C.S. 1967. Fans and pediments. American Journal of Science, 265, 81-105.

Dunne, T. 1977. Intensity and controls of soil erosion in Kajiado District, Kenya. U.N.F.A.D., Nairobi, 151 pp.

Dunne, T. 1978. Rates of chemical denudation of silicate rocks in tropical catchments. Nature, 274, 244-246.

Ellis, F. and Lasmbrechts, J.J.N. 1986. Soils. In:Cowling, R.M., Roux, P.W. and Pieterse, A.J.H. (eds.). The Karoo biome: a preliminary synthesis. Part 1 - physical environment, South African National Scientific Programmes, Report No. 124, FRD, CSIR, Pretoria, 18-38.

Elwell, H.A. and Stocking, M.A. 1976. Vegetal cover to estimate soil erosion hazard in Rhodesia. Geoderma, 15, 61-70.

Ferrar, A.A. (ed.) 1983. Guidelines for the management of large mammals in African conservation areas. South African Scientific Programmes, Report No. 69.

Graf, W.L. 1977. The rate law in fluvial geomorphology. American Journal of Science, 277, 178-191.

Graf, W.L. 1979. The development of montane arroyas and gullies. Earth Surface Processes, 4, 1-14.

Heede, B.H. 1974. Stages of development of gullies in Western United States of America. Zeitschrift für Geomorphologie, 18, 260-271.

Holling, C.S. 1973. Resilience and stability of ecological systems. Annual Review of Ecology and Systematics, 4, 1-23.

Hugo, W.J. (ed.) 1968. The small Stock Industry in South Africa. Gov. Press, Pretoria. Ch. XIX Principles of Veld Management in the Karoo and the Adjacent Dry Sweet-grass Veld, 318-340.

Ireland, H.A., Sharpe, C.F.S. and Eagle, D.H. 1939. Principles of gully erosion in the Piedmont of South Carolina. U.S. Department of Agriculture, Technical Bulletin, 633, 142 pp.

King, L. 1963. South African Scenery. Oliver and Boyd, London.

Kirkby, M.J. 1980. The problem. In: M.J. Kirkby and R.P.C. Morgan (eds.). Soil Erosion, Wiley, Chichester, 17-62.

Langbein, W.B. and Schumm, S.A. 1958. Yield of sediment in relation to mean annual precipitation. Transactions of the American Geophysical Union, 39, 1076-1084.

Leopold, W.B. and Miller, J.P. 1956. Ephemeral streams, hydraulic factors and their relationship to the drainage net. U.S. Geological Survey, Professional Paper, 282-A, 37 pp.

McCabe, K. 1987. Veld management in the Karoo. The Naturalist, 31, 8-15.

Meadows, M. 1985. Biogeography and Ecosystems of South Africa. Juta, Cape Town.

Melton, M.A. 1965. The geomorphic and paleoclimatic significance of alluvial deposits in southern Arizona. Journal of Geology, 73, 1-38.

Patton, P.C. and Schumm, S.A. 1975. Gully erosion, northern Colorado: a threshold phenomenon. Geology, 3, 88-90.

Patton, P.C. and Schumm, S.A. 1981. Ephemeral stream processes: implications for studies of Quaternary valley fills. Quaternary Research, 15, 24-43.

Ross, J.C. 1947. Land utilisation and soil conservation in the Union of South Africa. State Information Office, Pretoria.

Roux, P.W. 1981. Interaction between climate, vegetation and runoff in the Karoo. In: Maaren, H. (ed.), Workshop on the Effect of Rural Land Use and Catchment Management on Water Resources. TR 113, Department of Water Affairs, Forestry and Environmental Conservation, Pretoria, 90-106.

Roux, P.W. and Opperman D.P.H. 1986. Soil erosion. In: Cowling, R.M., Roux, P.W. and Pieterse, A.J.H. (eds.). The Karoo biome: a preliminary synthesis. Part 1 - physical environment. South African National Scientific Programmes Report No. 124. FRD, CSIR, Pretoria, 92-111.

Roux, P.W. and Vorster, M. 1983. Vegetation change in the Karoo. Proceedings of the Grassland Society Southern Africa, 18, 25-29.

Schumm, S.A. 1980. Geomorphic thresholds: the concepts and its applications. Transactions of the Institute of British Geographers, New Series, 4, 485-515.

Schumm, S.A. and Hadley, R.F. 1957. Arroyos and the semi-arid cycle of erosion. American Journal of Science, 225, 161-174.

Schumm, S.A., Harvey, M.D. and Watson, C.C. 1984. Incised Channels: Morphology, Dynamics and Control. Water Resources Publications, Littleton, Colorado. 200 pp.

Selby, M.J. and Hosking, P.J. 1973. The erodibility of Pumice soils of the North Island, New Zealand. Journal of Hydrology (N.Z.), 12, 32-56.

Snyman, H.A., Van Rensburg, W.L.J. and Opperman, D.P.H. 1986. Run-off, soil loss and water use efficiency of natural veld on various slopes and plant covers. Journal of the Grassland Society of Southern Africa, 3, 153-158.

Stocking, M.A. 1980. Examination of factors controlling gully growth. In: De Boodt M. and Gabriels D. (eds.), Assessment of Erosion. Wiley, Chichester, 505-520.

Thornes, J.B. 1976. Semi-arid erosional systems: case studies from Spain. London School of Economics, Geography Department, Paper No. 7, 88 pp.

Thornes, J.B. 1985. The ecology of erosion. Geography, 70, 222-235.

Thompson, W.R. 1935. Rainfall, soil erosion and run-off in South Africa. University of Pretoria, Series 1, 29.

Venter, J.M., Mocke, C. and de Jager, J.M. 1986. Climate. In: Cowling, R.M, Roux, P.W. and Pieterse, A.J.H.(eds.), The Karoo biome: a preliminary synthesis. Part 1 - physical environment. South African National Scientific Programmes Report No.

124. FRD, CSIR, Pretoria, 39-52.

Wilson, L. 1973. Variations in mean annual sediment yield as a function of mean annual precipitation. American Journal of Science, 273, 335-341.

Womack, W.R. and Schumm, S.A. 1977. Terraces of Douglas Creek, northwestern Colorado: an example of episodic erosion. Geology, 5, 72-76.

Zimmerman, R.C. and Thom, B.G. 1980. Physiographic plant geography. Progress in Physical Geography, 6, 45-59.

SOIL EROSION FORMS IN SOUTHERN AFRICA

GEORGE F. DARDIS, HEINRICH R. BECKEDAHL
Department of Geography, University of Transkei
TANYA A.S. BOWYER-BOWER
School of Geography, University of Oxford
PATRICIA M. HANVEY
Department of Geography, University of Transkei

1. INTRODUCTION

Soil erosion is becoming an acute problem in many parts of southern Africa, exemplified by, for example, an extension of badland areas, particularly in eastern and central southern Africa (Stocking, 1968; Darkoh, 1987; Beckedahl et al., 1988). A better understanding of the nature and factors responsible for soil erosion is needed if an adequate management of erosion phenomena is to be achieved. In order to gain such an understanding, it is necessary to identify the range of soil erosion end-forms present in the landscape, which indicate the effective resistance of the soil surfaces, underlying sediments and parent materials (cf. Graf, 1979), and may be used to provide a classificatory framework within which factors promoting soil erosion can be identified.

Soil erosion features are often known by a number of different terms, which may even be colloquial to the region (Table 1). As a result, it may be difficult to interpret their scientific meaning and extrapolate information from one area to another. Also, few attempts have been made to differentiate between forms and degrees of erosion grouped under one term. Taking the example of gullies, Watson et al.(1984) (after Hooker, 1984) have distinguished two characteristic gully forms in Swaziland and Natal; (1) ravine gullies (Fig. 1), and (2) (organ-pipe) dongas (or gullies)(Fig. 2). Ravine-type gullies are described as linear, flat-walled features developed in soil, colluvium and weathered bedrock, whereas organ-pipe dongas are described as being more dendritic gully forms, with fluted walls, cut in colluvium (Fig. 1).

Ireland et al.(1939) differentiated six characteristic gully forms in the Piedmont region of South Carolina. From these they recognised four distinct stages of gully growth; (1) V-shaped gullies developed in B-horizon material, (2) gully penetration

Geomorphological Studies in Southern Africa, G.F.Dardis & B.P.Moon (eds)
© *1988 Balkema, Rotterdam. ISBN 90 6191 831 6*

Table 1. Terminology of gully forms

TERM	MORPHOLOGICAL CHARACERISTICS	LANGUAGE OF ORIGIN*
Gully	Open channel	French
Ravine	Steep-sided, u- or v-shaped open channel	French
Arroyo	Steep-sided, open channel	Spanish
Donga	Open channel	Bantu
Lavakas	Ravine-type open channel, often	Malagasy**
Gulch	Ravine-type open channel	English (North American)

* After The Concise Oxford Dictionary of Current English, Seventh Edition, Oxford University Press (1982)

** After Darkoh (1987)

into the C horizon, (3) headcut development, with formation of a u-shaped channel, and (4) slope remoulding and re-establishment of vegetation. Leopold and Miller (1956) differentiated "continuous" and "discontinuous" gullies on the basis of their surface morphology. Heede (1974, 1976), however, has argued that it is insufficient to simply identify separate forms, and that transition states should instead be defined, as landforms evolve continuously through time, for example, in the Davisian interpretation, from young to old stages.

Imeson and Kwaad (1980), instead of concentrating on form or states of transition, looked at the influence of surface and subsurface flow, climatic conditions, and material eroded, in relation to erosion form. Four main gully types were defined as a result. This means of classification was applied to gullying, identifying four distinct types of gully based on the above criteria, and thus relating more to an understanding of their nature and formation than simply distinuishing end-forms.

On another level, the SARCCUS Subcommittee for Land-Use Planning and Erosion Control put forward a simplified classification of soil erosion phenomena for the southern African region. This attempted to provide a common approach to the iden-

Fig. 1. "Ravine-type" gully, Swaziland.

Fig. 2. "Organ-pipe" donga (gully) in the Mkondvo valley, Swaziland, showing heavily fluted surfaces.

189

tification and assessment of existing erosion, on the basis of surface morphology (SARCCUS, 1981). The classification was not designed to be comprehensive but merely to provide a simple practical system for rapid aerial photographic assessment of the degree of erosion severity. Thus, it did not assess the influence of subsurface processes on soil erosion, for example, which is possibly one of the more important agents of soil erosion in the southern African region (cf. Downing, 1968; Stocking, 1968; Beckedahl, 1977; Beckedahl et al., 1988).

This paper outlines a morphogenetic means of classifying a wide range of features of accelerated soil erosion, and is based on observations made in the southern African region.

2. BASIS OF THE CLASSIFICATION

The system of classification of soil erosion forms presented in this paper is based on the following criteria; (1) flow type, (2) flow regime, (3) geometry of erosion feature, (4) nature of the host material and (5) dominant processes acting on the particular soil erosion form.

2.1. Flow type

Three main flow types are defined here; (1) unconfined flow, (2) confined (closed conduit) flow, and (3) confined (open conduit) flow.

Unconfined flow refers to Horton overland flow (Horton, 1945), where flow is distributed relatively evenly over a planar surface, as a thin, irregular sheet of water. For the purpose of this classification, flows concentrated by turbulence on minor topographic irregularities are considered as unconfined flows (i.e. topographically unconfined). Other processes (e.g. rainsplash erosion) play an important role in unconfined flow (Dunne, 1980).

Confined (closed conduit) flow refers to flow in soil pipes and related subsurface tunnel forms (Jones, 1981). Flow conditions vary widely and in many instances the soil pipes or tunnels may be largely hydraulically inactive, but nevertheless contribute to soil erosion by other processes (e.g. roof and sidewall collapse). Confined (open conduit) flow refers to flow in open channels (cf. Chow, 1959).

The combinations of open and closed conduit confined flow paths (B/C, Type 6) in Table 2 overcomes the problem of having to define discontinuous gullies any further. Continuous gullies are designated type C flow paths. Type A/C flow paths refer to combinations of unconfined and confined flow in a degraded gully environment,

190

where sheetwash and related processes may operate on the sidewalls and floor (see Figs. 7-9).

2.2. Flow regime

Soil erosion occurs in two major flow regimes; (1) hydrodynamic and (2) aerodynamic. Most attention has focused on hydrodynamic aspects of soil erosion, despite considerable evidence of sand and dust transport in arid and semi-arid environments in Africa (cf. Goudie, 1978; Morales, 1979; McTainsch, 1980; Middleton, 1985; Tsoar and Pye, 1987). In many instances both regimes may operate to form soil erosion features (flow regime A/B, Soil Erosion type 9, Table 2).

2.3. Geometry of erosion feature

Geometry is defined in terms of cross-sectional form (i.e. planar, u-shaped, v-shaped) and according to the plan form of the soil erosion feature (linear, channel, cavity, network etc.). Geometry has been considered an important factor in destinguishing between gully types (cf. Ireland et al., 1939; Leopold and Miller, 1956; Heede, 1970; Imeson and Kwaad, 1980). In southern Africa, important relationships between soil erosion and parent material tend to be reflected in morphological variations, particularly in cross-sectional profile (Hanvey et al., 1988; Bowyer-Bower et al., 1988).

2.4. Host materials

The nature of the host materials is probably the single most important parameter influencing soil erosion type. The categories used (Table 2) are modified after Imeson and Kwaad (1980).

Three categories of host materials are identified; (1) a relatively resistant layer or near-surface layer of soil, colluvium or bedrock, (2) homogeneous, poorly resistant weathering products, colluvium or sand, and (c) heterogeneous weathering products, colluvium or sand, with variable resistance between individual beds or layers. The differences between (2) and (3) are described by Bowyer-Bower et al.(1988).

2.5. Dominant processes

The main processes associated with the different soil erosion forms are indicated. It should be noted, however, that only erosional and depositional processes operating

Table 2. Soil erosion types in southern Africa

TYPE	FLOW* PATH	FLOW** REGIME	GEOMETRY	HOST*** MATERIAL	DOMINANT PROCESSES
1	A	A	Sheet;horizontal or gently inclined planar surface; weak anastomosing or braided channelling	A	Overland flow; sheetwash; sheetflooding
2	A	B	sheet;horizontal or inclined planar, undulating or sinusoidal surface; no channelling	A/B	Eolian; saltation; deflation
3	B	A	linear channels (rills)	A/B/C	Overland flow; sheetwash
4	B	A	sub-surface cavity;vein; crack;isolated cavity; non-linear	C	Subsurface flow; solution; defloculation
5	B	A	conduit;linear;meandering; bedrock-cut;V-shaped	A/C	Subsurface flow; piping; roof and sidewall collapse
6	B/C	A	linear closed-open conduit; single channel or channel network	C	Overland and sub-surface flow; roof and sidewall collapse;headcut erosion
7	C	A	linear open conduit;single channel or channel network: u-shaped; flat-bottomed;	C	Overland flow; piping;sidewall collapse;headcut erosion; sidewall rilling
8	C	A	open conduit; single channel; or weakly developed channel network;v-shaped	A/B	Overland flow; sidewall rilling; headcut erosion; general absence of piping
9	A/C	A/B	planar surface;degraded	A	Overland flow; sheetwash; rill erosion;eolian activity

*   A:unconfined
B:Confined (closed conduit)
C: Confined (open conduit)

**   A: Hydrodynamic
B:Aerodynamic

*** A: Relatively resistant surface or near-surface soil, colluvium or bedrock layer
B: Homogeneous, poorly resistant weathering products, colluvium or sand
C: Heterogeneous weathering products, colluvium or sand with variable degrees of resistance between individual beds of layers

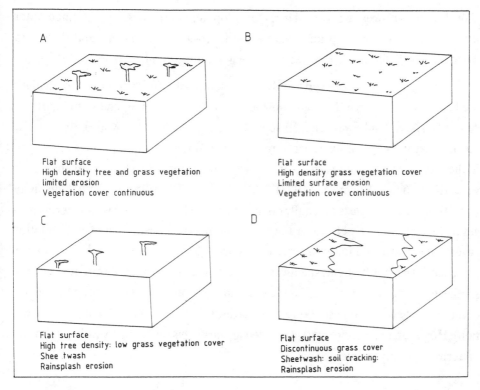

Fig. 3. Type 1 soil erosion forms.

on the soil erosion form are shown. Factors which are responsible for the operation of a particular process (e.g. climate, vegetation density and distribution, overgrazing, soil texture etc.) are not considered here (cf. Weaver, 1988; Rowntree, 1988).

3. SOIL EROSION TYPES

Nine major soil erosion types are described (Table 2). While the classification may suggest cyclical development of features (cf. Heede, 1976; Aghassy, 1973), no such emphasis is intended. Similarly, some soil erosion forms may be composite features, showing, for example, evidence of two or more soil erosion forms superimposed on each other (e.g. rills occurring on the sidewalls of a v-shaped gully).

3.1. Type 1 soil erosion forms

Relatively little erosion may occur where the vegetation cover is of high density (except where it is associated with subsurface processes) (Fig. 3a,b). Water-activated unconfined erosion (Fig. 3c,d) is generally associated with areas where vegetation has been removed or where topsoil or surficial sediment characteristics inhibit infiltration. The main processes involved in the development of regions of unconfined water-activated soil erosion are primarily, rainsplash erosion, soil cracking and sheetwash. Topographically-confined flow features are generally absent.

This erosion form occurs commonly in the southern African interior (particularly the Karoo, Orange Free State and northeastern Cape Province) and in the northern Transvaal, Natal and Swaziland. In semi-arid areas, such as the Karoo, this form of erosion may not necessarily be accelerated, as the long-term equilibrium state may be in the form of a continuous cycle of soil erosion triggered by short-term changes in vegetation cover (Rowntree, 1988) and seasonal variability in rainfall erosivity. Rowntree (1988) suggests that gully erosion may also be triggered by short term changes in vegetation cover. However the most common response to reduced vegetation cover is sheetwash and rainsplash erosion (Figs. 4, 5). This is supported by geological evidence (Dardis, 1987) that sheetwash processes have operated throughout eastern southern Africa over the past 25000 years, while gully erosion is a recent phenomena, occurring largely in the past 300 years (Dardis, submitted; Dardis et al., in preparation). However, much more work is needed to clarify this point and to evaluate erosion patterns over longer time periods.

3.2. Type 2 soil erosion forms

Unconfined soil erosion resulting from aeolian activity (Fig. 6) can cause relatively large-scale soil loss. This is particulary true in the southern African interior and southwest Africa and in coastal dune systems (Lancaster, 1979; 1986; Tinley, 1985; Goudie and Thomas, 1985). A considerable degree of unconfined erosion may be associated with transport of dust (cf. Tsoar and Pye, 1987). Despite adequate source areas for dust in southern Africa (e.g. Namib Desert, Kalahari, Karoo) little work has been devoted to dust production, transport and deposition. Contemporary dust storms occur frequently in arid and semi-arid areas (cf. Goudie, 1983) and occur frequently in southern Africa. It is likely, therefore that a relatively high percentage of present day erosion may be associated with aeolian activity, though few quantitative measurements have yet been made. Areas on the fringe of the Namib and Kalahari are likely depocentres for wind-transported materials, as most Quaternary dust deposition seems to have been concentrated in desert marginal areas (Yaalon and Dan, 1974; McTainsh, 1984; Smith and Whalley, 1981). Despite the likelihood that dust deposition may have been important during the Quaternary, no recognisable aeolian ele-

Fig. 4. Type 1 soil erosion, Tsolo district, Transkei.

Fig. 5. Type 1 soil erosion in bushveld terrain in the middle Lowveld region of Swaziland, associated with overgrazing. Note the discontinuous vegetation cover.

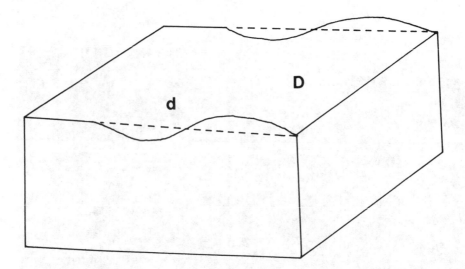

Fig. 6. Type 2 soil erosion forms associated with eolian activity. They are characterised by zones of net erosion, formed by deflation (d), and zones of net accumulation, where dune development (D) may occur.

ments have yet been identified in Quaternary colluvial sequences in southern Africa (Bull and Goudie, 1984). Other source areas for contemporary dust storms are fallow agricultural land and spoil heaps from mining activity.

The geomorphic effects of aeolian activity (e.g., loess, coversand, dune development, deflation) are well known and will not be considered further here. However, it should be pointed out that it is often difficult to establish to what extent the forms that develop in association with aeolian activity reflect accelerated erosion (cf. Hanvey et al., 1988). This requires further detailed studies before accelerated aeolian erosional and depositional forms can be adequately distinguished.

3.3. Type 3 soil erosion forms

Flow, by its very nature, will impose a shear stress on the soil surface and on the vegetation cover (if present). Where boundary shear stress exceeds the soil resistance, sheet erosion will occur (Dunne, 1980). The magnitude of boundary shear stress (t) at any point on a hillslope is

$$t = PgdcosXsinX \qquad (1)$$

where P is the fluid density, g is the acceleration due to gravity, d is the mean flow

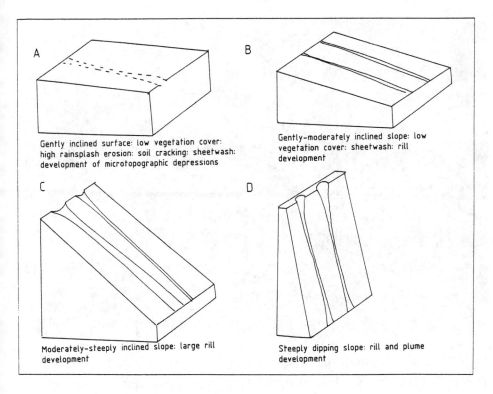

A

Gently inclined surface: low vegetation cover:
high rainsplash erosion: soil cracking: sheetwash:
development of microtopographic depressions

B

Gently-moderately inclined slope: low
vegetation cover: sheetwash: rill
development

C

Moderately-steeply inclined slope: large rill
development

D

Steeply dipping slope: rill and plume
development

Fig. 7. Type 3 soil erosion forms

depth measured vertically and X is the local angle of slope (Horton, 1945; Dunne, 1980). From Equation 1 it is clear that boundary shear stress will increase with bed slope angle. The convergance of sheetflows coupled with variable soil resistance may result in the development of micro-channels (or rills) and micro-depressions (Smith and Bretheron, 1972; Dunne, 1980). Rill development may occur on any unvegetated, inclined surface (Fig. 7). As a result they are commonly found on the sidewalls of gullies. The nature and extent of rill erosion depends on slopes and soil texture, and is highly variable (Bowyer-Bower and Bryan, 1986). The nature of rills developed within gullies is determined by a number of edaphic factors which are largely independent of gully form (cf. Bowyer-Bower and Bryan, 1986). Rill (Fig. 8) and fluting (Fig. 9) patterns, for example, may vary considerable between stratigraphic units. For this reason

organ-pipe rill patterns cannot be used to signify a distinctive morphogenetic gully form (in the sense of Hooker, 1984). Flutes commonly develop at high bed slope angles (Figs. 9, 10) and commonly show evidence of considerable downslope movement of material (Figs. 9, 10). Flutes appear to be higher order bedforms than rills (Fig. 10). They may be up to 3 m in width at the top, narrowing to 2-50 cms. Rills may occur in

Fig. 8. Variations in patterns of rill development between different stratigraphic units, Cedarville Flats area, Natal, South Africa. (Type 3 erosional form).

Fig. 9. Fluting in colluvium, Cedarville Flats area, Natal, South Africa. Note the abrupt development of flutes below a buried soil and downslope movement of materials. In this example, rills have developed within the flutes. (Type 3 erosional form).

Fig. 10. Flutings developed at high slope angles, Mkondvo valley, Swaziland. Note the downward tapering of the flutes resulting from increased channelling of flow. (Type 3 erosional form).

the erosional part of the flute. The lower part of a flute may consist of a pediment, varying in size from 50-500 cms (Fig. 10). Flutes are generally isolated but may in some instances converge at depth. The depth of the flutes is largely dependent on the thickness of the material in which the flute has formed and the nature of the host material.

3.4. Type 4 soil erosion forms

Type 4 soil erosion forms (Fig. 11) are mainly subsurface cavities, such as veins and

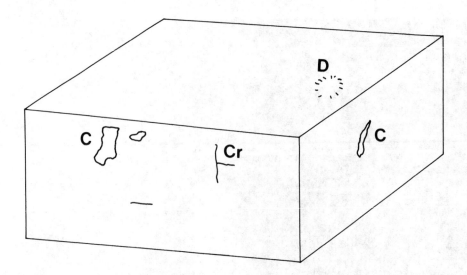

Fig. 11. Type 4 soil erosion forms, characterised by cavities (C), surface depressions (D), and linear seepage cracks (Cr).

pedotubules. These forms are generally only visible in section, and only where large cavities arise (Fig. 12). They develop mainly in heterogeneous sediments with variable degrees of resistance (Table 2). They are largely solutional pseudo-karstic features, most commonly found in dispersive sediments where deflocculation has occurred, and in calcareous sediments subject to chemical solution. Clays are removed in suspension along veins, cracks, pedotubules and by groundwater flow.

The role of these erosion forms in badland development in southern Africa is well documented. They have been reported in Natal (Downing, 1968; Beckedahl, 1977) and Lesotho (Faber and Imeson, 1982; Schmitz and Rooyani, 1987). Dardis and Beckedahl (1988) have found that they play a considerable role in the development of soil pipes and gully systems in Transkei. In advanced stages of development, they propogate towards the surface and form surface depressions (Fig. 11), which ultimately develop into sinkholes.

3.5. Type 5 soil erosion forms

Type 5 soil erosion forms (Fig. 13) are probably the most widespread form of confined soil erosion in southern Africa. They include soil pipes, tunnel erosion forms and linking surface cavities. They are characterised by closed

200

Fig. 12. Cavity developed in colluvium as a result of defloculation, Mqanduli, Transkei. (Type 4 erosional form)

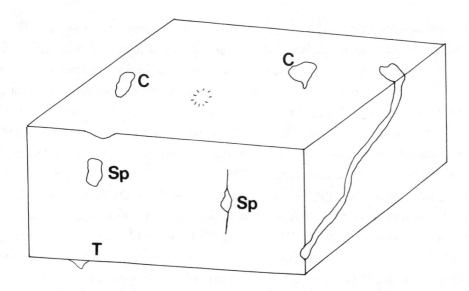

Fig. 13. Type 5 soil erosion forms, characterised by soil pipes (Sp) and tunnel erosion forms (T). They are commonly linked to surface cavities (c).

Fig. 14. Tunnel erosion (type 5 soil erosion form) at Inxu Drift, Transkei.

conduit flow (or pipe flow) and form mainly in heterogeneous materials of variable resistance. They vary in size, from small-scale cracks which are normally vertical or subvertical propagating from the surface or near-surface horizons, and between 1 and 25 mm in diameter, to pipes of up to 3 metres in diameter. They commonly penetrate to bedrock as tunnel erosion forms (Fig. 14). Detailed studies of soil pipe morphology in Natal, the Orange Free State, Transkei and Zimbabwe demonstrate that soil pipe networks are relatively complex, showing both straight and dendritic subsurface drainage networks of variable gradient (Downing, 1968; Stocking, 1976; Beckedahl, 1977; Beckedahl et al., 1988; Beckedahl and Dardis, 1988; Dardis and Beckedahl, 1988a). Sediments in southern Africa with variable dispersion ratios appear to be particularly susceptible to piping (Beckedahl, 1977; Faber and Imeson, 1982; Rooyani, 1985; Watson et al., 1987). Soil pipe networks are dynamic (i.e. unstable), and are continually enlarging by processes of subsurface cavitation, side-wall erosion and roof collapse (cf. Blong et al., 1982; Dardis and Beckedahl, 1988b). Rates of development are not well documented in southern Africa.

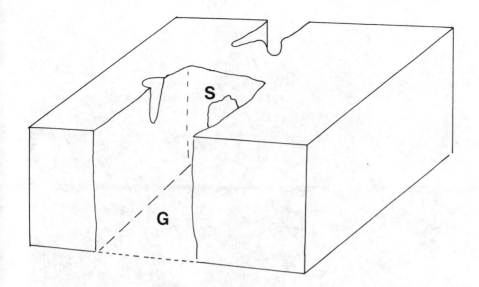

Fig. 15. Type 6 soil erosion forms, showing open conduits (G) linked by soil pipes (S).

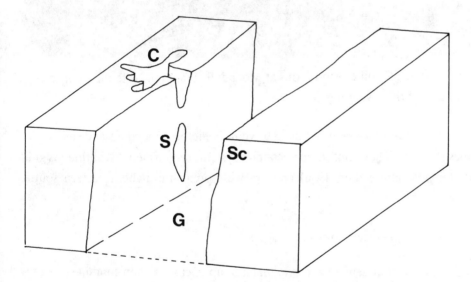

Fig. 16. Type 7 soil erosion forms, characterised by flat-bottomed, steep-walled open conduits. They may have associated soil pipes (S), Cavities (C) and evidence of sidewall collapse(Sc).

3.6. Type 6 soil erosion forms

Types 4 and 5 soil erosion forms commonly develop into type 6 soil erosion forms, through cavity enlargement by water flow, roof collapse etc, to produce discon-

Fig. 17. Type 6/7 soil erosion forms at Mqanduli, Transkei. Note the steep sidewalls and sidewall rill development.

tinuous closed-open conduits, linked by vein-like interconnections and soil pipes (Fig. 15). These closed-open conduit systems are termed discontinuous gullies (cf. Leopold and Miller, 1956) and are widespread throughout eastern southern Africa.

3.7. Type 7 soil erosion forms

The continual enlargement and eventual link-up of closed-open conduit systems will ultimately result in continuous gullies (Figs. 16,17). They tend to have steep side-walls and are generally flat-bottomed. They are generally associated with soil piping, and may form as a result of the collapse of the roofs of soil pipe networks. They may develop from type 6 soil erosion forms. These forms are associated with heterogeneous materials of variable resistance (Table 2). They most commonly occur in stratified colluvium. The lateral extension of these erosion forms occurs either by merging with collapsed cavities or soil pipes, or by rotational slumping of the steep sidewalls.

Type 7 soil erosion forms occur throughout the middle lowveld region, in Swaziland,

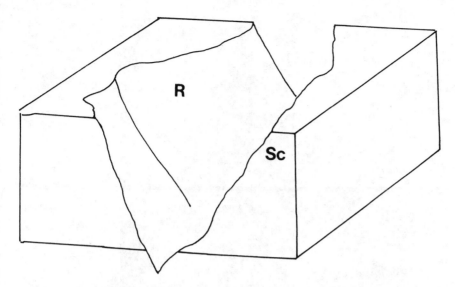

Fig. 18. Type 8 soil erosion form, characterised by a v-shaped profile, with sidewall rill development (R) and sidewall collape (Sc).

Natal, Transkei, Ciskei, the Orange Free State, the eastern Cape, and Lesotho (Watson et al., 1984; Beckedahl et al., 1988).

3.8. Type 8 soil erosion forms

Type 8 soil erosion forms are characterised by open, v-shaped conduits (Fig. 18), with generally steep ($10°$-$60°$) sidewalls. During the early stages of their development they often taper into interlocking spurs on the floor of the conduit. In advanced stages, the floors of the conduit may be flat-bedded, undulating or step-like, as a result of channel widening and down-cutting. They are generally linear (Figs, 19a,b) or bulbous (Fig. 20) in plan form and tend not to develop beyond second order drainage networks (Fig. 19). This is most likely a function of slope steepness (cf. Phillips and Schumm, 1987).

They form mainly in homogeneous materials (Table 2), such as weathered regolith (Figs. 19), coastal sands (Fig. 20) (Dardis et al., 1988) and soft bedrock forms (cf. Dardis and Beckedahl, 1988b). These forms may also develop in massive or poorly stratified colluvium.

The main processes acting on these soil erosion forms are runoff-related, including rill erosion, sidewall elluviation and head-cutting. Subsurface processes, such as soil piping or tunnel erosion, has not yet been identified in type 8 soil erosion forms (except in secondary talus deposits formed on the sidewall of the conduit by rill erosion).

Fig. 19. Oblique aerial photograph of type 8 soil erosion forms developed on weathered regolith, Mtata River, Transkei. Note the poor development of tributary gullies, the linear nature of the network and inter-locking spurs on the conduit floor. (Photograph taken: October 1987).

This soil erosion form occurs mainly in areas with high relief ratios (e.g. mountain slopes). It commonly occurs in the Middleveld and Highveld regions of Swaziland (Watson et al., 1984; Hooker, 1984), in the eastern Transvaal and Lebowa, and the coastal regions of Natal, Transkei and the eastern Cape. It also occurs in the Drakensberg (particularly on alluvial fans).

The depth and length of type 8 soil erosion forms is determined by slope gradient

Fig. 20. Oblique aerial photograph of type 8 soil erosion forms developed in massive, poorly stratified coastal red sands (of eolian origin), Mbolompo Point, Transkei. The depth of the sands is c.10 - 15 m. Note the bulbous form of the conduits.

and the depth of the weathered mantle. In Transkei and Ciskei, these forms seldom exceed 5-10 m depth. In the Highveld region of Swaziland, Lebowa, and the eastern Transvaal, in granodiorite terrain, they can reach considerable depths (up to 70 m in places) (Bowyer-Bower et al., 1988). Where the eroding material is deep the forms tend to be bulbous (see Fig. 20), reflecting the influence of sidewall collapse on lateral extension of the conduits.

Ravine-type gullies (Fig. 1) and organ-pipe gullies (Fig. 2)(cf. Watson et al., 1984; Hooker, 1984) both fall into this category of soil erosion form.

3.9. Type 9 soil erosion forms

In advanced stages of soil erosion, conduit widening may result in stabilisation of soil erosion, when a new equilibrium state has been achieved (cf. Rowntree, 1988). Where two or more juxtaposed channels coalesce, they may result in formation of isolated sidewall remnants (Fig. 21a).

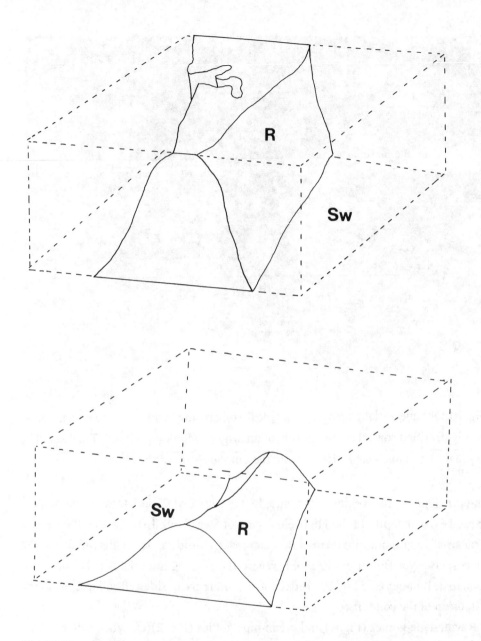

Fig. 21. Type 9 soil erosion forms, developed from (a) type 7 and (b) type 8 soil erosion forms. Rill erosion (R) and sheetwash (Sw) are the dominant erosional processes in each case.

At this stage of development, rill and sheet erosion are the dominant processes, regardless of the nature of the initial soil erosion form (e.g., type 7 or 8 conduits). Erosion processes may cease to operate entirely, resulting in re-establishment of vegetation.

Forms of this type do not occur widely in southern Africa, except perhaps in areas such as the Karoo, where sheet erosion may represent advanced (rather than initial) stages of soil erosion. This may reflect the relatively recent development of badland topography in the region, such that equilibrium stages have not been widely achieved.

4. FACTORS INFLUENCING THE GENESIS OF SOIL EROSION FORMS IN SOUTHERN AFRICA

Factors influencing the cessation of erosion in southern Africa are not clear. Factors influencing the initiation of erosion, such as base-level controls on soil erosion are however apparent in a number of studies (cf. Dardis and Beckedahl, 1988a,b), as is ecosystem imbalance (Rowntree, 1988).

It is often difficult to distinguish quantitatively between the initiation and further development of an erosion feature (Parker and Schumm, 1982). Very often observations are taken over such a short time span that it is difficult to understand the preconditions for the onset of accelerated erosion. Much of the erosion occurs, for example, in deposits (i.e. colluvium) which are relatively young (c.25000 years) in geological terms, and which may have formed under a unique set of circumstances (i.e. initiated by largescale climatic change during the late Pleistocene period). Until further data are forthcoming regarding the recurrence interval of particular erosion forms, and of the controls on their development, it will remain difficult to predict likely responses resulting from ecosystem imbalances (caused by human, climatic and edaphic factors).

In the short-term, some factors contributing to soil erosion can be readily identified. Whether water flow commences as surface or subsurface flow is determinable. The factors responsible for this thus provide a primary control on the nature of subsequent processes operating in, or contributing to, accelerated erosion. These include climate (influencing the effectiveness of hydrodynamic or aerodynamic processes, rainfall energy), parent (or host) material (the primary influence being the presence or absence of erodible material), vegetation cover (influencing runoff, infiltration, interception, peak discharge), nature of surface or subsurface sediment texture (influencing infiltration capacity, infiltration rate, patterns of interflow), presence or absence of a surface, or near surface, impermeable layer (influencing infiltration, seepage, solution or dissolution of surface or subsurface materials, and hence the potential for development of surface cracking, piping, tunnels or cavities), human interference (overgrazing, building practices, land-use), ecological balance (a loose term to describe the extent to which vegetation, land use etc. contribute to disequilibrium con-

209

ditions), and rock strength (influencing the penetration depth and rate of operation of erosion processes). These many factors contributing to soil erosion make it relatively difficult to establish and predict the genesis of soil erosion forms.

The classification presented here is aimed at establishing a relatively simple morphogenetic classification of soil erosion forms, to help provide a basis for establishing the main factors responsible for the development of particular soil erosion forms.

5. SUMMARY

A classification of soil erosion forms found in southern Africa is presented. The classification is based on flow path, flow regime, geometry of feature, nature of the host/parent material and dominant processes. Nine major morphogenetic soil erosion forms have been identified on the basis of these parameters. It is hoped that the classification will remove some ambiguities which arise in describing soil erosion types and provide a framework within which factors promoting soil erosion can be readily identified.

6. ACKNOWLEDGEMENTS

We thank Peter Storey, Tescor, for helicopter assistance in the course of fieldwork.

7. REFERENCES

Aghassy, J. 1973. Man-induced badlands topography. In: Coates, D.R. (ed.), Environmental Geomorphology and Landscape Conservation, Vol. 3, Non-Urban. Dowden, Hutchinson and Ross, 124-136.

Beckedahl, H.R. 1977. Subsurface erosion near the Oliviershoek Pass, Drakensberg. South African Geographical Journal, 59, 130-138.

Beckedahl, H.R. and Dardis, G.F. 1988. The role of artificial drainage in the development of soil pipes and gullies: Some examples from Transkei, southern Africa. (this volume).

Beckedahl, H.R., Bowyer-Bower, T., Dardis, G.F. and Hanvey, P.M., (1988). Geomorphic effects of soil erosion. In: Moon, B.P. and Dardis, G.F. (eds.), The Geomorphology of Southern Africa, Southern Book Co., Johannesburg.

Bowyer-Bower, T.A.S. and Bryan, R.B. 1986. Rill initiation: Concepts and experimental evaluation on badland slopes. Zeitschrift für Geomorphologie, Supplement Band 60, 161-175.

Bowyer-Bower, T.A.S., Dardis, G.F. and Beckedahl, H.R. 1988. Lithological controls on gully morphology in southern Africa. Paper presented at the Workshop on Soil Erosion and Parent Materials, Symposium on the Geomorphology of Southern Africa, University of Transkei, April 1988.

Bryan, R.B. and Yair, A. 1982. Badland Geomorphology and Piping. Geobooks, Norwich.

Chow, Ven Te 1959. Open Channel Hydraulics, McGraw-Hill, New York.

Dardis, G.F. and Beckedahl, H.R. 1988a. Drainage evolution in an ephemeral soil pipe-cavity system, Transkei, southern Africa. (this volume).

Dardis, G.F. and Beckedahl, H.R. 1988b. Gully formation in Archaean rocks at Saddleback Pass, Barberton Mountain Land, South Africa. (this volume).

Darkoh, M.B.K. 1987. Socio-economic and institutional factors behind desertification in Southern Africa. Area, 19, 25-33.

Downing, B.M. 1968. Subsurface erosion as a geomorphological agent. Transactions of the Geological Society of South Africa, 71, 131-134.

Dunne, T. 1980. Formation and controls of channel networks. Progress in Physical Geography, 4, 211-239.

Faber, T. and Imeson, A.C. 1982. Gully hydrology and related soil properties in Lesotho. In: Recent Developments in the Explanation and Prediction of Erosion and Sediment Yield, IAHS Publication 137, 135-144.

Goudie, A.S. 1978. Dust storms and their geomorphic implications. Journal of Arid Environments, 1, 291-310.

Goudie, A.S. and Thomas, D.S.G. 1985. Pans in southern Africa, with special reference to South Africa and Zimbabwe. Zeitschrift für Geomorphologie,

Graf, W.L. 1979. The development of montane arroyos and gullies. Earth Surface Processes, 4, 1-14.

Hanvey, P.M., Beckedahl, H.R. and Dardis, G.F. 1988. Soil erosion on a late Pleistocene coastal dune complex, Transkei, Southern Africa. Paper presented at the Workshop on Soil Erosion and Parent Materials, Symposium on the Geomorphology of Southern Africa, April, 1988.

Heede, B.H., 1967. The fusion of discontinuous gullies. Bulletin of the International Association of Scienctific Hydrology, 15, 79-89.

Heede, B.H., 1971. Characteristics and processes of soil piping in gullies. UDSA, Forest Research Paper RM-68, 15pp.

Heede, B.H., 1974. Stages in the development of gullies in western United States of

America. Zeitschrift fur Geomorphologie, NF 18, 260-271.

Heede, B.H., 1976. Gully development and control: the status of our knowledge. USDA Forest Service, Rocky Mountain and Range Experimentation Station, Fort Collins, Research Paper RM-169.

Hooker, R.M. 1984. Gully (donga) erosion in Swaziland - A distributional analysis. Unpublished manuscript.

Imeson, A.C. and Kwaad, F.J.P.M. 1980. Gully types and gully prediction. KWAG Geografisch Tijdschrift, 5, 430-441.

Ireland, H.A., Sharpe, C.F.S. and Eargle, D.H. 1939. Principles of gully erosion in the Piedmont of South Carolina. United States Department of Agriculture Technical Bulletin, 63, 143pp.

Jones, J.A.A. 1981. The Nature of Soil Piping - A Review of Research. Geobooks, Norwich.

Jones, J.A.A., 1987. The initiation of natural drainage networks. Progress in Physical Geography, 11, 207-245.

Leopold, L.B. and Miller, J.P., 1956. Ephemeral streams - hydraulic factors and their relation to the drainage net. US Geological Survey Professional Paper 282-A, 37pp.

McTainsh, G. 1980. Harmattan dust deposition in northern Nigeria. Nature, 286, 587-588.

Middleton, N.L. 1985. Effect of drought on dust production in the Sahel. Nature, 316, 431-434.

Morales, C. (ed.) 1979. Saharan Dust - Mobilization, Transport, Deposition. Wiley, Chichester, 316pp.

Parker, R.S. and Schumm, S.A. 1982. Experimental study of drainage networks. In Bryan, R.B. and Yair, A. (eds.) Badland Geomorphology and Piping, Geobooks, Norwich, 153-168.

Phillips, L.F. and Schumm, S.A. 1987. Effect of regional slope on drainage networks. Geology, 15, 813-816.

Rooyani, F. 1985. A note on soil properties influencing piping at the contact zone between albic and argillic horizons of certain duplex soils (Aqualfs) in Lesotho, southern Africa. Soil Science, 139, 517-522.

Rowntree, K.M. 1988. Equilibrium concepts, vegetation change and soil erosion in semi-arid areas: Some considerations for the Karoo. (this volume).

SARCCUS, 1981. A System for the Classification of Soil Erosion in the SARCCUS Region. Department of Agriculture and Fisheries, Pretoria.

Seginer, I. 1966. Gully development and sediment yield. Journal of Hydrology, 4, 236-253.

Stocking, M. 1976. Tunnel erosion. Rhodesian Agricultural Journal, 73, 35-40.

Tsoar, H. and Pye, K. 1987. Dust transport and the question of desert loess formation. Sedimentology, 34, 139-153.

Watson, A., Price Williams, D. and Goudie, A.S. 1984. The palaeoenvironmental interpretation of colluvial sediments and palaeosols of the Late Pleistocene Hypothermal in southern Africa. Palaeogeography, Palaeoclimatology, Palaeoecology, 45, 225-249.

Watson, A., Price Williams and Goudie, A.S. 1987. Reply to 'Is gullying associated with highly sodic colluvium? Further comment to the environmental interpretation of Southern African dongas' Palaeoclimatology, Palaeogeography, Palaeoecology, 58, 123-128.

FACTORS AFFECTING THE SPATIAL VARIATION IN SOIL EROSION IN CISKEI: AN INITIAL ASSESSMENT AT THE MACROSCALE

ALEX VAN BREDA WEAVER
Department of Geography, Rhodes University

1. INTRODUCTION

The southern African region is designated as an area which is susceptable to soil erosion by rainfall (Hudson, 1981). The extent and spatial variation in soil erosion is dependent on both anthropogenic and natural environmental factors. One of the most important practical contributions that can be made by the geomorphologist to society lies in the identification of sites that are free from the threat of geomorphic hazards (Scheidegger, 1975). In the field of soil erosion, this entails the identification of the anthropogenic and environmental factors which determine the severity and spatial extent of soil erosion.

Both Garland (1982) and Rooseboom (1983) point out that very little published work exists on the true nature and extent of the soil erosion problem in southern Africa. This is partcularly true for Ciskei where it is estimated that 47 % of the area is moderately to severely eroded (Ciskei Commission, 1980). A detailed study of erosion rates in the Roxeni Basin in Ciskei, shows that soil loss occurs at a rate exceeding 100 $t.km^{-2}yr^{-1}$ (Weaver, 1988). Stocking (1984) makes a plea for the development of erosion hazard assessment methods which are both simple in design and practical in implementation yet encompass as wide a range of evidence as possible.

This paper presents the results of a preliminary investigation of the relationship between 23 possible erosion hazard indices and soil erosion in four Ciskei catchments. The investigation forms the precursor to a more detailed multivariate analysis which is currently nearing completion.

Four catchment areas were selected in central Ciskei (Fig. 1) with as wide a range as possible in both the degree of erosion and environmental characteristics. The four catchments chosen were the Yellowwoods, Amatole, and Roxeni Basins, and the Middledrift portion of the Keiskamma Basin. The aim of the study was to identify the individual factors which most efficiently account for variations in soil erosion. An important constraint imposed on the study is that readily available data be used to describe the independent variables.

Geomorphological Studies in Southern Africa, G.F.Dardis & B.P.Moon (eds)
© 1988 Balkema, Rotterdam. ISBN 90 6191 831 6

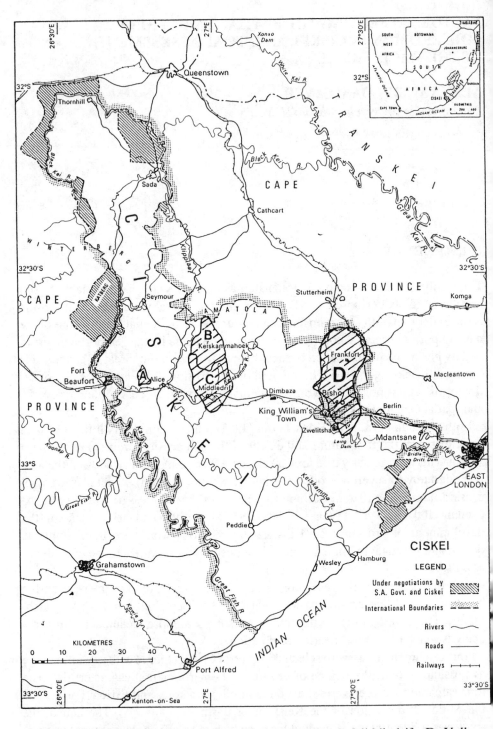

Fig.1. Location of study area (A: Roxeni; B: Amatole; C: Middledrift; D: Yellow-woods).

216

2. METHODS OF DATA COLLECTION

Independent variables were selected on the basis of the available literature and data on soil erosion promoting factors. The erosion promoting factors selected as independent variables for the study as well as data sources are outlined in Table 1. Soil erosion severity was mapped using the modified version of the SARCCUS (1981) erosion mapping system as described by Weaver (1987). Five levels of soil erosion are used (no erosion, sheet and rill erosion only, sheet and rill with evidence of gullies, intricate gullies and severe gullies). These five classes are made up of 16 possible SARCCUS classes. The 1:30000 aerial photographs, flown in 1984 by Aircraft Operating Company, were used for the stereoscopic analysis of soil erosion in the area in conjunction with 1:10000 orthophotos. Detailed groundtruth studies were undertaken for all the units mapped.

All data were mapped at a 1:50000 scale. Data on soil erosion and factors promoting soil erosion were extracted from the original maps using a 500 m square grid point sampling system. These data were then read into a computerised geographic information system.

3. ANALYSIS OF DATA

The relationship between the various independent variables and actual soil erosion was tested using the BMDP statistical software package (Dixon, 1985). As the scale of measurement of the independent variable (erosion) is ordinal, the use of statistical tests is limited to non-parametric statistics. Chi-square tests are used to test the relationship between erosion and variables measured at the nominal and ordinary scale whilst Spearman Rank correlation are used to test relationships which involve variables measured at the ordinal or the interval scale. The result of the statistical analysis of the data are outlined in Table 2.

4. DISCUSSION OF RESULTS

The results show that the one variable which appears to have the most marked effect on spatial variations in soil erosion is land use. Fig. 2 shows three frequency histograms depicting the distribution of various erosion classes for forestry, grazing and arable lands. The relatively low proportion of soil erosion in areas covered by forest is well in keeping with the literature reviewed. Gerlach (1976) for example showed soil loss from cultivated lands to be 15 to 30000 times greater than forested

Table 1. Summary of Data Sources used in the Study.

DATA	SOURCE
SOIL EROSION	Field survey and 1:30 000 aerial photographs (Aircraft Operating Company, 1984)
RAINFALL	South African Weather Bureau
ALTITUDE	1:50 000 topocadastral series, South Africa
EROSIVITY	Derived from original breakpoint and daily rainfall data (South African Weather Bureau) (Methods of EI(subscript 30) computation discussed in Weaver and Hughes, 1986)
SOIL ERODIBILITY	SLEMSA (1976) values applied to maps of Hill, Kaplan and Scott (no date 1, 2 and 1977) and ARDRI (1981)
SOIL TYPE	Hill, Kaplan and Scott (no date 1, 2 and 1977) and ARDRI (1981)
GEOLOGY	1:50 000 geological sheets, Geological Survey of South Africa
SLOPE ANGLE	1:50 000 topocadastral series, South Africa
SLOPE LENGTH	1:50 000 topocadastral series, South Africa
COMBINED SLOPE LENGTH AND ANGLE (LS)	1:50 000 topocadastral series, South Africa (Using the method described by Wischmeier and Smith, 1978)
PROFILE SLOPE SHAPE	1:50 000 topocadastral series, South Africa
LATERAL SLOPE SHAPE	1:50 000 topocadastral series, South Africa
COMBINED SLOPE SHAPE	1:50 000 topocadastral series, South African (Using the method described by Ruhe, 1975)
SLOPE ASPECT	1:50 000 topocadastral series, South Africa
LAND USE	1:30 000 aerial photographs (Aircraft operating company, 1984)
VELD TYPE	Hill, Kaplan and Scott (1977), Gibbs-Russel and Robinson (1982), Loxton, Hunting and Associates (1979) and field survey
VELD CARRYING CAPACITY	Gibbs-Russel and Robinson (1982), Willis and Trollope (1983) and field surveys
VELD CONDITION SCORE	Gibbs-Russel and Robinson (1982), Willis and Trollope (1983) and field surveys
POPULATION DENSITY	Ciskei Directorate of Planning (1984)
DISTANCE TO NEAREST VILLAGE	1:50 000 topocadastral sheets, South Africa
FOOTPATH DENSITY	1:50 000 topocadastral sheets, South Africa
LIVESTOCK DENSITY	Ciskei Department of Agricultures's Vetinary services dipping tank statistics

218

Table 2. Results of Statistical Tests.

SAMPLE POPULATION	INDEPENDENT VARIABLE	CHI-SQUARE	SPEARMAN RANK
1334	MEAN ANNUAL PRECIPITATION		-0,48
1290	ALTITUDE		-0,34
1334	RAINFALL EROSIVITY		-0,44
1309	SOIL ERODIBILITY		0,02(2)
1304	SOIL TYPE	301,2(1)	
1334	GEOLOGY	48,7	
1288	SLOPE STEEPNESS		-0,30
1288	SLOPE LENGTH		0,12
1288	L.S. FACTOR		-0,29
1288	PROFILE SHAPE	58,9	
1288	LATERAL SHAPE	48,0	
1324	THREE DIMENSIONAL SHAPE	129,4	
1334	SLOPE ASPECT	53,2	
1321	LAND USE	114,1	
1180	VELD TYPE	435,4	
859	VELD CARRYING CAPACITY		0,06(2)
485	VELD CONDITION SCORE		0,20
481	INCREASER II. BENCH		0,09(3)
1130	POPULATION DENSITY		0,14
1332	DISTANCE FROM VILLAGE		-0,21
1329	FOOTPATH INTENSITY		0,11
1237	LARGE STOCK UNIT DENSITY		0,29
1233	SMALL STOCK UNIT DENSITY		-0,31
1236	MATURE LIVESTOCK UNIT DENSITY		0,12

(1) Expected chi-square value <1 (possibly too many categories)

(2) Not significant at the 5 % level

(3) Not significant at the 1 % level

lands. Zachar (1982) estimates a 10000 fold difference between fallow land and forested lands. Even though forested areas received more attention than other land use areas in terms of ground truthing, it could be argued that litter or dense vegetation cover might obscure erosion evidence. It was therefore decided to rerun the analysis procedure excluding forested land. The results of the rerun are given in Table 3. It is evident from Fig. 2 and Table 3 that grazing land shows a higher incidence of erosion and arable land. Difficulties involved in distinguishing between arable land and grazing land meant that all non-forested land not currently under cultivation was classified as grazing land. This means that land previously cultivated and abandoned due to excessive erosion could be included in the grazing category.

Table 3. Results of Statistical Tests excluding forested regions

SAMPLE POPULATION	INDEPENDENT VARIABLE	CHI-SQUARE	SPEARMAN RANK
1165	MEAN ANNUAL PRECIPITATION		-0,35
1129	ALTITUDE		-0,16
1165	RAINFALL EROSIVITY		-0,32
1141	SOIL ERODIBILITY		0,04(2)
1135	SOIL TYPE	204,2(1)	
1165	GEOLOGY	55,2	
1129	SLOPE STEEPNESS		-0,10
1129	SLOPE LENGTH		0,06(3)
1129	L.S. FACTOR		-0,09
1129	PROFILE SHAPE	61,6	
1129	LATERAL SHAPE	22,3	
1156	THREE DIMENSIONAL SHAPE	112,5(1)	
1165	SLOPE ASPECT	6,1(2)	
1165	LAND USE	185,0	
1015	VELD TYPE	435,4	
790	VELD CARRYING CAPACITY		0,06(2)
428	VELD CONDITION SCORE		0,20
428	INCREASER II. BENCH		0,09(3)
988	POPULATION DENSITY		0,14
1164	DISTANCE FROM VILLAGE		-0,21
1161	FOOTPATH INTENSITY		0,11
1076	LARGE STOCK UNIT DENSITY		0,29
1073	SMALL STOCK UNIT DENSITY		0,31
1076	MATURE LIVESTOCK UNIT DENSITY		0,12

(1) Expected chi-square value <1 (possibly too many categories)

(2) Not significant at the 5 % level

(3) Not significant at the 1 % level

For the sake of brevity, the remainder of the discussion will focus on statistically significant relationships between soil erosion and the various independent variables in non-forested areas.

Three "climatic" indices were used in the study: mean annual precipitation (MAP), altitude (a surrogate measure of MAP due to the strong orographic influence of rainfall in the area) and rainfall erosivity (as defined by Wischmeier and Smith, 1978). The results show a statistically significant inverse relationship between the degree of soil erosion and MAP. As expected, altitude shows a similar, yet weaker relationship with soil erosion. The inverse relationship between erosivity and soil erosion is contrary to the expected. Although erosivity values are likely to be important controls on soil loss through time at a given point in space,

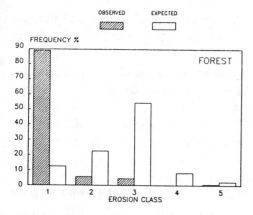

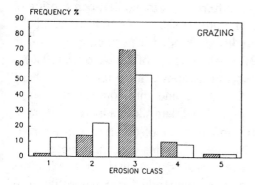

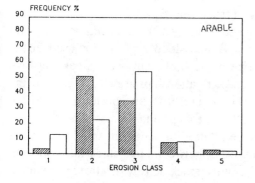

Fig. 2. Relationships between soil erosion and landuse.

the effect of spatial variation in soil erosion due to variations in erosivity might be complicated by the effect of rainfall on vegetation. Zachar (1982) emphasises the fact that soil erosion does not always occur where the erosivity is highest, but may occur where ecoclimatic conditions are unfavourable. Langbein and Schumm (1958), Douglas (1967) and Kirkby (1980) show that on a world scale, areas with a mean annual precipitation of less than 300 mm show an increase in erosion with increasing mean annual precipitation. At precipitation levels above 300 mm the protective effect due to increased vegetation cover increases and soil erosion decreases.

The relationship between soil erosion and underlying material was investigated in terms of underlying geology, soil type and soil erodibility. No significant relationship could be found between the SLEMSA (1976) erodibility ratings and soil erosion. Significant differences were however found in soil erosion on different soil types. Soils with a high clay content and a well structured A-horizon (e.g. Milkwood, Mayo, Arcadia and Shortlands) tend to show relatively low levels of erosion. Duplex soils with restrictive horizons (e.g. Valsrivier and Sterkspruit) and shallow soils with a low permeability (Mispah and Glenrosa) exhibit higher levels of erosion. These findings are in keeping with those of Eloff (1973) and d'-Huyvetter and Laker (1985). A significant difference is found between erosion of soils underlain by dolerite and those underlain by shales and mudstones of the Beaufort Group, the latter being the more heavily eroded. This finding occurs with the field observations of Mountain (1952) and the findings of Berzak et al. (1986).

Although statistically significant differences exist between the three slope shape parameters selected (lateral shape, profile shape and three-dimensional slope shape), no logical explanation could be found to explain these apparent differences. There is a strong possibility that the low number of occurrences of certain slope classes invalidates any conclusive statements on this data. A comparison of the two most common lateral slope shapes reveals no significant difference (Fig. 3). The same applies when comparing the two most frequently occurring profile slope shape classes (Fig. 4). A comparison of the four most common three-dimensionally described slope type shows a similar lack of trend (Fig. 5).

Slope length and slope aspect show no statistically significant relationship with soil erosion. LS (the combined slope length and slope angle factor used in the Universal Soil Loss Equation) and slope angle both show very weak negative relationships with soil erosion severity. These relationships are contrary to the expected (Morgan, 1980 and Hudson, 1981) and are probably spurious, due to the positive correlation between MAP and both slope angle and LS.

The results show that there is an inverse relationship between small stock units and soil erosion yet a direct relationship exists between large stock units and soil erosion. (Because of the units of measure of livestock density [hectares per animal unit] this means that the higher the number of small stock per unit area, the higher the erosion and vice versa for large stock units). The question as to whether there is a high amount of erosion because of the high density of small stock units or vice versa could only be adequately answered with more detailed historical records of livestock populations. Animals classified as large stock (cattle and equines) are predominantly grazing animals. It follows that the high densities of large stock units will be restricted to areas with good grazing (i.e. a high grass cover). These areas would also have good protection against soil erosion. Small stock units, on the other hand, include a large proportion of goats which are mainly browsers and are suited to areas with a relatively low grass cover and where bush encroachment

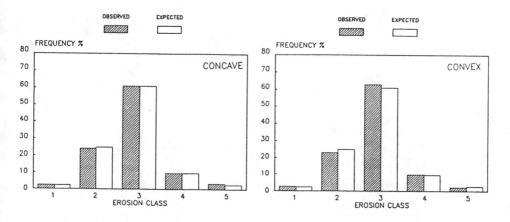

Fig. 3. Relationship between soil erosion and lateral slope shape.

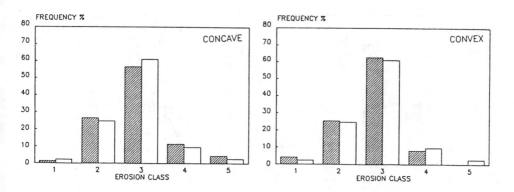

Fig. 4. Relationship between soil erosion and profile slope shape

(mainly Acacia Karoo) has occurred. Although significant relationships exists between stock density and soil erosion, difficulties justifying any cause-and-effect relationships make the usefulness of this index doubtful.

Of the four vegetation indices selected (veld type, veld condition score, the ratio of increaser 2 species to benchmark and veld carrying capacity), only veld type showed statistically significant differences. Dohne Sourveld comprises a dense grassveld with a basal cover exceeding 30 % (Acocks, 1975). It is not surprising that this veld-type shows the lowest susceptibility to soil erosion (Fig. 6). The other three vegetation types (False Thornveld, Valley Bushveld and Eastern Province Thornveld) exhibit less distinctive patterns in the frequency distribution of erosion (Fig. 6).

Perhaps the most noticeable feature in these vegetation types is that the Eastern Province Thornveld in the study area has no class one erosion (no erosion) and that most of the class 5 erosion occurs in areas of False Thornveld. The invasion of

223

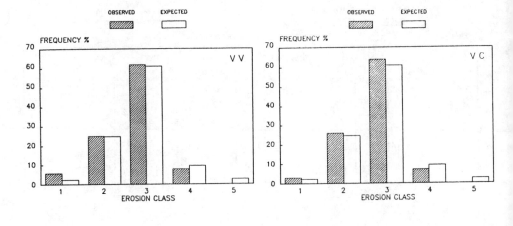

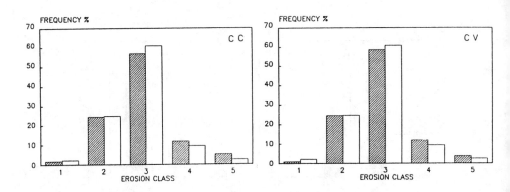

Fig. 5. Relationship between soil erosion and most frequently occurring two-dimensional slope shape. (VV = Convex in both lateral and profile dimension; VC = Convex in profile and concave in lateral dimension; CC = Concave in both lateral and profile dimension; CV = Convex in lateral and concave in profile dimension).

grassveld by Acacia Karoo in the False Thornveld areas has been linked to overgrazing (Acocks, 1975). According to Trollope and Coetzee (1978) False Thornveld areas are covered mainly by inferior vegetation and are amongst the most badly eroded areas in Ciskei.

Three anthropogenic variables were included in the study, population density, distance from the nearest village and footpath intensity. None of these variables showed any meaningful relationship with soil erosion.

5. CONCLUSIONS

It must be emphasised that the study described in this paper represents an initial

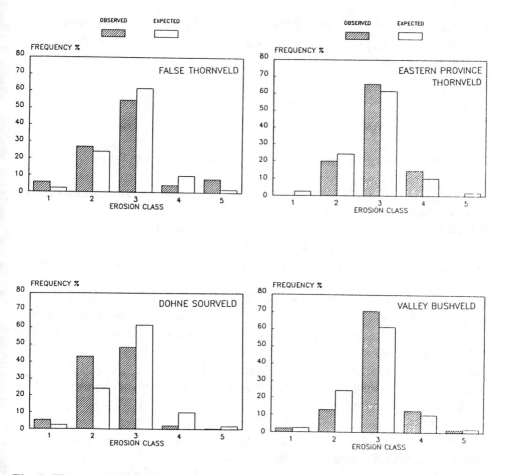

Fig. 6. The relationship between soil erosion and veld type.

investigation into the factors influencing spatial variations in soil erosion in Ciskei. A further point which needs emphasis is that, although the choice of indices was based largely on a view of the literature, the final choice of indices was governed to a large extent by the availability of data.

The study has shown that, on the regional scale, the individual variables which are significantly related to the spatial distribution in soil erosion in the areas in Ciskei selected for study are landuse, MAP, soil type, geology and veld type. The relationship between these individual variables and soil erosion is relatively weak. This is to be expected when one considers the complex and multi-dimensional nature of the soil erosion problem. Stocking (1972) points out that although individual variables might be weakly related to soil erosion, grouped variables often reveal stronger relationships. The results of the study described in this paper will be used as guidelines for the selection of variables for use in a multivariate analysis of soil erosion. The applicability of these results to other areas in Ciskei will be tested in other catchments. The final aim of the study is to develop an erosion hazard assessment technique applicable at the regional scale to as wide a range of conditions in Ciskei as possible.

6. REFERENCES

Acocks, J.P.H. 1975. Veld types of South Africa. Memoir of the Botanical Survey of South Africa, 40, Government Printer, Pretoria.

Aircraft Operating Company, 1984. 1:30 000 Aerial Photographic Survey of Ciskei. Johannesburg.

Adri, 1981. Unpublished 1:10 000 Soils Map of the Amatole Basin. Alice.

Berzak, M., Finchman, R.J., Liggit, B. and Watson, H.K. 1986. Temporal and spatial dimensions of gully erosion in Northern Natal, South Africa. Proc. Symp.I.S.P.R.S. 26, 4, 583-593.

Ciskei Commision 1980. The Quail Report. Conference Associates, Pretoria.

D'Huyvetter, J.H.H. and Laker, M.C. 1985. Determination of threshold slope percentages for the identification and delineation of Arable land in Ciskei. Final Report to the CSIR.

Dixon, W.J. (ed.) 1985. BMDP Statistical Software. U. California, Los Angeles.

Douglas, I. 1967. Man, vegetation, and the sediment yield of rivers. Nature, 215, 925-928.

Eloff, J.F. 1973. Erosie Kwesbaarheid en-indekse van S.A. Gronde.

Garland, G.G. 1982. An appraisal of South African research into runoff-erosion. South African Geographical Journal, 64, 138-143.

Gerlach, A. 1976. Splashing and its influence on the translocations of the soil on slopes. Studia gemorphologica Carpatho-Balcanica, X, 125-137.

Gibbs-Russel, G.E. and Robinson, E.R. 1982. Buffalo River Basin. Plant communities. 1:50 000 unpublished map, ARDRI, Fort Hare.

Hill, Kaplan, Scott and Partners 1977. Keiskamma River Basin study. Zwelitsha. Ciskei Government Report.

Hill, Kaplan, Scott and Partners 1979. Fish/Kat River Basins Study Zwelitsha. Ciskei Government Report.

Hill, Kaplan, Scott and Partners no date. 1:10 000 Soil Code Map for the Roxeni Basin. Unpublished map.

Hudson, N. 1981. Soil Conservation. Cornell University Press, New York.

Kirkby, M.J.1980. The problem. In: Kirkby, M.J. and Morgan, R.P.C. (eds), Soil Erosion. Wiley, London.

Langbein, W.B. and Schumm, S.A. 1958. Yield of sediment in relation to mean annual precipitation. Transactions of the American Geophysical Union, 39, 1076-1084.

Morgan, R.P.C. 1980. Soil Erosion. Longmans, London.

Mountain, E.D. 1952. Geology of the Keiskammahoek District, In: Mountain, E.D.(ed.), The Natural History of the Keiskammahoek District, 1, Keiskamma Rural Survey. Shuter and Shooter, Pietermaritzburg.

Rooseboom, A. 1983. Sediment studies, In: Maaren, H.(ed.), Proceedings of the First South African National Hydrological Symposium, TR119, Pretoria, 346-351.

SARCCUS 1981. A system for the classification of soil rosion in the SARCCUS region. Department of Agriculture and Fisheries, Pretoria.

Scheidegger, A.E. 1975. Physical Aspects of Natural catastrophes. Elsevier, Amsterdam.

Slemsa 1976. Soil loss estimator for Southern Africa. Natal Agricultural Research Bulletin, 7.

Stocking. M.A. 1972. Relief analysis and soil erosion in Rhodesia using multi-variate techniques. Zeitschrift für Geomorphologie N.F., 16, 432-443.

Stocking, M.A. 1984. Rates of erosion and sediment yield in the African environment, In: Walling, D.E., Foster, S.S.D. and Wurzel, P.(eds.), Challenges in African Hydrology and Water Resources. I.A.H.S., Publ. 144, 285-294.

Trollope, W.S.W. and Coetzee, P.G.F., 1978. Vegetation and veld management, In: Laker, M.C. (ed.), The Agricultural Potential of the Ciskei, University of Fort Hare, Alice, 5.

Weaver, A. Van B. 1987. Changes in landuse and soil erosion in South African and Ciskeian portions of the Yellowwoods drainage basin between 1975 and 1984, In: Firman, J.B.,(ed.), Landscapes of the Southern Hemisphere, Elsevier, Amsterdam (in press).

Weaver, A. Van B. 1988. Soil erosion rates in the Roxeni Basin, Ciskei. South African Geographical Journal (in press).

Weaver, A. Van B. and Hughes, D.A. 1986. A preliminary study of the estimation of rainfall erosivity values for Ciskei, In: Schulze, R.E.(ed.), Proceedings of the Second South African National Hydrological Symposium, University of Natal, Pietermaritzburg, 229-243.

Willis, M.J. and Trollope, W.S.W. 1982. The grazing and browzing capacity of the Buffalo River Basin. ARDRI, University of Fort Hare, Alice.

Wischmeier, W.H. and Smith, D.D. 1978. Predicting rainfall erosion losses - a guide to conservation planning. United States Department of Agriculture, Handbook, 537.

Zachar, D. 1982. Soil Erosion. Elsevier, New York.

THE ROLE OF ARTIFICIAL DRAINAGE IN THE DEVELOPMENT OF SOIL PIPES AND GULLIES: SOME EXAMPLES FROM TRANSKEI, SOUTHERN AFRICA

HEINRICH R. BECKEDAHL, GEORGE F. DARDIS

Department of Geography, University of Transkei

1. INTRODUCTION

Increasing evidence is forthcoming to suggest that soil systems in the southern African region exist at thresholds within finely balanced states of dynamic equilibrium (Stocking, 1978; Beckedahl et al., 1988). This is controlled primarily by soil properties (Chakela, 1981; Faber and Imeson, 1982; Watson et al., 1982; Yaalon, 1987) and climate (particularly effective precipitation) (Douglas, 1967), which may render many parts of southern Africa prone to soil erosion. Extrinsic variables , such as human interference (e.g., poor farming practice, footpath erosion), are commonly cited as important factors contributing to soil erosion in southern Africa (Stocking, 1978; Garland, 1985; Beckedahl et al., 1988; Marker, 1988). Whereas much attention has been focused on human interference, such as overgrazing and poor farming practices, little attention has been focused on the effects of artificial drainage on land systems.

The term artificial drainage, as used in this paper, refers to unnatural, often accelerated, runoff, associated with roads, footpaths, buildings or reservoirs. Even though it is well known that erosion rates may increase dramatically where artificial drainage occurs (Wolman, 1967), little is known about its role in the development of badland topography or about the complexity of the effects of artificial drainage, whether positive or negative, upon the physical landscape. In certain situations, for example, greater runoff at a particular point may result in increased vegetation growth, with a concomitant increase in the retardation and retention capacity of the surface, damping erosional effects. In other cases, however, the increased retention of runoff has directly contributed to increases in subsurface erosion (cf. Stocking, 1976).

A number of examples from Transkei, southern Africa, are considered to demonstrate the complex effects of artificial drainage on relatively unstable land sys-

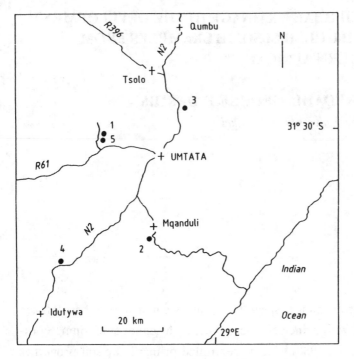

Fig. 1. Location map of sites referred to in the text

tems. This area provides excellent examples of these effects in view of the potentially unstable nature of the land systems, where small extrinsic changes may result in
major geomorphic responses within relatively short time periods.

Preliminary observations of the role of artificial drainage in the development of soil
pipes and gullies at a number of sites in Transkei are outlined (Fig. 1). The sites are
in colluvium-mantled terrain, which characterises much of the lowveld region of eastern southern Africa (Watson et al., 1984; Dardis, 1987a). The colluvium is considered
to be mainly late Pleistocene in age (30000-12000 years BP), though recent evidence
from Transkei suggests that it may, in places, be much younger (Dardis, submitted).

2. ARTIFICIAL DRAINAGE AND GULLY DEVELOPMENT: INITIAL
   STAGES

Where runoff is artificially directed off a man-made surface (e.g. a road or culvert)
into natural terrain, it may have a number of effects, in terms of piping and gullying.
The initial stages of piping and gullying resulting from drainage of this type is illustrated at sites 1 and 5 (Fig. 1).

Fig. 2. The road culvert at site 1 (see Fig. 1) with the first of the depressions in which water is concentrated (a). Pipe intakes occur at point b.

2.1. Site 1 (Mabeleni Dam Region)

The Mabeleni Dam site is located 35 km southwest of Umtata (site 1, Fig. 1). It lies at 880 m a.s.l., in a region of steep terrain in the foothills of the Drakensberg Mountains. The site shows extensive soil piping and soil erosion developed in close association with a culvert under the tarred road, which has been constructed since 1976 (Figs. 2,3). A series of micro-scale topographic depressions lead from the culvert into grassland (Fig. 2). These depressions are densely vegetated (point b, Fig. 2; points b,c, Fig. 3). Soil pipe intakes formed by soil pipe roof collapse (Fig. 4) occur at several points in the densely vegetated areas. These are c.25-40 cm wide and are up to 1.5 m deep.

The development of pipes (rather than open gully forms) at the site seems to be due to enhanced growth of vegetation at the culvert outlet, leading to increased retention of surface water and a higher infiltration capacity (Graf, 1979), which contributes to higher resistance to runoff erosion but in turn lowers resistance to seepage erosion, associated with cracking and/or dispersible soils (cf. Gibbs, 1945; Grouch, 1976; Stocking, 1976; Beckedahl, 1977). The morphology of incipient pipes suggests that they formed from seepage cracks (Fig. 5).

Enlargement of the pipes has resulted in roof collapse and "discontinuous gully" development (cf. Gibbs, 1945) (Fig. 6). The resulting gully is steep-walled and v-

231

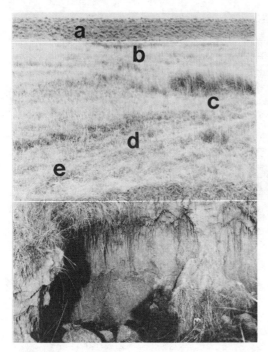

Fig. 3. View from the head of the pipe-gully system to the culvert (a), showing dense-ly-vegetated micro-topographic depressions (b,c,d) acting as pipe intakes and sub-sidence of the soil (e) above a subsurface drainage line. Note the sharp junction of the vegetation cover on the sidewall of the gully.

Fig. 4. Initial stage of pipe-roof collapse (taken at point d, Fig. 3).

shaped, with dense grass vegetation up to the edge of the sidewall (Fig. 5). The gully meanders, with pronounced spurs, reflecting the primitive subsurface drainage net-

Fig. 5. Incipient soil pipe associated with seepage cracks at site 1. Note the rectilinear cracks (above left of pipe) and rectilinear nature of the pipe. Pipe is c.10 cm deep.

work of the soil pipes from which it developed. The gully has not developed a branching network (though sidewall block breakage and undercutting has commenced) and is quasi-linear (cf. Ireland et al., 1939). Soil arches (point b, Fig. 6A), associated with soil pipe roof collapse, occur in the discontinuous gully system. Soil pipes (Fig. 6B) measuring 50-70 cm diameter link the gully network. In its lower reaches, the discontinuous gully system parallels an older, continuous gully for some 150 m, before breaching the narrow sidewall and following the old, partially stabilised, gully down to the valley bottom.

It is clear, therefore, that the culvert has acted principally to concentrate runoff, and has not acted directly to cause gully erosion. In fact, there is no evidence of the development of natural surficial channelling at the culvert outlet, suggesting that much of the runoff is dissipated by seepage or subsurface flow (cf. Jones, 1978). These observations are in accordance with other findings which suggest that soil piping and subsurface erosion are frequently related to regions where surface impedence of runoff and ponding are concomitant with a high infiltration rate (Stocking, 1976; Beckedahl,

Fig. 6. A Section in the main gully, showing soil arches (b) developed beneath a surficial resistant layer of colluvium. B. Soil pipe, occurring at point a, Fig. 6A.

1977; Jones, 1981). The actual subsurface transport path cannot be established on the basis of present observations. Exposures in the gully (Fig. 6A), however, suggest that the transport path may have been determined by the position of relatively impermeable layers within the stratified colluvium (Fig. 6A) and by the depth of crack propagation into the colluvium.

2.2. Lugxogxo

More complex initial effects of artificial drainage are evident where drainage trenches have been constructed to divert runoff away from road surfaces. Some of these effects can be seen at Lugxogxo in Transkei (Site 5, Fig. 1).

The Lugxogxo site (2 km east of site 1) has a drainage ditch leading away from the tarred road surface (Fig. 7). This is one of the most common forms of artificial drainage in Transkei; a trench is constructed initially parallel to the road and then the accumu-

Fig. 7. The Lugxogxo site (site 5, Fig. 1), showing the drainage ditch (x) leading away from the tarred road surface.

Fig. 8. The drainage trench at Lugxogxo, showing rills and gullying on the diversion channel sidewalls.

lated water is diverted onto the natural terrain and away from the road.

A trench has been incised into bedrock (Burgersdorp Formation mudstones, Karoo Sequence) at the point of diversion. The trench is 1.5-2.0 m deep and varies from 0.75-2.4 m in width. Channelling at this point has resulted in local base-level lowering, in-

Fig. 9. Extensive down-trench gullying within and on the sides of the trench at Lug-xogxo. Note the dendritic channel forms developing on the floor of the eroded area.

itiating shallow gullying on the sidewalls of the trench (Fig. 8). The gullying has developed in a thin discontinuous layer of colluvium, up to 1 m thick, which overlies bedrock at this site. There is rill development and incipient gullying within 1-2 m of the sides of the trench at the start of the diversion channel (Fig. 8) but extends up to some 5 m beyond the trench in a down-channel direction, where shallow piping also becomes evident (Fig. 9.). The sidewall piping and gullying has extended the trench to form a shallow dendritic gully system, characterised by narrow, v-shaped tributary gullies (Fig. 9). This downcutting effect is similar to that observed in bedrock-cut gullies at Saddleback Pass, in the eastern Transvaal (Dardis and Beckedahl, 1988), where base-level lowering associated with road-cutting has resulted in accelerated erosion. This process has been described in many alluvial channel systems (Schumm and Hadley, 1957; Nordin, 1964; Pickup, 1975) and has been noted in association with gully development in colluvium in southwestern Australia (Conacher, 1982).

3. ARTIFICIAL DRAINAGE AND GULLY DEVELOPMENT: ADVANCED STAGES

Advanced stages of gully development associated with artificial drainage are evident at many sites in Transkei. One site at KuLozulu, near Mqanduli (Site 2, Fig. 1), shows the effects of drainage related to two culverts in moderately steep terrain.

Fig. 10. The KuLozulu site, Transkei.

3.1. KuLozulu

The KuLozulu site is located 5 km southeast of Mqanduli (Site 2, Fig. 1). It lies at 665 m a.s.l. on a slope varying between 10 and 25 degrees in undulating terrain. The site has a relatively thick (0.5-8.0 m) veneer of stratified colluvium and is underlain by soft mudstones of the Beaufort Group of Permo-Triassic age.

An extensive, complex pipe-gully system is exposed at KuLozulu (Fig. 10). The pipe-gully system has many morphological attributes similar to those seen at Lugxogxo, but is at a more advanced stage of development.

The pipe-gully system is associated with two culverts on the main tar road to Coffee Bay. The tarred surface was added only recently and the culverts appear to have been emplaced during this construction phase. They are thus less than 5 years old. Both culverts have seepage zones, characterised by a heavily saturated area, with ponded water on the down-culvert side and occasional runoff between interconnected ponds, extending some 4-5 m downstream from the culvert (Figs. 11 and 12). At the time of investigation (November-December, 1987), both zones were water-logged, with runoff

Fig. 11. View of the main culvert (middle foreground) at KuLozulu, showing the seepage zone downstream of the culvert, with standing water and the pipe-gully system which commences c.10 m downstream of the culvert.

flowing through the main culvert, and extending into the main gully (Fig. 13).

The headwall of the main gully is c.2.5 m high and commences abruptly. No pipe or other subsurface drainage features were observed in the immediate vicinity of the headwall zone.

A series of soil pipes and cavities are well exposed on the gully sidewalls downstream of the headwall zone. These are mainly tributary features. Deep, narrow gullies predominate (Fig. 14). Their morphology and evidence of discontinuous gullies, with linking soil pipes (Fig. 15), clearly indicate that the gullies are remnants of collapsed

238

Fig. 12. The second tributary culvert at KuLozulu, showing a seepage zone on the down-culvert side, with ponded water, and discontinuous cavities occurring approximately 25 m downslope of the culvert. A soil pipe is exposed in the main cavity (point x). The culvert is shown at point y.

soil pipes. In some instances the gullies are incised into bedrock, which is mainly Burgersdorp Formation mudstones.

A tributary system of the main pipe-gully system, extends some 70 m up-slope, to the second culvert (Fig. 12). In its upper reaches, it consists of discontinuous cavities some 2-3 m in diameter, associated with pipe roof collapse (Fig. 15), immediately down-slope of the second culvert (Fig. 12). This shows downslope transition from discontinuous gullies with interconnecting soil pipes into narrow, deep meandering gul-

Fig. 13. Runoff over the gully headwall at KuLozulu.

Fig. 14. Deep narrow linear gullies at KuLozulu. Note the rill development on the sidewalls.

Fig. 15. Discontinuous gully with interconnecting soil pipe leading from the second culvert at KuLozulu.

Fig. 16. Meandering gully system at KuLozulu with soil arches, representing remnants of a soil pipe roof. The gully follows the path of the soil pipe.

lies (Fig. 14). Gullies of this type exist throughout the central part of the pipe-gully system (Fig. 16).

This sequence appears therefore to be broadly similar to the pattern of gully development observed in the pipe-gully system at Lugxogxo, although the KuLozulu site demonstrates a more advanced stage of development. Field evidence suggests that the subsurface drainage systems are complex features, characterised by relatively continuous drainage lines. The gully pattern largely reflects the nature and direction of the subsurface drainage.

The drainage lines associated with the second culvert have only been exposed in the past four months. The gully system was not evident in reconnaisance studies undertaken in February and August of 1987, and is evidence that the pipe-gully system as a whole is developing very rapidly.

4. DISCUSSION

The above observations suggest that the artificial concentration of runoff from road surface drainage is having a profound effect on badland development in certain areas of Transkei. Runoff does not appear to be of sufficient magnitude to result in direct erosion by particle detachment. The principal erosion processes operating in these areas are soil piping and collapse. These can be related to the occurrence of saturated zones in the down-culvert zone. Hydraulic pressure associated with seepage in these zones, combined with high rates of dispersion or deflocculation appears to be the main cause of cavitation and piping at these sites (cf. Terzaghi, 1922; Sherard et al., 1972; Beckedahl, 1977; Cedergren, 1977).

Seepage and subsurface dissolution of the soil is enhanced by soil cracking, and the presence of sodic and duplex soils, which are characterised by high dispersion rates and are widespread in southern Africa (Stocking, 1976; 1978; Beckedahl, 1977; Faber and Imeson, 1982; Watson et al., 1984).

Once established, a pipe system will be enlarged by several processes (fluvial erosion, sidewall collapse, roof collapse etc.). When roof collapse ultimately occurs, a well developed gully system is exposed, which may not be easily controlled by conventional methods. This is particularly applicable within the southern African context, where indications are that pipe diameters rate among the largest in the world, with a mean value of some 1.5-2.0 m.

5. CONCLUSIONS

The artificial channelling of surface runoff associated with a road network has been shown to contribute significantly to soil erosion at a number of sites in Transkei. This type of soil erosion is probably widespread in badland areas of southern Africa, and is associated with both tarred and untarred road surfaces. A greater understanding of the problem and of the factors contributing to the high erodibility of the soil in certain areas implies that adequate precautions can then be taken in the future design and construction of road drainage.

6. SUMMARY

Artificial channelling of overland flow through road culverts has resulted in the rapid development of soil pipes, micro-topographic depressions associated with sub-surface cavities, and gullies. Concentration of surface runoff as a result of channelling has enhanced rates of infiltration in discrete parts of the soil system. This in turn has resulted in the development, initially, of subsurface cavities and soil pipes, rather than unconfined soil erosion forms such as gullies. This suggests that subsurface dissolution processes may play a major role in the initial development of gullies in colluvium-mantled terrain in southern Africa. Some problems and implications for road construction, and the potential effects in terms of soil erosion in sensitive terrain, are discussed.

7. REFERENCES

Beckedahl, H.R. 1977. Subsurface erosion near the Olivershoek Pass, Drakensberg. South African Geographical Journal, 59, 130-138.

Beckedahl, H.R. Bowyer-Bower, T.A.S.,Dardis, G.F. and Hanvey, P.M., 1988. The geomorphic effects of soil erosion. In: Moon, B.P. and Dardis, G.F. (eds.), The Geomorphology of Southern Africa, Southern Book Co., Johannesburg, (in press).

Cedergren, H.R. 1977. Seepage, Drainage and Flow Nets. John Wiley and Sons, New York, 534pp.

Chakela, Q.K. 1981. Soil Erosion and Reservoir Sedimentation in Lesotho. Report 54, Department of Geography, Uppsala University, 152pp.

Conacher, R.J. 1982. Salt scalds and subsurface water: a special case of badland development in southwestern Australia. In: Bryan, R.B. and Yair, A. (eds.), Badland Geomorphology and Piping, Geobooks, Norwich, 195-219.

Crouch, R.J. 1976. Field tunnel erosion - a review. Journal of Soil Conservation Service, New South Wales, 30, 98-111.

Dardis, G.F., 1987a. Quaternary erosion and sedimentation in badland areas of Transkei, Southern Africa. Paper presented at Workshop on Erosion, Transport and Deposition in Arid and Semi-Arid Areas, Hebrew University of Jerusalem, March-April 1987.

Dardis, G.F. (submitted). Age of Quaternary colluvium in eastern southern Africa.

Dardis, G.F. and Beckedahl, H.R. 1988. Gully formation in Archaean rocks at Saddleback Pass, Barberton Mountain Land, South Africa. (this volume).

Douglas, I. 1967. Man, vegetation and the sediment yield of rivers. Nature, 215, 925-928.

Downes, R.G., 1956. Conservation problems on sodic soils in the State of Victoria (Australia). Journal of Soil and Water Conservation, 11, 228-232.

Downing, B.M., 1968. Subsurface erosion as a geomorphological agent in Natal. Transactions of the Geological Society of South Africa, 71, 131-134.

Faber, T. and Imeson, A.C., 1982. Gully hydrology and related soil properties in Lesotho. In: Recent Developments in the Explanation and Prediction of Erosion and Sediment Yield, IAHS Publication, 137, 135-144.

Garland, G.G. 1985. Erosion Risk from Footpaths and Vegetation Burning in the Central Natal Drakensberg. Draft Report for the Natal Town and Regional Planning Commission.

Gibbs, H.S. 1945. Tunnel-gully erosion on the Wither Hills, Marlborough. New Zealand Journal of Science and Technology, 27, 135-146.

Graf, W.L. 1979. The development of montane arroyos and gullies. Earth Surface Processes and Landforms, 4, 1-14.

Henkel, J.J., Bayer, A.W. and Coutts, J.R.H., 1938. Subsurface erosion on a Natal Midlands farm. South African Journal of Science, 35, 236-241.

Hóly, M., 1980. Erosion and Environment. Pergamon Press, Oxford.

Ireland, H.A., Sharpe, C.F.S. and Eargle, D.H. 1939. Principles of gully erosion in the Piedmont of South Carolina. United States Department of Agriculture Technical Bulletin, 63, 143pp.

Jones, J.A.A., 1981. The Nature of Soil Piping: A Review of Research. Geobooks, Norwich.

Jones, J.A.A. 1987. The effect of soil piping on contributing areas and erosion patterns. Earth Surface Processes and Landforms, 12, 229-248.

Marker, M.E. 1988. Soil erosion in a catchment near Alice, Ciskei, southern Africa. (this volume)

Sherwood, J.L., Decker, R.S. and Ryker, N.L. 1972. Piping in earth dams of dispersive clay. Proceedings, Specialty Conference on the Performance of Earth and Earth-Supported Structures, Purdue University, June 1972, vol. 1, part 1.

Stocking, M.A., 1976. Tunnel erosion. Rhodesian Agricultural Journal, 73, 35-40.

Stocking, M.A., 1978. The prediction and estimation of erosion in sub-tropical Africa. Problems and prospects, Geo-Eco-Trop, 2, 161-174.

Terzaghi, K. 1922. Der Grundbruch an Stauwerken und seine verhütung. Die Wasserkraft, 17, 445-449.

Watson, A., Price-Williams, D. and Goudie, A.S., 1984. The palaeoenvironmental interpretation of colluvial sediments and palaeosols of the late Pleistocene hypothermal in southern Africa. Palaegeography, Palaeoclimatology, Palaeoecology, 45, 225-249.

Wolman, M.G. 1967. A cycle of sedimentation and erosion in urban river channels. Geografiska Annaler, 49A, 385-395.

Yaalon, D.H. 1987. Is gullying associated with highly sodic colluvium? Further comment to the environmental interpretation of Southern African dongas. Palaeogeography, Palaeoclimatology, Palaeoecology, 58, 121-123.

DRAINAGE EVOLUTION IN AN EPHEMERAL SOIL PIPE-GULLY SYSTEM, TRANSKEI, SOUTHERN AFRICA

GEORGE F. DARDIS, HEINRICH R. BECKEDAHL

Department of Geography, University of Transkei

1. INTRODUCTION

Even though soil pipes are considered a major process in soil erosion (Jones, 1981) and drainage network development (Jones, 1987), many aspects of soil pipe development remain poorly understood. This is due largely to difficulties in examining the geometry and long profiles of subsurface drainage networks, particularly those of small diameter (Jones, 1978). As a result, little is known about the stages of development of soil pipes, the nature of soil pipe systems, erosion and deposition processes operating within pipes, and the nature of soil pipe development over time. Morphometric aspects of pipe networks have been considered in a number of studies, aimed at examining their general evolution (cf. Jones, 1971; 1975; Beckedahl, 1977) and modelling hydrological response (cf. Gilman and Newson, 1980). Studies of the pattern of pipe networks have been reviewed in some detail by Jones (1981) and it appears that pipe networks tend to be of dendritic or anastomising form and are often discontinuous, with greater continuity and dendritic form indicative of more advanced development. Actual patterns depend to some degree on the scale of mapping (Jones, 1981). Most studies have concentrated on two-dimensional patterns of soil pipe networks, examined in terms of hydrological response, with pipe size and continuity taken as an indication of discharge (cf. Jones, 1975; 1978). This relationship is probably reasonable in temperate environments, but is less applicable in semi-arid and arid environments, where flow is largely ephemeral and other factors determining geometry, such as seepage erosion or sidewall and roof collapse attain a relatively greater significance (Beckedahl, 1977; Beckedahl and Dardis, 1988). This complexity has been highlighted by Jones (1981, p.150) who states that "....there is very little direct evidence on which to build a general model for the evolution of pipe networks....it is clear that no one single model could totally encompass all possible (soil pipe network) developments, largely because of differing initiating processes....".

Geomorphological Studies in Southern Africa, G.F.Dardis & B.P.Moon (eds)
© *1988 Balkema, Rotterdam. ISBN 90 6191 831 6*

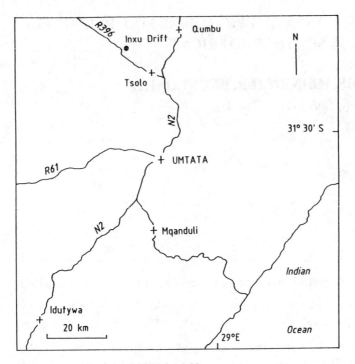

Fig. 1. Location map of Inxu drift.

The geometry of soil pipe systems has been described in South Africa (Beckedahl, 1977) and Zimbabwe (Stocking, 1976). Beckedahl et al.(1988) and Dardis et al.(1988) have briefly reviewed the geomorphic effects of soil erosion and referred to the occurrences of simple and composite soil pipe systems.

In this paper, the geometry of a collapsed ephemeral soil pipe system in Transkei, southern Africa, is described and the main stages of development of the system are outlined. Processes acting to enlarge the system are identified from detailed mapping.

2. STUDY AREA

The pipe system occurs at Inxu Drift (Lat:$31°15'$S; Long: $28°40'$E; Alt: 1020 m a.s.l.) in Transkei (Fig. 1.). It forms part of an extensive gully system (Fig. 2), approximately 800 m long and up to 250 m wide. The gully is incised into stratified colluvium varying in thickness from 2-10 m.

The pipe system is incised into stratified colluvial sediments, with individual beds of colluvium varying in thickness from 20-200 cm, interbedded with clast-supported pebbly gravels, and parallel-laminated sands. The sands and gravels are of fluvial origin and are found mainly at the base of the sequence. There is a buried organic-rich palaeosol near the top of the colluvial sequence. Though the palaeosol has not

248

Fig. 2. The gully system at Inxu Drift, Transkei. Note the flat valley-bottom topography and steep sidewalls of the gully system.

Fig. 3. The two metre grid network superimposed on the partially collapsed soil pipe system at Inxu Drift, Transkei.

been dated, it occurs in a similar stratigraphic position to palaeosols at Cedarville and Mbolompo Point in Transkei, which have been radiocarbon dated to 2100 + -80 years BP (Beta-20437) and 1990 + -60 years BP (Beta-20436) respectively (Dardis, submitted).

A partially collapsed composite soil pipe-gully system (Figs. 3,4) on the flank of the main gully network at Inxu Drift was selected for study. These features are relatively common in gully networks in Transkei and constitute one of the major ways in which

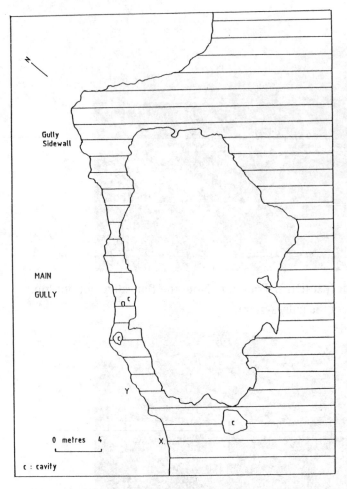

Fig. 4. The collapsed pipe system at Inxu Drift. The horizontally shaded area marks the gully and collapsed pipe system sidewalls. X and Y mark the location of the main and secondary pipe outlets from the system.

gully expansion takes place (Beckedahl et al., 1988).

The pipe system was surveyed in November 1987, by constructing a 2 metre square grid network over the site (Fig. 3). The extent of the pipe system was mapped in detail and a series of cross-profiles were taken across the system. Preliminary mapping of soil pipes was undertaken within the system and on the sidewall of the adjacent gully (Fig. 4). The position of rill development, sidewall collapse features, debris mounds and areas of fluvial activity within the collapsed pipe system were recorded.

Pegs used to construct the grid network have been left in place to facilitate future studies of the rate of expansion of the system.

Fig. 5. View of the central part of the collapsed pipe system at Inxu Drift. Detailed mapping was undertaken within the system using a 5 metre staff along transects marked by the grid network.

3. MORPHOLOGY OF THE SOIL PIPE-GULLY SYSTEM

3.1. General morphology

The partially collapsed pipe system covers are area of c. 600 sq. m. It comprises a major collapsed zone or sinkhole, 26 m long and 12 m wide, with three ancillary sinkholes developed on the sidewall ajoining the main gully (Fig. 4). The ancillary sinkholes are small scale features, the largest being 2 m in diameter.

The main sinkhole has a highly variable relief, making high resolution mapping difficult (Fig. 5). Contouring of the system (Fig. 6), however, has shown a number of important features within the system. The main feature of the system is a deep trench, varying from 2.0-6.5 m in depth, and extending from the main pipe outlet along the length of the collapsed pipe system (trench y, Fig. 6). The trench has an irregular

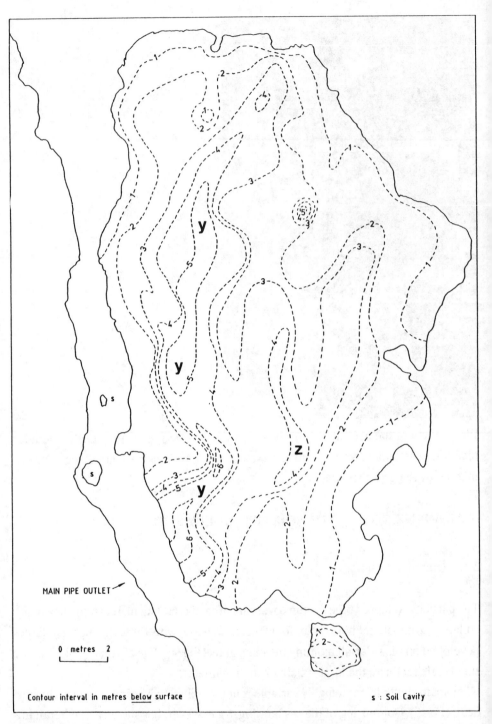

Fig. 6. Morphometry of the pipe-gully system at Inxu Drift, Transkei. Y and Z mark the positions of the two main trenches in the pipe-gully system. The contour interval is in metres below an artibrary datum taken on the gully sidewall surface.

252

topography, but deepens generally towards the main pipe outlet. It is relatively straight in the shallower part and meanders close to the main pipe outlet, where the trench also bifurcates. The deeper trench segments are linked by a series of arched soil pipes. The long profile of the trench is generally flat-bottomed. Thin (10 cm) beds of fluviatile sands, silts and clays occur in places in the lower parts of the trench. The slope of the long profile is c.9 degrees. This is comparable to slope gradients recorded in soil pipes by Drew (1972 - 20°), Jones (1975 - 6° and 7°) and Morgan (1977 - 10°). A second major trench occurs in the system (trench z, Fig. 6). This is up to 4.6 m deep and, like the main trench, is discontinuous, with inter-linking soil pipes. The trench is of crude dendritic form, with two first order branches. It deepens towards the main trench but is not connected directly to it by an open trench.

The deepest parts of the system are mainly associated with the trenches, although other isolated deep sections occur, associated mainly with soil pipes. The sidewall separating the collapsed system from the main gully (Fig. 6) is the steepest in the sinkhole, with near-vertical faces extending down from the surface to depths ranging from 0.5-6.0 m, reaching their greatest extent in the vicinity of the main pipe outlet (see Figs. 5,6). The other sidewalls are less steep, with gradients ranging from 20°-60°. The system is therefore assymmetric, steepening in the direction of the hydraulic gradient.

3.2. Soil pipe geometry

Soil pipes occur within the two main trenches of the system, their outlet into the main gully, and in association with the ancilliary sinkholes (X, Y, and Z, Fig. 7). They form remnants of what was an extensive soil pipe network.

The soil pipes in the trenches vary in diameter from 0.15-1.2 m (Fig. 7). Several of the pipes are composite features (i.e. with two or more pipes stacked one above the other)(Fig. 8). Composite pipes were found at pipes E and G (Fig. 7). The pipes are small, varying from 0.1-0.7 m in diameter, and are difficult to trace. However, it is clear that pipes A-F form part of a complex, three dimensional network. There is no clear relationship between pipe diameter and position within the network, except that in composite pipes, pipe diameters appear to increase successively with depth (see Fig. 8). This pattern in cross-sectional area has been noted in a number of other pipe systems (cf. Gilman and Newson, 1980; Jones, 1975).

It is interesting that only relatively small pipes have been found in the main sinkhole. This suggests a stepwise enlargement of the pipe diameter, with the successive collapse of a vertical sequence of pipes. This process can be seen in the main pipe out-

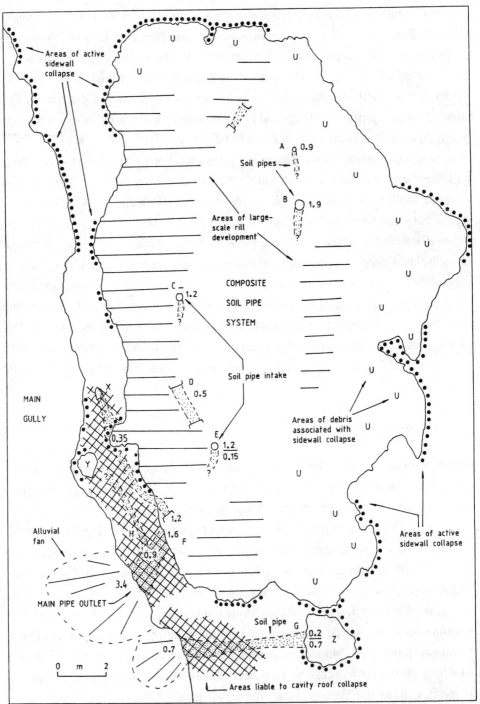

Fig. 7. Morphology of the pipe-gully system at Inxu Drift, Transkei. The locations of soil pipes (stippled, with diameters recorded in metres), areas susceptible to rill development (horizontal shading), sidewall collapse (dots) and areas liable to roof collapse (cross-hatched) are shown.

254

Fig. 8. Composite pipes occurring in sinkhole Z (see Fig. 7). Note the smaller pipe (A) overlying the larger pipe (B).

let.

3.3. Morphology of the main pipe outlet

The main pipe outlet (Fig. 9) is up to 3.4 m in diameter, and is irregularly shaped. It is partly associated with linear cracks and seepage lines formed within the sidewall of the main gully (Fig. 10). It is a composite pipe, with several soil pipes, in addition to the two main trenches, feeding into the main pipe.

The pipe is of variable gradient. The main pipe outlet dips at 10°-15° where it debouches into the main gully. It is highly irregular in shape, with smoothed, water-sculpted sidewalls (Figs. 11,12). The roofs of the pipes are generally irregular, reflecting block breakage and detachment (Figs. 11,12,14). The actual pattern of particle detachment from the roof is determined in part by the sedimentary character of the

Fig. 9. View of the main pipe outlet (A) from the pipe-gully system.

Fig. 10. View of the main and secondary pipe outlets (A and B respectively) from the main gully at Inxu Drift. Note the associated cracks and seepage lines within the gully sidewall.

Fig. 11. View inside the main pipe, Inxu Drift. Note the irregular pipe geometry, sculpted sidewalls and irregular roof, reflecting limited effects of pipe discharge.

Fig. 12. Water sculpted forms on the sidewalls of the main pipe at Inxu Drift. Note the pronounced etching of seepage cracks.(Width of section 80 cm).

Fig. 13. Elongate tributary soil pipe on the sidewall of the main soil pipe. The tributary pipe is associated with one of many seepage lines which penetrate through the roof of the main soil pipe.

colluvium (e.g. massive, stratified, homogeneous or heterogeneous). Observations of the system in November 1986 showed that the pipe had two main soil pipe channels, one overlying the other. Present observations (Fig. 9) indicate that the pipes have since merged as a result of collapse of the floor of the upper soil pipe.

The roofs show no evidence of direct water sculpture, though more than 30 per cent of the soil pipe roof shows evidence of "mulching" (i.e. the formation of a highly viscous slurry composed of fine grained material and water), associated with sedimentation and remobilisation of material by water penetrating through cracks in the roof. Mulching is also evident on sidewalls, and contributes considerably to propagation of the tributary soil pipes, most of which have developed along seepage lines (Figs. 12, 13).

The outlet of the main soil pipe is triangular in shape (Fig. 12) at the base of the main gully. This shape appears to be related partly to the seepage lines within the sidewall of the main gully. The development of the collapsed pipe system therefore

Fig. 14. Block detachment from the roof of a soil pipe feeding into the main soil pipe at Inxu Drift. Note the irregular roof profile and irregular blocks lying on floor of the soil pipe. Length of section c.2 m.

appears to be linked to the rate of downcutting of the main gully, which is 8 m deep at the pipe outlet (Fig. 2).

3.4. Hydrodynamic activity of the soil pipe

Water flow in the soil pipe is ephemeral. Observations at the start of the 1987 wet season showed that, following high intensity storms (c. 25 mm precipitation in 12 hours), the percentage of pipe cross-section filled by flows 12 hours
after the rainfall event was low (5-15 %), with the pipe floor exhibiting channels-within-channels (cf. Jones, 1981). The flow pattern was characterised by braiding, reflecting the low slope angle of the floor in the vicinity of the pipe outlet. Sands, silts and clays were deposited across 70 per cent of the pipe floor. The complex, inter-bedded nature of the stratified deposits suggests a relatively low but highly variable discharge (cf. Tunbridge, 1982). Few of the tributary pipes had any discharge at the time of observation.

3.5. Sidewall processes

Marked differences were observed in the nature of sidewall processes operating within and adjacent to the pipe-gully system. Rill development commonly occurs on the south- and southeast-facing sidewalls within and adjacent to the pipe-gully system

Fig. 15. Rill development in stratified colluvium on the sidewall of the collapsed pipe system. Note the relative absence of rills in the uppermost horizons of the colluvium.

(Figs. 7, 15). On the same material, they tend to be poorly developed (though still apparent) on north-facing sidewalls (see Fig. 10).

The rills are characterised by tributary rills joining a main rill downslope (Fig. 15), and are similar to the "organ-pipe" rill forms found in some Swaziland gullies (cf. Watson et al., 1984). They are developed on steep sidewall slopes (60°-90°) and commonly commence 1.2-1.6 m below surface. Rills are not developed above this level. It is interesting to note that these well developed rill patterns have formed where the runoff-contributing area is relatively small. In contrast, the southern side of the pipe-gully system, with a considerably larger contributing area, shows little rill development. It appears therefore that an inverse relationship may exist in this instance between surface runoff and rill development, although other factors such as aspect and geotechnical properties, may also be significant in their development (Engeln, 1973; Bowyer-Bower and Bryan, 1986). The relative absence of sidewall collapse debris adjacent to rills (see Fig. 7) suggests that the rills are effective in removing the debris. It is also likely that the higher runoff over the north and west facing sidewalls results to some extent in sidewall undercutting, thus accelerating the collapse, resulting in more debris in these areas.

The second dominant sidewall process is sidewall collapse, which is presently active on up to 50 per cent of the pipe-gully system (Fig. 7). It commonly results in concave collapse scars developed along the sidewalls (Fig. 7). Debris mounds associated with

Fig. 16. A break formed in a collapsed pipe system lying adjacent to the system under study, formed by roof collapse, Inxu Drift, Transkei.

sidewall collapse infill those parts of the pipe-gully system affected by sidewall collapse which are not affected by rill development.

The third dominant process is pipe roof collapse, although this process is not only found on the gully sidewalls. It is occurring principally in the area adjacent to the main pipe outlet, where the main concentration of minor sinkholes is found (Fig. 7). The hatched region (Fig. 7) is unstable and liable to collapse. When collapse occurs the pipe-gully system will be fully incorporated into the main gully network (Fig. 16).

4. DRAINAGE EVOLUTION

4.1. Pipe initiation

The observations show that the lateral extension of the main gully network is effected by piping. The position of pipe initiation in the system is likely to be complicated by a number of factors; (1) soil water table fluctuations which may result in new pipes forming either above or below older pipes, (2) variations in the rate of downcutting within the drainage network, in response to extrinsic changes (e.g. elevation of base level associated with reservoir development), and (3) the varying characteristics of the individual colluvial horizons.

261

4.2. Soil pipe network development

The occurrence of piping at different levels makes it difficult to map the horizontal structure of pipe networks. As a result few maps exist which show networks at two or more levels (Jones, 1981). The preliminary studies at this site have not facilitated detailed mapping of multi-storied networks. It is clear, however, that they existed in the system and that their development was not simply unidirectional (i.e. downward). More detailed studies of variations in grain size, porosity and permeability of individual beds within the colluvium, together with detailed tracing of soil pipes are needed to reveal their structure. Future work at this site will concentrate on these aspects of the pipe-gully system.

The morphology of the channels does indicate, however, that a soil pipe network of dendritic form developed. The actual pattern is complicated by the development of a three-dimensional drainage network, with stacked soil pipes forming at various points in the system. There is no clear evidence to suggest age relationships between pipes on the basis of their relative vertical positions. For example, the upper pipe exposed in sinkhole Z (Fig. 8) formed after the lower pipe during the past year.

The system is analogous to a karst cave network, with the development of a master conduit sequence (cf. Drew, 1972). Drew (1972), in investigations of piping in the Big Muddy Valley, Saskatchewan, suggested an evolutionary sequence from initial throughflow in many channels, eventually leading to turbulent flow in an integrated dendritic pipe network. It is not entirely possible to verify the initial stages at this site, though it seems likely, given the morphology of the system, that a subsurface dendritic soil pipe network was established at an early stage. Neither is it clear, from observations of other pipe systems in Transkei (see Beckedahl and Dardis, 1988), whether the master conduit hypothesis will apply, even in advanced stages of soil pipe development.

It seems highly unlikely that this type of pipe-gully system would have developed without an increase in hydraulic gradient. There is little evidence to suggest high flow velocities at the phreatic surface to cause appreciable pipe development (cf. Woodward, 1961). At Inxu Drift, the development of the pipe-gully system is associated with downcutting in the main gully network, punctuated by seasonal watertable fluctuations.

The clustering of outlets at Inxu Drift may indicate initial development of the network at locations which had prior concentration of drainage (e.g. percolines)(cf. Jones, 1971). The observations indicate a well integrated network similar to those reported in Arizona (Jones, 1968), Nant Gerig, Wales (Pond, 1971; Gilman, 1971; Davis, 1972; Newson and Harrison, 1978), Gerrig-yr-Wyn, Wales (Newson, 1976; Morgan, 1976; Gilman and Newson, 1980), and the Lake District (Humphreys, 1978). Though detailed analysis of the network at Inxu Drift has not been undertaken, it appears to be of higher drainage density, with greater occurrences of multi-storeyed pipes, than soil pipe networks developed in temperate environments. (see Jones, 1981, pp.135-151). More detailed studies of drainage density, pipe geometry, long profiles and pipe hydrology at this site are planned.

5. CONCLUSIONS

The observations at this site suggest that soil pipe and drainage network development in semi-arid environments is controlled by a number of edaphic factors in addition to hydraulic regime. The pattern of soil pipe development is apparently determined primarily by sediment texture, stratigraphy and soil erodibility, with seepage lines determining the overall pattern of the network. The dominant processes affecting the internal geometry of the pipe-gully system appear to be non-hydraulic, while the extent of downcutting is determined both by intrinsic factors (with pipes developed at perched soil-water tables, whose position is determined by soil texture and stratigraphy) and extrinsic factors (e.g. local base-level lowering). Future work at this site will focus on aspects of the geotechnical properties, soil permeability and soil geochemistry, as contributing factors to soil pipe network development.

6. SUMMARY

Detailed mapping of a psuedo-karstic pipe-gully system at Inxu Drift, Transkei, southern Africa shows that it developed from a well-integrated soil pipe network of dendritic form. Multi-storied soil pipes are common in the network, forming at different levels at different times, in response to edaphic conditions such as water table fluctuations, and changes of local base-level. Drainage was directed to a master conduit, part of which is intact. Detailed studies within the master conduit indicate that it seldom, if ever, achieved full pipe flow conditions. Pipe propagation occurred along seepage lines while pipe cross-sectional enlargement was facilitated by roof collapse, sidewall mulching and pipe floor fluvial incision. The main pipe was originally multi-storied but has developed into a large single conduit, up to 3.4 m in size, with arterial pipes and seepage lines.

Subaerial exposure of the system by pipe roof and subsequent sidewall collapse, has resulted in rill development, and sidewall undercutting and further collapse. These processes may operate separately or in conjunction with one another.

7. REFERENCES

Beckedahl, H.R. 1977. Subsurface erosion near the Oliviershoek Pass, Drakensberg. South African Geographical Journal, 59, 130-138.

Beckedahl, H.R. and Dardis, G.F. 1988. The role of artificial drainage in the development of soil pipes and gullies: Some examples from Transkei, southern Africa (this volume).

Beckedahl, H.R., Bowyer-Bower, T.A.S., Dardis, G.F. and Hanvey, P.M. 1988. The geomorphic effects of soil erosion. In: Moon, B.P. and Dardis (eds.), The Geomorphology of Southern Africa, Southern Book Co., Johannesburg (in press).

Bowyer-Bower, T.A.S. and Bryan, R.B. 1986. Rill initiation: Concepts and experimental evaluation on badland slopes. Zeitschrift fur Geomorphologie, Supplement Band 60, 161-175.

Dardis, G.F. (submitted) Age of Quaternary colluvium in eastern southern Africa.

Dardis, G.F., Beckedahl, H.R., Bowyer-Bower, T.A.S. and Hanvey, P.M. 1988. Soil erosion forms in southern Africa. (this volume).

Davis, R.A. 1972. Subsurface hydrological features in the Nant Gerig. Institute of Hydrology, Unpublished Report, 6pp.

Drew, D.P. 1972. Geomorphology of the Mig Muddy Valley area, Southern Saskatchewan, with reference to the occurrence of piping. In: Paul, A.H., Dale, E.H. and Schichtmann, D. (eds.), Southern Prairies Field Excursion Background Papers, Department of Geography, Regina Campus, University of Saskatchewan, 197-212.

Drew, D.P. 1982. Piping in the Big Muddy badlands, Southern Saskatchewan, Canada. In: Bryan, R.B. and Yair, A. (eds.), Badland Geomorphology and Piping, Geobooks, Norwich, 293-304.

Engeln, G.B. 1973. Runoff processes and slope development in Badlands National Monument, South Dakota. Journal of Hydrology, 18, 55-79.

Gilman, 1971. A semi-quantitative study of the flow of natural pipes in the Nant Gerig sub-catchment. NERC Institute of Hydrology, Subsurface Section, Internal Report No. 36, 16pp.

Gilman, K. and Newson, M.D. 1980. Soil Pipes and Pipe-flow - A Hydrological Study in Upland Wales. Geobooks, Norwich, 114pp.

Humphreys, B. 1978. A Study of Some of the Geomorphological and Hydrological Properties of Natural Soil Piping. Unpublished Dissertation, University of East Anglia, 35pp.

Jones, J.A.A. 1971. Soil piping and stream channel initiation. Water Resources Research, 7, 602-610.

Jones, J.A.A. 1975. Soil Piping and the Subsurface Initiation of Stream Channel Networks. Unpublished Ph.D. thesis, University of Cambridge, 467pp.

Jones, J.A.A. 1978. Soil pipe networks: distribution and discharge. Cambria, 5, 1-21.

Jones, J.A.A. 1981. The Nature of Soil Piping - A Review of Research. Geobooks, Norwich, 301pp.

Jones, J.A.A. 1987. The initiation of natural drainage networks. Progress in Physical Geography, 11, 207-245.

Jones, N.O. 1968. The Development of Piping Erosion. Unpublished Ph.D thesis, University of Arizona, 168pp.

Morgan, L.A. 1976. An Investigation of the Location, Geometry and Hydraulics of Ephemeral Soil Pipes on Plynlimon, Mid-Wales. Unpublished dissertation, University of Manchester.

Morgan, R.P.C. 1977. Soil erosion in the United Kingdom: field studies in the Silsoe area, 1973-1975. National College of Agricultural Engineering, Silsoe, Bedfordshire, Occasional Paper No. 5.

Newson, M.D. 1976. Soil piping in upland Wales: a call for more information. Cambria, 1, 33-39.

Newson, M.D. and Harrison, J.G. 1978. Channel studies in the Plynlimon experimental catchments. NERC Institute of Hydrology, Report 47, 61pp.

Pond, S.F. 1971. The occurrence and distribution of subsurface pipes in upland catchment areas. NERC Institute of Hydrology Internal Report 47.

Price-Williams, D., Watson, A. and Goudie, A.S. 1982. Quaternary colluvial stratigraphy, archaeological sequences and palaeoenvironments in Swaziland. The Geographical Journal, 148, 50-67.

Stocking, M.A., 1976. Tunnel erosion. Rhodesian Agricultural Journal, 73, 35-40.

Tunbridge, I.P. 1982. Sandy, high-energy flood sedimentation - some criteria for recognition, with examples from the Devonian of south-west England. Sedimentary Geology, 28, 79-95.

Woodward, H.P. 1961. A stream piracy theory of cave formation. National Speleological Society (America) Bulletin, 23, 39-58.

SOIL EROSION IN A CATCHMENT NEAR ALICE, CISKEI, SOUTHERN AFRICA

MARGARET E.MARKER
Department of Geography, University of Fort Hare

1. INTRODUCTION

Soil erosion is a serious environmental problem in southern Africa (Stocking, 1978), and one which appears to be particularly prevalent in eastern southern Africa, in areas such as Ciskei. Many catchments in Ciskei appear to be undergoing rapid changes in soil erosion intensity, with concomitant diminution in crop yields. Sheet erosion affects old cultivated lands and gullies have incised into many valley fills. These changes have been attributed to the prevalence of soils in these areas which are susceptible to soil erosion, where long and short-term climatic variability, population increase, and associated grazing pressures render the soils as a considerable erosion hazard.

Most investigations into soil erosion have concentrated on parameters associated with soil erosion distribution, erosion characteristics and erosion intensity or with the development of soil erodibility indices (Stocking, 1978) or with rates of soil loss (Weaver, 1988). All these studies indicate that soil erosion is a very serious and growing economic problem. Far fewer studies have, however, attempted to monitor changes in soil erosion at one site over a longer time period.

The aim of this paper is to present preliminary data on changes in rates of soil erosion over time in a small catchment in Ciskei. It is a third order catchment, 3.6 sq. km in area, tributary to the Tyume river. It is situated 3 km northwest of Alice between the villages of Mavuso and Nkobonkobo. Soil erosion in this catchment has been monitored regularly over the past 10 years. Aerial photo coverage is available for specific years from 1949 to the present, facilitating examination of the changes in soil erosion intensity and location over the past four decades. Considerable resettlement and population increase has occurred over this period, with concomitant increase in grazing pressure and destruction of the natural vegetation cover, and therefore provides an opportunity to assess the impact of these changes on rates of soil erosion.

2. DATA COLLECTION METHODS

The empirical data have been derived from analysis of air photos of the catchment for the years 1949, 1963, 1972, 1975, 1976, 1980 and 1984. Soil erosion , evidenced as gullies and rill wash, is easily mapped from these air photographs and sheet erosion shows up as tonal differences. In order to obviate distortion and problems of variable scales a base map was constructed from 1:50000 map sheet 3226 DD with additional channel detail taken from the air photographs. For each specific year channel gullies, rill erosion, and sheet erosion were recorded onto base maps. The distribution of settlements and the extent of cultivated area were also mapped. The same information for 1987 was collected in the field.

The actual types of soil erosion present were established from field investigation. Sheet erosion is widespread but is replaced by rill wash over time and on less permeable B horizons or bed rock. Channel gullies with spreading heads are the most spectacular form of soil erosion. Tunnelling occurs at the base of the soil A horizon with subsequent collapse to form gullies. Slumping or slope collapse is restricted to gully walls. On shale, gully incision may continue into bedrock. Slope profiles, from which slope angles were recorded, were measured in the field.

3. THE CATCHMENT

The catchment has an altitudinal range of only 150 m. It is incised below a level plateau maintained by a dolerite sill intruded into Beaufort Group shales which form the lower slopes of the valley. The plateau slopes from 720 m above sea level (a.s.l.) in the south towards the Tyume valley. The Nkobonkobo interfluve at 660 m a.s.l is both broader and higher than that on the eastern, Mavuso side (Fig. 2 b). This shallow valley has relatively low angle slopes. A bimodal slope angle distribution occurs; 47 % of all slopes lie between 5 and 8 degrees and 25% fall between 12 and 13 degrees (Fig. 1). The only slopes steeper than 14 degrees are channel banks which are often vertical, where incision has been vigorous.

The valley floor is broad, infilled by colluvial and alluvial fan deposits. Two valley floor terraces exist separated by a step of 2-3 m and profiles exposed in the gully indicate that infilling has been episodic. At least three buried soil profiles can be identified. On the valley floor, soil depths often exceed 3 m but on the slopes most untruncated soils are under 1 m in depth. Many of the upper slopes are now devoid of soil as a result of severe soil erosion.

4. CHANGES IN SOIL EROSION OVER TIME

In 1949 all settlements were traditional open clusters with kraals and were sited on

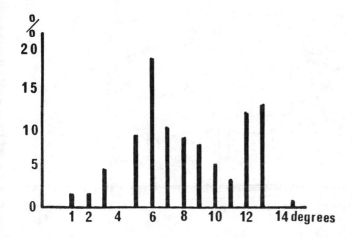

Fig. 1. Slope angle variation.

interfluves around the valley. The valley floor was cultivated and soil erosion was limited to intermittent gullies, isolated rill development and patches of sheet erosion (Fig. 2a.). Shallow, intermittent gullies occurred along the main channel, probably indicative of tunnel erosion in the catchment but only downstream of the road was the main channel deeply incised.

By 1963 the traditional settlements had extended and shifted. The main channel gully had cut headwards to the position of the old road and an upstream segment had incised back along tributary B and cut up to the track. Some other tributary gullies had also extended. Rill erosion was more pronounced everywhere on the sites where sheet erosion had previously been visible. The area upstream of the present dam was particularly badly eroded (Fig. 2a.).

By 1972 traditional interfluve settlements with kraals had been resettled as two planned nucleated villages, Mavuso and Nkobonkobo, as part of the Land Betterment Policy recommended by the Tomlinson Commission. At that time the main dam was built and a new road aligned north of the old road possibly because extension of the main gully was causing it to collapse. All gullies and rill areas had extended and areas under cultivation in 1963 had been abandoned. Sheet erosion had become noticeable on the eastern slope below the track south of Mavuso village (Fig. 2a.).

The incidence of soil erosion in 1975 had declined (Fig. 2a.). Only tributary C and the lower portion of B were badly gullied. Some recovery even along the main channel is apparent. The erosion scarred areas of the western slopes and revegetated and rill scars showed only on the lower slopes. Even the severe erosion site upstream of the new dam had largely healed. Sheet erosion remained serious on the eastern side. However only a year later some of these sites were again showing signs of erosion (Fig. 2b.).

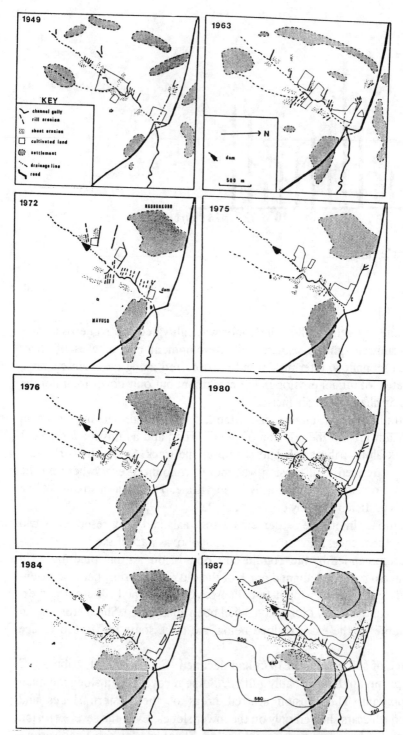

Fig. 2. The soil erosion distribution and settlement pattern for specific years 1949, 1963, 1972, 1975   1976, 1980, 1984, 1987. (Note that contours are given on the 1987 map).

By 1980 soil erosion had again increased (Fig. 2b.). Gullies had extended up many side channels. There was sheet erosion in many areas and badland rill sites below the Mavuso track had become a network of gullies above tributary B. Mavuso main settlement had increased in density and spread southwards as an area of shack settlement. Sheet erosion was visible in many places on the Mavuso ridge in association with the new settlement area.

The situation by 1984 showed less total area of soil erosion but with extended rill and gully systems and more sheet erosion associated with Mavuso Extension and in the area east of the dam. Another dam had been built on the interfluve beyond Nkobonkobo and a holding pond close to the road where a water pump had been installed (Fig. 2b).

In 1987 soil erosion areas were similar in location and extent to 1984 but gullies were longer and more continuous and finger extensions were incising back into the fan terrace in many places (Fig. 2b). Some attempts had been made to reduce runoff by filling some gullies with stones, and aloes had been planted, so the lower western slopes below Nkobonkobo are revegetating despite the loss of virtually all soil. The worst area is now the eastern slope associated with Mavuso Extension. This area adjacent to tributary B is a badland of gullies incising into the shale. All soil has been lost. The main channel is one almost continuous gully incised about 3 m below the former surface. Gully depth is controlled by fan gradient and bedrock steps.

From the base maps (Fig. 2) measurements could be taken to quantify approximately the area of soil erosion in any one year. Measurements were taken to the nearest 100 m for gully length and to the nearest 100 sq m for areas. The main channel gully which is the base level for the whole catchment has shown an intermittent increase overall (Fig. 3D). When aggregate gully length is considered, it can be seen that headwards erosion was at its most rapid in the 5 year period between 1975 and 1980. Other types of erosion show a similar jump and again a further increase during the past 3 years since 1984. An improvement in soil erosion incidence occurred between 1972 and 1976. The area under cultivation has remained more or less constant, showing a decline following resettlement in 1972 and again in 1984, a bad drought period. The cultivated areas have not however always occupied the same sites. These figures must be taken as indicative only, as air photo mapping is dependant on the scale and quality of the photograph. Nevertheless within these limitations they depict the changes that have occurred in this catchment (Fig. 3).

In addition to measurement of soil erosion, the number of dwellings, first in the traditional settlements and then in two villages, were recorded where possible for each sample time. A survey in Mavuso has shown that 65 % of households occupy a single dwelling. Twenty per cent have two dwellings and only 15 % occupy three dwelling units. The average household size is 5 persons (Boqwana 1984). The number of dwelling units showed a steady increase, more rapid in recent years (Fig. 3B).

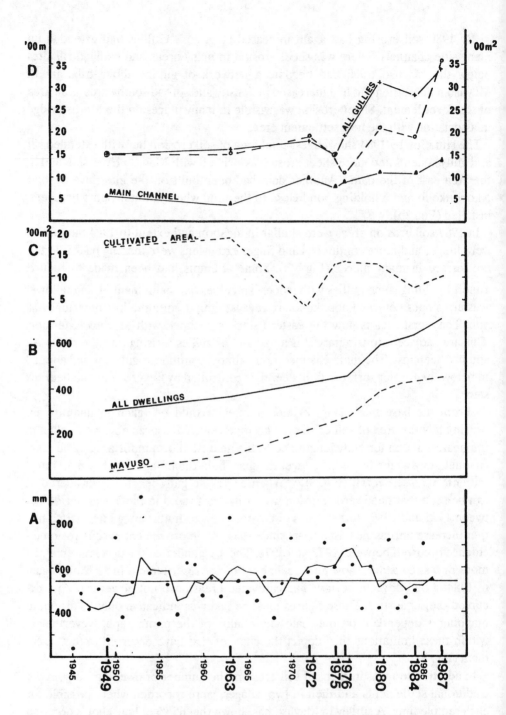

Fig. 3. Soil erosion incidence change over time (D) in relation to   cultivated area (C), population increase (B), and rainfall (A). Actual annual totals is plotted in dots, and mean annual rainfall (544 mm) as a straight line; rainfall fluctuation curve is based on a four year running means.

5. DISCUSSION

The distribution of soil erosion in the catchment can related to soil type, soil depth and to slope angle. With two exceptions all areas of soil erosion have occurred on soils derived from Beaufort Group shales. Dolerite soils have incurred erosion only in the headwaters of gully C, adjacent to the road and in the headwaters of the main catchment. The former site has probably been exaccerbated by runoff from the road itself and the photos suggest that in the headwater area erosion was initiated by runoff along tracks leading to water sources.

Before 1976 the worst eroded areas were associated with deeper colluvial soils, areas utilised for cultivation. Most other slopes should then have been stable because there was only limited incision along the main channel.

A distinction has to be made between channel gullies and sheet erosion that develops into rill erosion. Channel gullies extend by headwards erosion and once initiated will work back upslope unless harder bedrock is encountered or a dam is built. Gully C demonstrates this clearly. Its catchment had recovered and was not eroded in 1972 when a dam had been installed. By 1975 when this had been washed out, erosion was reactivated. Similarly the headwater catchment of the main channel recovered from serious sheet erosion after the main dam was installed. Only since 1980 has erosion again manifested itself in this area. Clearly the extension of the main channel gully which has a depth exceeding 3 m over much of its length, has affected the entire catchment through creation of a new base level that effectively steepens the entire slopes.

Many of the gullies on the lower slopes are associated with tunnel erosion. The grey silty soils are particularly prone to water movement above the B horizon with removal of the silt fraction in suspension. Gullying seems also to be initiated along the break of slope between the two terraces and then to have extended up-slope.

Sheet erosion is caused by increased surface runoff due to soil compaction or reduction in vegetation density. Much of the early sheet erosion can be correlated with abandoned cultivated lands. Later erosion is associated with thinning of the woody vegetation component. It is not possible to trace changes in the grass component. By 1980 the entire Mavuso interfluve had been cleared of woody vegetation with the exception of two large trees. Prior to 1972 the area had supported open bushveld. Thinning of bush cover has also occurred above the main dam, particularly in the past few years. A temporary settlement was sited there from 1985-6 but the area also serves as a source of fuel. Clearance to this extent has not occurred on the western slopes. The extension of Mavuso along the interfluve has also resulted in increased runoff with increased sheet erosion.

Sheet erosion becomes transformed into rill erosion when the A horizon has been lost and the less permeable B horizon is exposed. Where harder bedrock occurs, such areas may vegetate after soil loss but by then have totally lost their agricultural potential. This has happened on parts of the western slope. However where friable shales are the bedrock, rill erosion continues to incise vertically creating badland topography such as that developed below Mavuso extension.

The increase in gully length parallels that of sheet and rill erosion. The areal extent of erosion increased slowly until 1972. An improvement then occurred following resettlement but a marked deterioration is visible since 1976 with only a slight slowing in the deterioration between 1980 and 1984 (Fig. 2D).

The causes of the improvement between 1972 and 1976 and for the subsequent rapid deterioration needs further investigation. The area under cultivation has been relatively constant and has declined overall since resettlement. The recent decline in cultivated areas is also the result of the serious drought in the early 1980s.

There is little or no observable correlation between changes in the incidence of erosion and fluctuations in rainfall (Fig. 3). Although cultivated area fluctuates with rainfall it shows no relationship to soil erosion incidence. It is possible however that runoff associated with flooding in 1976 and 1979 caused renewed gully incision. Water is the agent of erosion and if the major variable causing changes in erosion incidence were moisture availability, an inverse relationship with rainfall could be anticipated. Rainfall records for Alice (Fort Hare University campus) are available since 1944. Annual rainfall is highly erratic; annual average rainfall being 544 mm/yr. To establish a pattern for comparative purposes a smoother curve was obtained from the use of 4-year running means (Fig. 3A). No relationship between above average or drought periods and changes in soil erosion incidence can be recognised (Fig. 3).

Population increase with associated grazing pressure is a further factor. As far as can be established from a count of dwelling units, growth has been steady but not excessive in this area. It has not exceeded the rate of natural increase except immediately prior to 1980 in anticipation of Ciskei independence. A population increase at this time is well documented in other villages and is associated with resettlement of farm labour from South Africa as a consequence of the 1976 and 1977 Farm Labour Acts. Mavuso, virtually a suburb of Alice, bore the brunt of the influx. Nkobonkobo has remained largely unchanged since 1972 except for infilling of garden plots. A survey of Mavuso has, however, shown that only 10 % of its population have settled since 1975 and of these, 60 % came from the white farms (Boqwana, 1984).

Nevertheless the surrogate population measure, dwelling units, demonstrates a rise that closely parallels the incidence of erosion. The extension of Mavuso southwards along the interfluve has had as important influence on the shift of serious soil erosion from the western to the eastern slopes of the catchment. It is directly related to population increase. Nevertheless the Mavuso slope had a naturally higher potential erodibility. It faces west, a drier micro environment more susceptible to drought damage to vegetation. It is underlain by friable Beaufort Group shales overlain by silty soils with a high erodibility index. Its slopes range between 9 and 12 degrees. The increase of population in Mavuso resulted in clearance of woody vegetation and extension of settlement along the interfluve was the final trigger, through increased runoff, to accentuate erosion on that slope.

274

This small catchment appears typical of most others in the Ciskei foothill zone where rainfall is highly variable and drought makes subsistence agriculture hazardous. Grazing of stock is virtually the sole mode of production and as wealth is assessed by numbers rather than by quality, overgrazing is a serious problem. Drought results in stock losses after vegetation has been decimated, and subsequent veld recovery.

Field work in late 1987 showed lower stock numbers than in previous years, attributed to stock losses during the drought, and attempts to improve the situation. Prickly pear was being planted on two mid slope sites and the cultivated lands had been effectively fenced. The main dam is also fenced but small stock are not inhibited.

Although no soil loss figures for this catchment are available, soil loss calculated from dam siltation has been assessed for a nearby Roxeni catchment at 113.7 t/km^2 /yr (Weaver, 1988). From the incidence of soil erosion, it is probable that similar amounts of soil loss would be recorded here also.

6. CONCLUSIONS

Using air photo analysis combined with field investigation, it has been possible to measure changes in soil erosion over time. The data obtained indicate a temporary improvement after resettlement which coincided with above average rainfall, and a serious deterioration since 1980 coincide with drought. It was also possible to compare soil erosion incidence with variables such as area under cultivation, rainfall periodicity and population increase. Only population increase showed any correlation with soil erosion incidence. Yet the increase in population as measured from dwelling unit numbers, is well below the Ciskei average and within the limits of natural increase, except immediately prior to 1980. The effect on the environment of population increase is both direct through removal of the woody component of the natural vegetation for fuel and building material and indirect through increased stock numbers leading to lower vegetation density and reduced cover through overgrazing.

The inescapable conclusion is that this catchment has reached a point where irreversible changes could occur. Attempts are presently being made to plant prickly pear on level areas and to fence the cultivated fields. Stock numbers are undoubtedly lower at present but whether this level, which may allow the veld to recover, will be maintained, remains to be seen.

7. SUMMARY

Changes in soil erosion in a small headwater catchment of the Tyume drainage system adjacent to Alice is the focus of this study. Air photographs for specific years in the period 1949 to 1984 are available so changes in the incidence of soil erosion distribution, intensity and types have been recorded from 1949 when little erosion was present, up to 1987. Variables such as settlement distribution, settlement increase as a surrogate for population growth, pattern of cultivated area, diminution of natural vegetation cover and rainfall variability have been applied in an attempt to explain the changes in soil erosion observed. There is evidence that soil erosion has increased dramatically since 1980. Gullies have deepened and extended headwards and the focus of badly eroded areas has moved from east-facing slopes. An increase in soil erosion incidence can be seen up to 1972. An improvement in the position was apparent by 1976 subsequent to resettlement villages away from the watershed.

8. REFERENCES

Boqwana, E.P. 1985. Housing problems in Alice town and neighbouring villages of Mavuso and Kwa Ntselamanzi. Honours dissertation, University of Fort Hare (unpublished).

Stocking, M. 1978. Prediction and measurement of erosion in subtropical Africa. Geog. Eco. Trop. 2, 161-74.

Weaver, A.V.B. 1988. Soil erosion rates in the Roxeni basin, Ciskei. South African Geographical Journal (in press).

SEDIMENT PROPERTIES AS A FACTOR IN SOIL EROSION

P.W. VAN RHEEDE VAN OUDTSHOORN
Department of Geography, University of Vista

1. INTRODUCTION

The potential erodibility of the colluvium which is present in the water courses and valleys of the Aliwal North area is extremely high. Although various factors such as topography, and human influence affect the erodibility of sediments, sediment charac-teristics are actually the most important factor (Faniran and Areola, 1978; Morgan, 1979). According to Morgan (1979) the erodibility of sediments varies according to texture, aggregate stability, infiltration capacity, organic content, chemical composi-tion and the shear strength of the sediments. These sediment characteristics are inter-dependent on one another and therefore they act as a single entity in the erosion process. Inspite of this, this study has attempted to consider specifically what the pos-sible influence of sediment characteristics on erosion is. To achieve this, the colluvial deposits of two third order sub-basins of the Bossieslaagte basin were examined. The Bossieslaagte basin is situated about 55 km east of Aliwal North and about 10 km south of Lady Grey (Fig. 1). The sedimentary and volcanic formations in the study area form part of the Karoo Sequence. Specifically they belong to Elliot (Shale and Mudstone), Clarens (sandstone) and Drakensberg (basalt) formations. The sedimentary deposits in one of the sub-basins comes largely from the Elliot and Clarens formations, while that in the other sub-basin comes largely from the Elliot, Clarens and Drakensberg formations.

Erosion scars about 15m deep and in some places as much as 35m wide have been cut into the colluvial deposits. The erosion scars in sub-basin A have a branching pat-tern while those in sub-basin B have a linear form.

2. CLIMATE AND VEGETATION

The climate of the area may be described in terms of the Koppen system of classifica-tion as subtropical monsoon with cool dry winters and warm wet summers. The

Geomorphological Studies in Southern Africa, G.F.Dardis & B.P.Moon (eds)
© *1988 Balkema, Rotterdam. ISBN 90 6191 831 6*

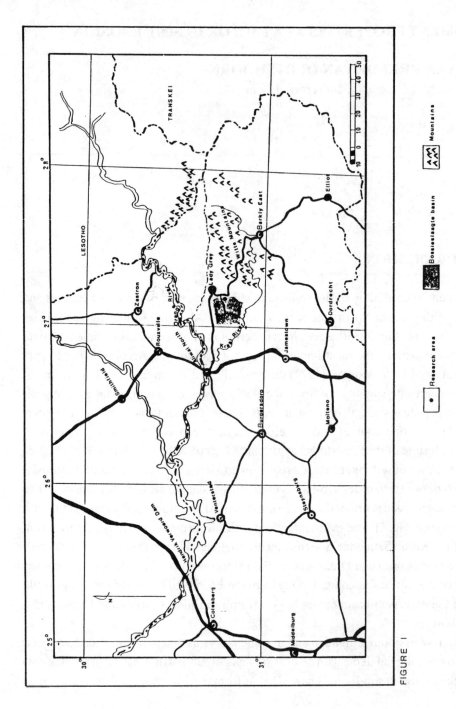

FIGURE I

Fig. 1. Location Map

278

average annual rainfall is 542 mm of which roughly 70 % occurs during the summer in the form of short thunderstorms.

According to Acock's (1953) classification of the vegetation of the Republic of South Africa, the Bossieslaagte catchment area has three types of vegetation, namely types 48, 50 and 58. The plants forming these kinds of vegetation are largely grasses which grow thickly in certain areas and in hardy (sparse) tufts in others. A few trees, aloes and low shrubs are present. In general the vegetation cover is moderate to poor. Consequently it offers little protection against surface and gully erosion.

3. PROCEDURES AND METHODS

Sediment samples of the colluvial deposits were taken at intervals of 0.3 m from the exposed erosion banks in sub-basins A and B. Since the thickness of the deposits varied at the observation points, forty samples were taken in sub-basin A and twenty in sub-basin B.

The granular composition of the sediment samples was determined, as was the presence of exchangeable cations (magnesium, calcium, potassium and sodium) in the sediment. Since no data on the rates or extent of erosion in the two sub-basins exists, an erosion index, which indicates the difference in the density of erosion scars in the two sub-basins, was determined. This erosion index is purely an indication of the length of the erosion scars per unit area and is not a measure of the amount of sediment per unit time which is carried away. The branching erosion pattern in sub-basin A and the more linear pattern in sub-basin B are reflected by the erosion index of 3.30 for sub-basin A and 1.62 for sub-basin B.

4. GRAIN-SIZE COMPOSITION OF THE SEDIMENTS

4.1. Sub-basin A

The granular composition of the sediments indicates that the deposits in sub-basin A consists largely of very fine material, especially in the upper seven metres of the deposit, which contains a very high proportion (roughly 90%) of fine material (Fig. 2). The percentage of clay decreases from the top to the bottom of the deposit profile. Up to a depth of about six metres the percentage of clay in the deposits is about 30 per cent. From six to twelve metres below the surface this decreases to roughly 17 per cent. The percentage of silt in roughly the upper seven and a half metres of the profile is about 20 %, after which it decreases until it reaches about 12% at about twelve

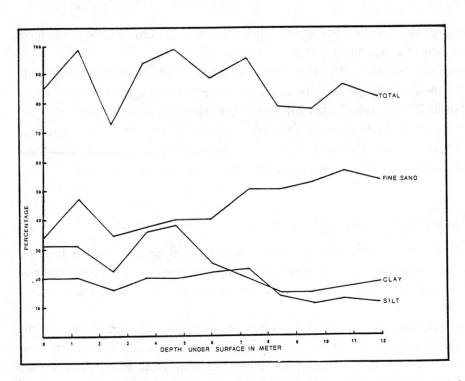

Fig. 2. Variations in grain size composition of colluvial deposits with depth, Sub-basin A

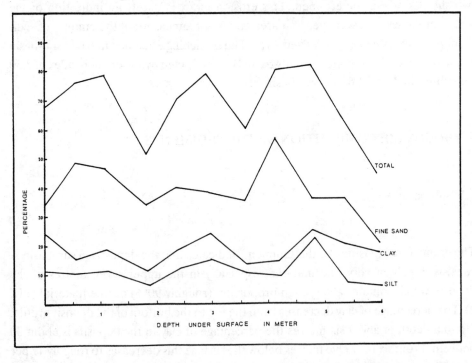

Fig. 3. Variations in grain size composition of colluvial deposits with depth, Sub-basin

metres. If the percentage of fine sand in the sediment is added to the percentages of silt and clay, the percentage of fine material in the upper seven metres increases to about 90 %, as compared with about 80 % in the lower five metres of the deposit. A further indication of the degree of fineness of the sediments in the deposits of sub-basin A is that the average size of the particles is roughly 0.09 mm.

4.2. Sub-basin B

The sediments of sub-basin B generally contain less fine material than the sediments in sub-basin A (Fig. 3). The percentage of clay in the sediments of sub-basin B is, with a few exceptions, fairly constant throughout the deposit at about 17 per cent. From the surface to a depth of about six metres the silt content is also fairly constant at roughly 11 per cent. If, as in the case of sub-basin A, the percentage of fine sand is added to the percentages of silt and clay, the sediment of sub-basin B can be described as containing about 70 % of fine material throughout. The sediment in sub-basin B is generally coarser than that of sub-basin A, with an average particle size of about 0.12 mm.

5. THE PRESENCE OF EXCHANGEABLE CATIONS AND EROSION

Various authors have described the effect of monovalent (Na^+ and K^+) and divalent (Mg^{2+} and Ca^{2+}) cations on the dispersivity of sediments. The effect of Na^+ on dispersivity is strongly influenced by the type of complimentary cation (Mg^{2+} or Ca^{2+}) which is present with Na^+ in the sediment. A higher percentage of exchangeable sodium (ESP) is needed when it occurs with Ca^{2+} than when it occurs with Mg^{2+} to achieve the same degree of clay dispersion (Emerson and Backer, 1973; Levy, 1985). According to Shainberg, Rhodes and Prather (1981), sediment with a low ESP content disperses very quickly when it is soaked in distilled water (simulated rainwater) and much slower when the sediments are soaked in water with a low electrolyte content. The exact role of K^+ is, however, not clear since contradicting findings on the effect of K^+ on the hydraulic conductivity are to be found in the literature of the subject.

   The analyses of the sediments of Sub-basins A and B indicated the following about the presence of exchangeable cations. Nearly twice as much exchangeable Na^+ is present in the sediments of Sub-basin A as in the sediment of Sub-basin B (2.4 % and 1.4 % respectively). The quantity of EPP in the sediments of sub-basins A and B is 3.0 % and 3.7 % respectively. The average amount of exchangeable magnesium in the sediment of the Sub-basin A is 14.1 % compared to 21.1 % in the sediment of Sub-basin B. Extremely high percentages of exchangeable calcium were found in the sediments of sub-basins A and B (99 % and 75 % respectively). These high levels of exchangeable calcium represent the adsorbed as well as the dissolved cations in the sediment.

6. DISCUSSION

The following conclusions may be drawn from the preceding results. Three inter-dependent factors may contribute to the high erosion rate in the Aliwal North area. The nature of the rainfall, the texture of the sediments and the presence of the monovalent Na + cation. The high percentage of fine material (clay, silt and fine sand) occurring in the upper half of the deposit in Sub-basin A as well as the fine particle size of the sediment may have the effect of decreasing the infiltration rate. According to Fitzpatrick (1971) and Foth (1978), sediments with a coarser texture have a higher infiltration tempo, while those with a fine texture and which contain dispersed clay allow practically no water to infiltrate. Precipitation in the form of brief thunderstorms, together with a low infiltration rate will result in an increased overland flow of water. In Sub-basin B, where the percentage of fine material (clay, silt and fine sand) in the sediment is lower and the average particle size is higher, the infiltration rate may be higher and thus the surface flow-off may be less.

Although the percentage of exchangeable calcium in the sediments of both sub-basins A and B is high, the 2.4 % of exchangeable sodium in the sediment of Sub-basin A is sufficient to disperse the clay through the addition of rainwater, which has a low electrolyte content. The dispersed particles have a two-fold effect on the sediment. Firstly, they block the openings in the sediment, which lowers the infiltration rate further. Secondly, together with the impact of raindrops, the dispersed particles form an almost impenetratable crust on the surface through which water does not penetrate. As a result, the surface flow-off of water is increased. In contrast to this, the ESP in the sediment of Sub-basin B averages 1.37 per cent. The lower ESP, coarser texture, lower percentage of fine material and high percentage of exchangeable calcium will not make the sediment in Sub-basin B disperse to such an extent as the sediment in Sub-basin A. In Sub-basin B the infiltration of water will consequently not be suppressed to the same extent as in Sub-basin A. According to Levy (1985), exchangeable potassium will contribute to crust formation in cases of sediments with a clay content of 20 % or more. In Sub-basin A, the potassium may possibly contribute to crust formation since the clay content in the upper section of the deposit is roughly 30 per cent. Equally, potassium is not likely to play a role in crust formation in sub-basin B, since the clay content in the deposit there is about 17 per cent.

In that part of the research area which is classified as semi-arid, basalt may have an effect on erosion rate in two different ways. Firstly, the sediment which contains basalt (Sub-basin B) has a higher average particle size than the sediment devoid of basalt (Sub-basin A) and consequently the infiltration rate is higher. Secondly, the sediment contains less exchangeable sodium. This, together with the lower fine material content means that dispersion does not occur to the same extent as in the sediment of Sub-basin A.

The branched erosion pattern in Sub-basin A and the more linear pattern in Sub-basin B is reflected in the erosion indices of 3.30 for Sub-basin A and 1.62 for Sub-basin B. The difference between the two erosion patterns and the erosion indices may be directly attributed to surface water flow. Since surface flow is the result of the texture characteristics of the sediment, the presence of monovalent cations and the nature of the rainfall, the erosion indices and erosion patterns of Sub-basins A and B may be the result of the above-mentioned three factors.

The influences of each of the three factors on the erosion pattern of the two sub-basins are inextricably intertwined. As a result of the stormy nature of the rainfall, large quantities of run-off water comes from the mountainous parts of the two sub-basins. In sub-basin A the crust layer on the top of the deposit prevents water infiltration from taking place, while in sub-basin B, the coarser sediment allows infiltration to take place more readily. The presence of the Na+ cation in the sediment of Sub-basin A leaves the surface sediment unprotected against the large quantities of highly energetic run-off water. Consequently a dendritic erosion pattern occurs in Sub-basin A, while higher infiltration and less Na+ in the sediment of sub-basin B results in a more linear erosion pattern.

7. SUMMARY

Apart from external factors which contribute towards soil erosion, physical and chemical characteristics of sediments can be regarded as the most important factors promoting soil erosion. A study of the severe surface and donga erosion in the Aliwal North region has been conducted. The grain size composition of the sediments and the presence of exchangeable cations (Na^+, K^+, $Ca2^+$ and Mg^+) were highlighted as the possible sediment characteristics which contributed the most towards erosion and the patterns of erosion dongas.

8. REFERENCES

Acocks, J.P.H. 1953. Veld types of South Africa. Memoir of the Union Botanical Sur-

28, Pretoria.

Emerson, W.W. and Bakker, A.C. 1973. The comparative effects of exchangeable calcium, magnesium and sodium on some physical properties of red-brown earth subsoils.II. The spontaneous dispersion of aggregates in water. Australian Journal of Soil Research, 11, 151-157.

Faniran, A. and Areola, O. 1978. Essentials of Soil Study. Heinneman, London.

Fitzpatrick, E.A. 1971. Pedology. A Systematic Approach to Soil Science. Oliver and Boyd, Edinburgh.

Foth, H.D. 1978. Fundamentals of Soil Science. John Wiley & Sons, New York.

Levy, G.J. 1985. The effect of exchangeable potassium on the permeability of soils. Proceedings of the K-symposium. Department of Agicultural Economics and Marketing, Pretoria.

Morgan, R.P.C. 1979. Soil Erosion. Longman, London.

Shainberg, I., Rhoades, J.D. and Prather, R.J. 1981. Effect of low electrolyte concentration on clay dispersion and hydraulic conductivity of a sodic soil. Journal of the Soil Science Society of America, 45, 273-277.

GULLY FORMATION IN ARCHAEAN ROCKS AT SADDLEBACK PASS, BARBERTON MOUNTAIN LAND, SOUTH AFRICA

GEORGE F. DARDIS, HEINRICH R. BECKEDAHL
Department of Geography, University of Transkei

1. INTRODUCTION

Many forms of soil erosion have been identified in southern Africa, ranging from unconfined (sheet and rill erosion, degraded gully systems) to confined (discontinuous and continuous gully systems) forms (Schmitz and Rooyani, 1987; Beckedahl et al., 1988; Dardis et al., 1988). Most of these forms are developed in unconsolidated materials, such as colluvium, though a number of distinctive gully forms are developed in weathered bedrock in a number of areas in southern Africa (Price-Williams et al., 1982; Watson et al., 1984; Dardis et al., 1988). Gullies may in some instances develop in bedrock. Dardis (1987) and Beckedahl and Dardis (1988) have briefly reported gullies (which are remnants of soil pipes or tunnel erosion forms) and soil pipes dissecting soft mudstones of the Burgersdorp Formation (Adelaide Sub-Group, Permo-Triassic Karoo Sequence) in Transkei, while Schmitz and Rooyani (1987) have reported gully formation in soft sandy shales of the Elliot Formation in Lesotho. Schmitz and Rooyani (1987, p.103) have also reported gullying associated with advanced stages of "rock stripping" on Karoo sandstones and basalts.

In this paper, a gully network cut into bedrock at Maid of the Mist Mountain (Lat: 25 50'S Long: 31 5'W Alt: 1480 m), 20 km southeast of Barberton (Fig. 1) is described. The site occurs in the Saddleback Pass, linking Barberton to Pigg's Peak (in Swaziland). The geomorphology of the site is described and factors associated with the development of the gully network are outlined.

2. STUDY AREA

The study area occurs in steep mountain terrain, with slope angles from 20-60 degrees. Debris slope material has been largely removed from the hillslopes at this site, exposing the bedrock (Fig. 2). Rock stripping (cf.Schmitz and Rooyani, 1987)

Geomorphological Studies in Southern Africa, G.F.Dardis & B.P.Moon (eds)

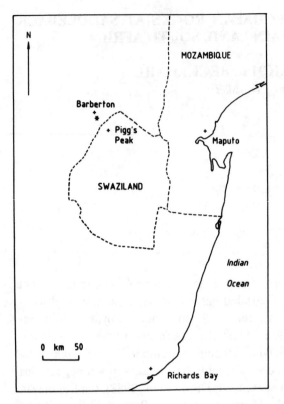

Fig. 1. Location of the study area in eastern southern Africa.

is localised, occurring in a zone 200 m long by 30 m in depth. The bedrock comprises quartzitic sandstones of the Archaean Moodies Group (Swaziland Supergroup), which is a relatively unmetamorphosed assemblege (Eriksson, 1978, 1980). The sandstones are coarse-grained, reflecting derivation from granitic rocks which were of widespread occurrence at the time of deposition of the Moodies Group. The rocks are drab or whitish in colour, reflecting a relatively high quartz content (Eriksson, 1978).

3. GEOMORPHOLOGY OF THE GULLIED BEDROCK ZONE

3.1. Gully morphology

The gullied bedrock zone consists of a series of deep (0.2-3.0 m) gullies of sub-parallel pattern, generally extending along the length of the area of exposed bedrock (Fig. 2). The depth of the incision into bedrock increases downslope, from

286

Fig. 2. Rock-incised gullies showing sub-parallel, v-shaped cross-sectional morphology.

Fig. 3. Open, v-shaped, asymmetric gully forms, showing potholing and waterfall erosional features. Note the development of grikes on gully sidewalls and adjacent slopes, and remnant debris mantling of gully interfluves.

0.3 m at the top to 3.0 m at the base of the slope.

The gullies in the lower part of the gullied zone are open (unconfined), continuous conduit forms (Figs. 2,3)(type 7 soil erosion forms; Dardis et al., 1988). Waterfall features are common within the gullies in this region (Figs. 3,4a,b), suggesting that they are mainly associated with overland flow processes. There is clear

Fig. 4. Asymmetric gully forms at Saddleback Pass, reflecting lithological or structural control on gully morphology.

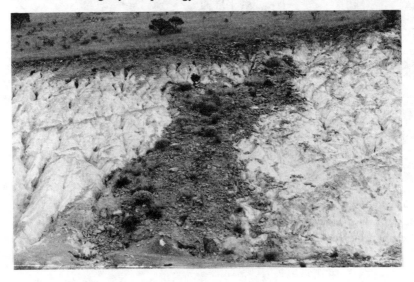

Fig. 5. Debris flow overlying gullied bedrock, Saddleback Pass site.

evidence, however that the gullies may underlie the debris slope materials in the upper part of the gullied zone (Figs. 5,6). The gullies underlying the debris slope materials at the top of the eroded zone are v-shaped. They vary in width and depth

Fig. 6. Erosion at the top of the gullied bedrock zone, Saddleback Pass site. Note the v-shaped forms penetrating beneath the regolith.

Fig. 7. Changes in the degree of incision associated with lithological/structural changes in the bedrock. Note the attempts to establish vegetation within the gullied zone.

up to 1.5 m (Fig. 6). The size of these features suggests that they extend for some distance upslope (possibly up to 50-100 m), beneath the regolith cover. These are distinctly different from the u-shaped, bedrock-incised tunnel erosion features seen at Mqanduli, Inxu Drift and at Ncise in Transkei (Beckedahl and Dardis, 1988; Dardis and Beckedahl, 1988a,b).

289

Fig. 8. Outlier of bedrock, overlain by up to 2 m of debris slope deposits, which has not been subject to gullying.

Fig. 9. Debris slope materials accumulating at the base of the gullied bedrock zone.

Distinct lateral variations in the severity of gullying are evident (Fig. 7).These appear to be related to changes in bed inclinations within the Moodies Group sandstones, which often results in the abrupt termination of gullying (Fig. 8). Small-scale gullying and grike formation also occurs throughout the eroded zone (Figs. 4,5a). Sub-vertical grikes are common and appear to be associated with a pronounced structural lineation within the sandstones. It also influences the asymmetry apparent in most of the gullies (Figs. 3,4,5a,b).

3.2 Debris slope deposits

The bedrock was apparently overlain by clast-supported debris slope materials. These deposits are exposed in a bedrock outlier within the eroded zone (Fig. 8). They consist of disorganised (cf. Walker, 1975), angular to sub-angular clasts set in a coarse granular matrix. The debris slope deposits vary in thickness from 0.4-2.0 m. Remanie exposures within the gullied zone (Fig. 8) suggest that the presently exposed gullied bedrock was at one time overlain by debris slope deposits of up to 2 m in thickness, and that the bedrock in that area was probably up to 2-3 m higher in the gullied zone than at present.

The debris slope materials appear have been removed from the eroded zone by two main processes; (1) fluvial activity associated with gullying, and (2) mass movement. Remnants of debris slope materials occur on the crests of interfluves in the eroded zone (Figs. 2,4a), while much of the debris slope materials have accumulated as talus at the base of the eroded zone (Fig. 9). Debris slope material at the upper boundary of the gullied bedrock zone appears to have been relatively unstable. This has resulted in debris flow into and across the eroded zone, infilling gullies in transit (Fig. 5). Vegetation growth on the debris slope accumulations is relatively poor (see Fig. 5) which suggests that they are either active at present or have been active in the recent past (Innes, 1985).

4. DISCUSSION

4.1. Gully forms

The type of gully forms developed at Saddleback Pass are similar to gullies reported from a number of areas. Ireland et al. (1939) has described gullies of this type, associated with channel erosion by downward scour (mainly into soil or weak or weathered bedrock). The gullies were typically shallow and strongly v-shaped, with potholing common in individual channels, whereas gully heads are pointed. The narrow v-shaped gully forms, when developed in bedrock, are normally restricted to soft mudstones, marls or silty-clay formations (Smith, 1958; Schumm, 1956; 1962; Bryan and Yair, 1982), or weathered bedrock (Watson et al., 1984). Although gully forms have been described from other lithologies (cf.Ruxton, 1958; Feininger, 1969), they differ markedly from the erosional forms encountered at Saddleback Pass.

It appears that the cross-sectional gully profile will be maintained while the gully is incising material of uniform consistency and at an approximately constant rate. The rate of downcutting is strongly dependent on slope angle.

4.2. Gully network generation

The spatial pattern of gullies in the eroded bedrock zone is quite unlike normal drainage lines. The drainage density appears to be abnormally high, which may reflect special structural control (for example, close correspondence between strike and slope directions) or unusual antecedent topography (Smart, 1972; Jones, 1978). Why such a dense network should develop is unclear. Lithology may play a role in determining the drainage pattern. Bryan et al.(1978) have found almost instantaneous runoff on a steep sandstone slope in the Dinosaur badlands of Alberta, with a slower runoff response on shale, which has a direct influence on drainage patterns. Drainage pattern may also be influenced by regional slope. The subparallel/parallel drainage pattern and high drainage density is similar to those generated experimentally by Phillips and Schumm (1987) on homogeneous slopes steeper than 2-3 degrees. These patterns reflect inequalities of surface slope and inequalities of rock resistance. They may also reflect to some degree the nature of the initial slope and structure of the rock surface (Zernitz, 1932). Given the high density of the network and tentative evidence of subsurface initiation (see 4.3. below), it seems likely that the gully pattern may be inherited in part from a pre-existing subsurface pipeflow or percoline network (cf. Jones, 1975;1978), possibly associated with the debris slope materials.

4.3. Genesis of the gullied bedrock zone

The gully network may have had two possible modes of development; (1) by subsurface processes, and (2) by subaerial processes.

Although there is clear evidence that the network developed to its present extent largely as a result of overland flow, the morphology of the upper boundary between the gullied zone and the debris slope materials suggests that initial development of the network was probably the result of subsurface processes, related, possibly, to erosion by percolating groundwater at the regolith-bedrock interface (cf. Jones, 1987).

It is feasible, given the magnitude of the erosional forms, that erosion has been underway for several years. It seems likely, however, that the rate of development of the gully system has been accelerated by road cutting. Similar bedrock erosional features occur in the Barkly Pass area in the northeastern Cape Province and also appear to be associated with road cutting.

Gullies of this type may develop as a result of a change of base level, resulting in a change from sheetflow (or pipe flow) to downward gullying, or enhanced sheetflow or pipeflow leading to stripping of debris slope materials (Fig. 10). Mass

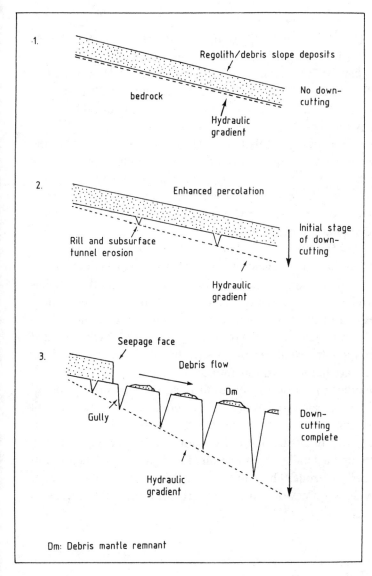

Fig. 10. A simplified model of gully development at Saddleback Pass. Downward migration of the hydraulic gradient as a result of road-cutting results in progressive enlargement of subsurface drainage networks, mass movements, exposure of bedrock and progrssive enlargement of rock-cut gullies.

movements may result from lateral subsurface eluviation of material along the seepage line and from build-up of water pressure within the regolith (cf. Campbell, 1975; Jones, 1987). Seepage faces (Fig. 10) form on the regolith, leading to further downslope migration of debris slope deposits (cf. Hadley and Rolfe, 1955). Removal of the debris slope materials would ultimately increase runoff generation (cf. Bryan et al., 1978), while changes in hydraulic gradient would accelerate downcutting.

The lowering of base-level (either naturally or by human interference) as a cause of accelerated erosion has been noted in alluvial channels (Schumm and Hadley, 1957; Nordin, 1964; Pickup, 1975) and in simulation experiments (Howard, 1971; Smart and Moruzzi, 1971; Parker, 1977; Begin et al., 1981; Begin, 1982). It has also been noted in association with gully development in unconsolidated sediments in southwestern Australia (Conacher, 1982).

In a number of these cases it was found that degradation (or "disturbance") of the original surface was uneven, with disturbances of different magnitude propogating at different rates. High magnitude disturbances generally propogated at slower rates than low magnitude disturbances (Begin, 1982). The net result of this was to produce a relatively complex drainage network, with local base-levels forming within sub-basins in the drainage network.

Observations in Transkei (Beckedahl and Dardis, 1988; Dardis and Beckedahl, 1988a) have demonstrated that gully development (regardless of the nature of the initial factors resulting in soil erosion) is promoted by changes in hydraulic gradient within gully networks which are incised into unconsolidated colluvium and related materials, and provides one of the most important mechanisms for the lateral extension of gully networks. It is clear, from the observations at Saddleback Pass, that changes in hydraulic gradient, by whatever process (in this instance road cutting), can result in considerable accelerated erosion, into consolidated bedrock.

It is not yet clear whether this process will result in accelerated erosion on all bedrock types subject to largescale changes in hydraulic gradient. Further work will focus on the physical properties (cf. Beckedahl, 1987a; 1987b; Hall, 1987) of lithologies exhibiting this form of accelerated erosion (Dardis et al., in prep), erosion pin experiments, and a detailed examination of variations in gully morphology within the eroded bedrock zone. This will allow a better assessment of rates of development and provide a framework to predict growth rates for this erosion form.

5. SUMMARY

Gully erosion at Saddleback Pass, near Barberton, appears to have been initiated by road cutting. Several stages of gully development have been identified. Gully morphology is determined by lithology and by migration patterns of the hydraulic gradient. Bedrock-cut gullies underlie the regolith and appear to have developed initially as subsurface tunnel erosion forms.

6. REFERENCES

Beckedahl, H.R. 1987a. Rock mass strength and landscape development on the

Vredefort Collar Zone. Unpublished M.Sc. thesis, University of the Witwatersrand, Johannesburg, 130pp.

Beckedahl, H.R. 1987b. Rock mass strength determinations as an aid to landscape interpretation. In: Gardiner, V. (ed.), International Geomorphology 1986 Part I, John Wiley and Sons, 393-405.

Beckedahl, H.R. (submitted). Approaches to rock mass strength data and landscape interpretation.

Beckedahl,H.R. and Dardis, G.F. 1988. The role of artificial drainage in the development of soil pipes and gullies: Some examples from Transkei, southern Africa. (this volume).

Beckedahl, H.R., Bowyer-Bower, T.A.S., Dardis, G.F. and Hanvey, P.M. 1988. Geomorphic effects of soil erosion. In: Moon, B.P. and Dardis, G.F. (eds.), The Geomorphology of Southern Africa, Southern Book Co., Johannesburg (in press).

Begin, Z.B. 1982. Application of "diffusion" degradation to some aspects of drainage net development. In: Bryan, R.B. and Yair, A. (eds.), Badland Geomorphology and Piping, Geobooks, Norwich, 169-179.

Begin, Z.B., Meyer, D.F. and Schumm, S.A. 1981. Development of longitudinal profiles of alluvial channels in response to base-level lowering. Earth Surface Processes and Landforms, 6, 49-68. Bryan, R.B., Yair, A. and Hodges, W.K. 1978. Factors controlling the initiation of runoff and piping in Dinosaur Provincial Park badlands, Alberta, Canada. Zeitschrift für Geomorphologie, Supplement Band 29, 151-168.

Campbell, R.H. 1975. Soil slips, debris flows and rainstorms in Santa Monica Mountains and vicinity, Southern California. U.S. Geological Survey Professional Paper 851.

Conacher, A. 1982. Salt scalds and subsurface water: a special case of badland development in southwestern Australia. In: Bryan, R.B. and Yair, A. (eds.), Badland Geomorphology and Piping, Geobooks, Norwich, 195-219.

Dardis, G.F. 1987. Quaternary erosion and sedimentation in badland areas of Transkei, southern Africa. Paper presented at Workshop on Erosion, Transport and Deposition in Arid and Semi-Arid Areas, Hebrew University, Jerusalem, March-April 1987.

Dardis, G.F. and Beckedahl, H.R. 1988a. Drainage evolution in a composite soil pipe-cavity system, Inxu Drift, Transkei, southern Africa. (this volume).

Dardis, G.F. and Beckedahl, H.R. 1988b. Soil Erosion, Quaternary Stratigraphy and Mass Movements in Transkei. Pre-Symposium Field Excursion Guidebook, Symposium on the Geomorphology of Southern Africa, April, 1988.

Dardis, G.F., Beckedahl, H.R., Bowyer-Bower, T.A.S. and Hanvey, P.M. 1988. Soil erosion forms in southern Africa. (this volume).

Dardis, G.F., Beckedahl, H.R. and Hall, K.J. in prep. The influence of rock properties in the development of bedrock-incised gullies.

Eriksson, K.A. 1978. Alluvial and destructive beach facies from the Archaean Moddies Group, Barberton Mountain Land, South Africa. In: Miall, A.D.(ed.), Fluvial Sedimentology, Canadian Society of Petroleum Geologists Memoir 6, 287-311.

Eriksson, K.A. 1980. Transitional sedimentation styles in the Moodies and Fig Tree Groups, Barberton Mountain Land, South Africa: evidence favouring an Archaean continental margin. PreCambrian Research, 12, 141-160.

Feininger, T. 1969. Pseudokarst on quartz diorite, Colombia. Zeitschrift für Geomorphologie, 13, 287-296.

Graf, W.J. 1979. The development of montane arroyos and gullies. Earth Surface Processes, 4, 1-14.

Hadley, R.F. and Rolfe, B.N. 1955. Development and significance of seepage traps in slope erosion. Transactions, American Geophysical Union, 36, 792-804.

Hall, K.J. 1987. The physical properties of quartz-micaschist and their application to freeze-thaw weathering studies in the maritime Antarctic. Earth Surface Processes and Landforms, 12, 137-149.

Howard, A.D. 1971. Simulation of stream networks by headward growth and branching. Geographical Analysis, 3, 29-50.

Imeson, A.C. and Kwaad, F.J.P.M. 1980. Gully types and gully prediction. KWAG Geografisch Tijdschrift, 5, 430-441.

Ireland, H.A., Sharpe, C.F.S. and Eargle, D.H. 1939. Principles of gully erosion in the Piedmont of South Carolina. United States Department of Agriculture Technical Bulletin, 63, 143pp.

Jones, J.A.A. 1975. Soil piping and the subsurface initiation of stream channel networks. Unpublished Ph.D thesis, University of Cambridge.

Jones, J.A.A. 1978. The spacing of streams in a random-walk model. Area, 10, 190-197.

Jones, J.A.A. 1987. The initiation of natural drainage networks. Progress in Physical Geography, 11, 207-245.

King, L.C. and Fair, D. 1944. Hillslopes and dongas. Transactions of the Geological Society of South Africa, 47, 1-4.

Nordin, C.F. 1964. Study of channel erosion and sediment transport. American Society of Civil Engineers, Journal of Hydraulics Division, 90, 173-192.

Parker, R.S. 1977. Experimental study of drainage basin evolution and its hydrologic implications. Unpublished Ph.D. thesis, Colorado State University, Fort Collins.

Phillips, L.F. and Schumm, S.A. 1987. Effect of regional slope on drainage networks. Geology, 15, 813-816.

Pickup, G. 1975. Downstream variations in morphology, flow conditions and sediment transport in an eroding channel. Zeitschrift für Geomorphologie, 19, 443-459.

Ruxton, B.P. 1958. Weathering and subsurface erosion in granite at The Piedmont Angle, Balos, Sudan. Geological Magazine, 95, 353-377.

Schmitz, G. and Rooyani, F. 1987. Lesotho Geology, Geomorphlogy, Soils. National University of Lesotho, 204pp.

Schumm, S.A. 1962. Erosion on miniature pediments in Badlands National Monument, South Dakota. Bulletin of the Geological Society of America, 73, 719-724.

Schumm, S.A. 1966. Evolution of drainage systems and slopes in badlands at Perth Amboy, New Jersey. Bulletin of the Geological Society of America, 67, 597-646.

Schumm, S.A. and Hadley, R.F. 1957. Arroyos and the semi-arid cycle of erosion.

American Journal of Science, 255, 161-174.

Smart, I.S. 1972. Quantitative characterization of channel network structure. Water Resoruces Research, 8, 1487-1496.

Smart, I.S. and Moruzzi, V.L. 1971. Computer simulation of Clinch Mountain drainage networks. Journal of Geology, 79, 572-584.

Smith, K.G. 1958. Erosion processes and landforms in Badlands National Monument, South Dakota. Bulletin of the Geological Society of America, 69, 975-1008.

Walker, R.G. 1975. Resedimented conglomerates of turbidite association. Bulletin of the Geological Society of America, 86, 737-748.

Watson, A., Price Williams, D. and Goudie, A.S. 1984. The palaeoenvironmental in-terpretation of colluvial sediments and palaeosols of the late Pleistocene hypother-mal in southern Africa. Palaeogeography, Palaeoclimatology, Palaeoecology, 45, 225-249.

Zernitz, E.R. 1932. Drainage patterns and their significance. Journal of Geology, 40, 498-521.

TERRACETTES IN THE NATAL DRAKENSBERG, SOUTH AFRICA

HELEN K. WATSON
Department of Geography, University of Durban-Westville

1. INTRODUCTION

As noted by Killick (1963) the slope microrelief of the Natal Drakensberg's Little Berg is clearly revealed following the removal of grass by burning. Steps in the surface soil, either parallel or sub-parallel to the contour or concentric in shape are a characteristic feature. Their abundant presence in the Drakensberg was first recorded by King (1944) who termed them terracettes. Clarke (1976) and Vincent and Clarke (1976) noted that the wide range of terminology used in the international literature to refer to these distinctive forms or microrelief is indicative of their polygenetic nature. These authors concluded that the origin and development of terracettes is dependent on a combination of topographic, soil and vegetation factors. Processes involved in their initiation include soil slippage, soil flow and soil creep. Once formed they are subjected to a range of erosional and denudational processes. Animal disturbance may be instrumental in triggering the initiating processes or in faciliating their subsequent development. Garland (1987) reviewed all literature dealing with Drakensberg terracettes, a total of nine references. With the exception of Vester et al. (1984), these references are limited to general description and speculative hypothesis. Most of them favour shear failure and solification as terracette initiating processes, and frost heave and wind ablation as secondary processes for their continued development. Measurements by Vester et al. (1984) of a number of terracette morphological and soil variables at three locations appear to support shear failure as the initiating process. Garland's (1987) study is the only rigorous study of Drakensberg terracettes carried out to date. Garland (1987) also concluded that Drakensberg terracettes are polygenetic in origin, and that their subsequent development is influenced by different geomorphological processes. His measurement of the stability of the terracettes and of a wide range of soil, site and morphometric variables at fourteen locations in the central Drakensberg however, suggest that   shear failure and sheet wash erosion are important initiating processes, that once formed further movement of the terracettes is restricted, and that the continued development of terracettes by scarp retreat is primarily due to rain splash and runoff processes. This study aimed to assess the influence of various environmental parameters on the distribution and morphology of terracettes at a location in the northern Drakensberg.

2. STUDY AREA

The Natal Drakensberg is an abrupt continuous scarp girdling the interior of the Province. Crest heights average 3000 m. a.s.l. A single cycle of erosion and a uniform geology throughout its length has resulted in a highly repetitious structural expression (King, 1972). Below the main escarpment is a basalt- capped terrace, termed the Little Berg. Its dissection by deep ravines has resulted in rounded spurs projecting into Natal. This study was conducted at the Cathedral Peak Forestry Research Station which is located at $29°00'S$ and $29°15'E$ and comprises fifteen gauged catchments at the head of three such isolated spurs. The soils of the Little Berg are of residual and colluvial origin and are derived entirely from basalt. They are generally acid, highly leached, highly weathered and structureless (Le Roux, 1972). The mean annual rainfall is 1406 mm. About 85 per cent of this falls during summer (Schulze, 1979). Bioclimatically the area is classified as Montane in which humid to humid-subhumid conditions, and cool to cold summer and cool-cold to very cold winter temperature range prevail (Phillips, 1973). The vegetation mainly comprises Themeda-Festuca Alpine Veld which is restricted from developing into a woody Montane Fynbos by recurrent fires (Granger, 1976).

A 306 ha area, comprising catchments IV, V, VI, VII, XI, XII and XIII, was selected for use in this study. The distribution of soil types and topographic factor classes within this area are presented in Table 1. All of these catchments are dominated by a consocies of the short grass *Themeda triandra*. Herbaceous geophytic perennials and dicotyledons although often locally conspicuous are a minor overall contributor to their floral composition. Woody communities dominated by *Leucosidea sericea,* Buddleia salviifolia and *Cliffortia linearifolia* are restricted to the percoline area in close proximity to their stream channels.

Table 1. Distribution of soil types and topographic factor classes as a percentage of the total study area.

SOIL TYPE	Hutton	Mispah-Glenrosa	Griffin			
	60	32	8			

SLOPE ASPECT	N	NW	NE	W	E	SW
	31	41	12	7	7	2

SLOPE ANGLE	<4	4,1-8	8,1-12	12,1-16	16,1-20	20,1-23	> 23,1	DEGREES
	6	23	29	22	13	5	2	

SLOPE LENGTH	<49,9	50-99,9	100-149,9	150-199,9	200-249,9	250-299,9	300-349,9	> 350	METRES
	22	23	20	16	9	4	4	2	

3. METHODS

The riser height, tread width and length of each of the 1123 terracettes in the study area were measured. The standard Product Moment Correlation Co-efficient was used to determine whether a significant degree of association existed between these terracette dimensions. The position of each terracette in each catchment was plotted on a 1:5000 map which was superimposed on a series of maps to obtain the combination of environmental factors present at the location of each terracette. The slope angle, aspect and length class maps were prepared for this study. A soil type and depth map by Tomlinson et al.(1980) and a vegetation map by Farrell (1986) were available. The latter map defines a total of fourteen different communities within the study area. The standard Product Moment Correlation Coefficient was used to determine whether a significant degree of association existed between each of these environmental factors and each of the terracette dimensions measured. An Analysis of Variance and a 0.05 confidence level was used to establish whether differences in the terracette dimensions between soil types and topographic factor were significant.

Table 2. Summary of the data on the terracette dimensions (in metres).

	RISER HEIGHT	TREAD WIDTH	TREAD LENGTH
MEAN	,33	,27	5,12
STANDARD DEVIATION	,13	,11	3,66
MAXIMUM	1,45	1,37	34,00

The correlations between the combinations of terracette dimensions were all significant at 0.001 confidence level. The strongest correlation was between the tread width and the riser height.

A comparison of terracette dimensions in the soil type   and each of the topographic factor classes revealed that terracettes on the Griffin soils differed significantly in morphology to those on either Hutton or Glenrosa soils. Slope angle was the only topographic factor where morphological differences in terracettes between classes was significant.

Table 3. Correlations between terracette dimensions and site factors that are significant at a 0.001 confidence level.

		Soil Type	Soil Depth	Slope Angle	Slope Aspect	Slope Length	Vegetation Type
RISER	HEIGHT					−	
TREAD	WIDTH	+	+	+		+	+
TREAD	LENGTH		−				−

4. DISCUSSION

4.1. Terracette morphology

There are striking differences between the terracette dimensions measured in this study and those measured by Garland (1987). The riser height and tread length of terracettes in this study were on average 0.6 m and 0.35 m respectively, of those measured in the central Drakensberg. The maximum tread width of terraces measured in this study was a third of the mean tread width reported by Garland (1987). All terracettes measured in this study occurred on volcanic soils. Garland (1987) did note that terracettes on such soils were narrower than those on sedimentary soils. The substantially narrower tread widths and the fact that they correlate well with tread length and riser height may indicate that the terracettes in this study area have not undergone considerable size modifications by secondary erosional and denudational processes including animal disturbance. The more extreme climate and the greater accessibility facilitated by the lower altitude may have served to encourage such secondary processes in the central Drakensberg. The terracette dimensions recorded in this study conform more closely to those cited by Clarke (1976) in her review of the international literature, and to estimates by Killick (1963) also for the Cathedral Peak area.

4.2. Soil factors

The density of terracettes on the three soil types present in the study area appears to be influenced by differences in the clay content and depth of the soil. The highest density was found on the xeric, clay loam Mispah-Glenrosa soils. These occur as shallow

Table 4. Distribution of terracette tread lengths as a percentage of the total number of terracettes within each soil type and topographic factor class.

		Terracette Tread Lengths in Metres				
		0 – 5	5,1 – 10	10,1 – 15	15,1 – 20	20,1
SOIL TYPE	HUTTON	65	26	7	1	1
	MISPAH–GLENROSA	55	37	6	1	1
	GRIFFIN	81	13	4	2	0
SLOPE ASPECT	N	72	24	3		1
	NW	55	33	8	2	2
	NE	72	24	4		
	W	59	38	3		
	E	74	23	3		
	SW					
SLOPE ANGLE	< 4 DEGREES	31	69	3		
	4,1 – 8	65	32	7		
	8,1 – 12	54	36	4	1	2
	12,1 – 16	70	24	6	1	1
	16,1 – 20	63	28	10	2	1
	20,1 – 23	54	33	2	2	1
	> 23,1	74	22		1	1
SLOPE LENGTH	< 49,9 METRES	65	28	7		
	50 – 99,9	61	33	4	2	
	100 – 149,9	59	30	8	2	1
	150 – 199,9	52	32	14	1	1
	200 – 249,9	59	26	3	1	
	250 – 299,9	75	25			
	300 – 349,9	71	19	10		
	> 350	67	29	4		1

Table 5. Percentage of terracettes present within each soil type and topographic factor class, and the mean number of terracettes per hectare for each soil type and topographic factor class.

SOIL TYPE	Hutton	Mispah-Glenrosa	Griffin				
%	51	45	4				
x̄ No./ha	2,9	4,8	2,0				

SLOPE ASPECT	N	NW	NE	W	E	SW
%	30	51	10	5	4	0
x̄ No./ha	3,8	4,9	3,3	2,8	2,0	0

SLOPE ANGLE	<4	4,1-8	8,1-12	12,1-16	16,1-20	20,1-23	>23,1	DEGREES
%	1	9	18	26	27	11	8	
x̄ No./ha	0,7	1,5	2,4	4,5	8,1	8,0	19,0	

SLOPE LENGTH	<49,9	50-99,9	100-149,9	150-199,9	200-249,9	250-299,9	300-349,9	>350	METRES
%	18	20	28	13	10	5	3	3	
x̄ No./ha	3,2	3,5	5,6	3,2	4,2	5,7	3,1	5,4	

305

soils in which the mean depth of the orthic A1 horizon is about 70 cm, or as a thin veneer covering rock outcrops. The lowest density was found on the Griffin soils. Both the Griffin and Hutton soils are sandy clay loams. The Griffin soils are, however, much deeper than the Hutton soils. Carson and Kirkby (1972) also reported that terracettes are most prevalent where the soil mantle is shallow. They believe that the binding effect of vegetation is relatively greater in shallow soils and that this imparts the necessary extra strength that while restricting landslides permits small scale instability manifested as terracettes. Clarke (1976) noted that terracettes are more prone to develop on soils with a high clay content where the surface layer is more mobile than the subsurface layer. The tread width and length of terracettes measured in this study are correlated with soil factors. Garland (1987) attributed the strong association he found between riser height and soil depth to the fact that the base of many of the terracette risers he surveyed rested directly on the bedrock.

Several authors, speculating about the origin of Drakensberg terracettes, have viewed soil factors as a major control of the initiating process. West (1951) suggested that terracettes formed as a result of water-saturated soil slippage over the underlying rock. De Villiers (1962) suggested that shear failure initiated terracettes in response to the over-steepened angle of repose of many of the Little Berg soils which he attributed to increasing climatic aridity. The shallow Mispah-Glenrosa are the most xeric in the study area. Granger (1976) suggested soil creep as an initiating process in response to the differential clay mineral proporties of the surface and sub-surface horizons, and also to the over-steepened slope angles. He suggested that the change in the angle of repose of these slopes was due to the replacement of deeper rooting vegetation (possibly fynbos), with shallower rooting grassland following an increase in the frequency of dry season fires associated with increased human influence in the Drakensberg. Clarke (1976) cites several references that support this view of terracette formation as a response to the transformation from woody vegetation to grassland associated with human habitation. Grazing animals are seen to constitute a critical added stress factor on the oversteepened grass slopes. However, she also cites several references that suggest terracettes presently found in grassland areas predated the clearance of the former woody vegetation. Terracettes are seen to have been formed by sediment derived from slope wash processes and deposited behind tree bases.

Whether different terracette initiating processes operate within the study area's different soil types, or if these processes are common to the different soil types, what factor favour their increased activity in the Mispah-Glenrosa soils, remain queries. Clarke (1976) and Vincent and Clarke (1976) maintain that such queries can only be resolved by obtaining information on the actual strength of the regolith. Garland (1987) found a relatively weak but significant association between tread angle and shear strength and has drawn attention to the possible influence of shear resistance on soil erodibility.

4.3. Topographic factors

4.3.1. Slope aspect

Slope aspect does not appear to be a factor of major influence on the process initiating terracette formation in this study. The density of terracette was substantially higher on the drier northern aspects. There was however, no degree of association between aspect and any of the measured terracette dimensions. These findings conform with those of several other researchers cited by Clarke (1976), most of whom considered that the influence of aspect on terracette formation was through its influence on soil development and vegetation growth. Fluctuations in pore water pressure, which reduce soil strength, have been shown to be more frequent and intense on drier slope aspects. Variations in soil and moisture characteristics attributed to aspect have been shown to influence the type, density and rooting habit of vegetation. These factors control the strength of vegetation and hence slope stability.

4.3.2. Slope angle

Of all the environmental factors considered in this study slope angle appears to exert the strongest influence on the processes initiating terracette formation. The density, overall size range and tread width of terracettes increased steadily with increased slope angle. These findings conform with those of numerous other researchers cited by Clarke (1976); however, most of them found that the terracette forming process became active at much steeper slope angles. Clarke (1976) favours solification as the initiating process of terracettes formed on lower slope angles such as those found in this study. She further suggested that where the overall terracette size range is related to slope angle as in this study, the shear strength of soil is implicated as an important factor effecting terracette morphology.

4.3.3. Slope length

Slope length does not appear to be a factor of major influence on the processes initiating terracette formation in this study. Terracettes were not concentrated at the top summit and basal area of slopes as has been found by several other researchers cited by Clarke (1976). Slope length did however, appear to exert a significant influence on the morphology of terracettes. Clarke (1976) has suggested that the combined trends associating riser height and tread width with slope length found in this study, indicate basal rotational slumping as an important process during terracette formation.

4.4. Vegetation factors

Clarke (1976) and Vincent and Clarke (1976) maintain that vegetation factors exert a major control over all stages of terracette formation and development. These factors exert control through their influence on vegetation strength, soil strength and angle of slope repose. In this study all the terracettes occurred in grass dominated communities and both tread length and width were significantly associated with vegetation type. The examination of vegetation factors was however not sufficiently rigorous to detect the nature and extent of this influence on terracette distribution and morphology. All the terracettes had bare risers and vegetated treads indicating that animal disturbance is both a minor initiating and secondary factor.

5. CONCLUSIONS

Clarke (1976) and Clarke and Vincent (1976) maintain that the methodological approach used in this study, viz: relating the distribution and morphology of terracettes to a range of environmental factors, cannot provide the information necessary to explain why terracettes are present in some localities and absent in others. Such an explanation is seen to be possible only with the addition of information on the strength of the soil and vegetation cover. The significance of the various associations found in this study, while not permitting identification of specific processes, do indicate that a variety of processes (i.e. soil slippage, soil flow and soil creep) may operate to initiate terracette formation. These processes, as well as the resultant terracette morphology, appear to be controlled by various combinations of environmental factors. These associations, however, failed to yield any information on the nature and intensity of secondary processes responsible for further development of the terracettes.

All terracettes occurred in grass dominated communities. Their densities were highest on the driest shallowest soils with the highest clay content, and increased in proportion to increases in slope angle. Slope aspect and length appeared to have a minor influence on the processes initiating terracette formation. The width and length of terracette treads were both significantly associated with soil depth, slope length and vegetation type. The width of terracette treads was additionally significantly associated with slope angle. None of the environmental factors considered in this study appeared to have any influence of terracette riser height.

6. SUMMARY

The position, riser height, tread width and length of a large number of terraces in the northern Drakensberg were measured and correlated with data on the soil type and depth, slope aspect, angle and length, and vegetation type present at each terracette location. All terracettes occurred in grass-dominated communities. Their densities were highest on the driest shallowest soils with the highest clay content, and increased in proportion to increases in slope angle. Slope aspect and length appeared to have a minor influence on the processes initiating terracette formation. The width and length of terracette treads were both significantly associated with slope angle. None of the environmental factors considered in this study appeared to have any influence of terracette riser height.

7. REFERENCES

Carson, M.A. and Kirkby, M.J. 1972. Hillslope, Form and Process. Cambridge University Press.

Clarke, J.V. 1976. The Origin and Development of Terracettes. Unpublished Ph.D. thesis, University of Salford.

De Villiers, J. 1962. A Study of Soil Formation in Natal. Unpublished Ph.D. thesis, University of Natal, Pietermaritzburg.

Farrell, V. 1986. The vegetation of the Cathedral Peak Research area. Unpublished map, Cathedral Peak Forestry Research Station.

Garland, G.G. 1987. Erosion Risk from Footpaths and Vegetation Burning in the Central Drakensberg. Natal Town and Regional Planning Supplementary Report, 20.

Granger, J.E. 1976. The Vegetation Changes, Some Related Factors and Changes in the Water Balance following Twenty Years of Fire Exclusion in Catchment IX, Cathedral Peak Forestry Research Station. Unpublished Ph.D. thesis, University of Natal, Pietermaritzburg.

Killick, D.J. 1963. An account of the plant ecology of the Cathedral Peak area of the Natal Drakensberg. Memoirs of the Botanical Survey of South Africa, 34.

King, L.C. 1944. Geomorphology of the Natal Drakensberg. Transactions Geological Society of South Africa, 47, 59-92.

King, L.C. 1972. The Natal Monocline. University of Natal Press, Pietermaritzburg.

Le Roux, J. 1972. Clay mineralogy of Natal oxisols. In: Proceedings of International Clay Conference, Madrid.

Phillips, J. 1973. The Agricultural and Related Development of the Tugela Basin and its Surrounds. Natal Town and Regional Planning Report, 19.

Schulze, R.E. 1979. Hydrology and Water Resources of the Drakensberg. Natal Town and Regional Planning Commission Report, 42.

Tomlinson, D.N., Clemence, B. and Jeffry, R. 1980. The soils of the Cathedral Peak Research area. Unpublished map, Cathedral Peak Forestry Research Station.

Verster, E., Van Rooyen, T.H. and Liebenberg, E.C. 1984. The Terracette morphology: relationships with soil properties and slope angle. Paper presented to Soil Science Society Congress, Bloemfontein.

Vincent, P.J. and Clarke, J.V. 1976. The terracette enigma - a review. Biuletyn Peryglac-jalny, 25, 65-77.

West, O. 1951. The vegetation of Weenen County, Natal. Botanical Survey of South Africa Memoir, 23.

MEASUREMENT OF SOIL MOVEMENT ON TWO HILLSLOPES DISPLAYING TERRACETTES IN HUMID SOUTH AFRICA

E. VERSTER and T.H. VAN ROOYEN

Department of Geography, University of South Africa

1. INTRODUCTION

The measurement of soil movement on steep, grassy hillslopes displaying terracette phenomena in humid South Africa is of great importance in explaining the formation of these tiered features. Likewise, terracettes are proof of mass movements occurring on hillslopes which, in turn, are regarded as partly responsible for the shaping of hillslopes. It seems evident, therefore, that quantitative description of such movements can be of great significance in determining the rate of hillslope evolution.

Soil movement due to the process of soil sreep is defined by Young (1972) as the slow downslope movement of the regolith as a result of the net effect of movements of its individual particles. Various factors, such as water and temperature changes, freeze-thaw, plants and soil fauna, are assumed to regulate, severally or in combination, total soil movement. In turn, soil creep is regarded as a necessary process in geomorphology to explain denudation on hillslopes that are too gentle for mass failure and that show no evidence of erosion by running water or surface wash (Finlayson, 1985). According to Young (1972), soil creep is predominant on hillslopes in temperate climates with dense grass cover. There seem to be several approaches to the study of soil creep: the engineering approach is directed, persistent and strong and is essentially a molecular process (Culling, 1983); geomorphologists think of soil creep as random, intermittent and weak and as a particulate phenomenon (Culling 1983); whereas Finlayson (1985) advocates a revised approach incorporating micro- and macropedological techniques and a plurality of causation and interaction.

From the wealth of literature it is evident that numerous studies of soil creep have been conducted. Information on field measurement methods and results is provided by Young (1972) and Finlayson (1985), while Saunders and Young (1983) summarized and collated rates from various studies. No extensive research has been conducted on rates of soil transport by creep in South Africa.

For this reason two areas, displaying visible soil movements in the form of ter-

Geomorphological Studies in Southern Africa, G.F.Dardis & B.P.Moon (eds)
© *1988 Balkema, Rotterdam. ISBN 90 6191 831 6*

Table 1. Location and environmental features of the two study areas.

Area	Location	Rain-fall (mm/y	Geology	Alt. (m)	Hill-slope Unit	Aspect	Vegetation
Long Tom Pass	25o09'S, 30o37'E	1700	Quartzite Shale*	2040	Mid-slope	East	Dense grass-veld
Mikes Pass	28o58'S 29o14'E	1350	Basalt+	1830	Mid-slope	East	Dense grass-veld

* Pretoria Group   + Lebombo Group

Table 2. Data on average mesoslope and terracette morphology for the seven sites.

Site	Meso-slope angle	Terracette morphological set			
		River Angle (o)	River Height (mm)	Tread Angle (o)	Tread Width (mm)
Long Tom 1	8.5	67.6	180	4.0	3830
Long Tom 2	11.0	65.1	242	4.6	3890
Long Tom 3	14.5	58.3	333	5.9	4150
Long Tom 4	21.0	57.4	413	7.3	4280
Mikes Pass 1	26.0	67.3	410	17.9	3980
Mikes Pass 2	15.0	-	-	-	-
Mikes Pass 3	20.0	72.3	374	14.2	3280

racettes were selected in order to measure the rate and direction of movement over a number of years and to make meaningful statements on the processes involved in soil movement and hence in hillslope evolution.

2. ENVIRONMENTAL FEATURES

Four sites at Long Tom Pass, Sabie (Eastern Transvaal) and three at Mike's Pass, Cathedral Peak (Natal), on slope angles varying from 8-26°, were investigated.

Table 3. Data on soil properties observed at the seven sites.

Site	Soil Form	Soil Depth	Soil Properties (Aver. for profile)					
			Clay	Silt	Stoni-ness	Liquid limit	Plasti-city index	Linear shrink-age
		mm	%	%	%			%
LT1	A	560	30	9	2	38	16	7.5
LT2	A	320	30	9	5	38	16	7.5
LT3	A	850	35	15	5	41	13	7.5
LT3	A	550	23	17	8	27	11	4.5
MP1	B	700	30	38	1	57	24	10.5
MP2	A	450	46	23	1	56	22	11.0
MP3	C	450	39	22	1	62	30	14.0

A: Clovelly B: Griffin C: Hutton

These areas cover a variety of environmental features (Table 1). With regard to climate, both areas can be classified according to Thornthwaite's system as humid and warm with no soil-water deficiences throughout the year (Schulze, 1947).

The terracettes observed at all these sites probably belong to the tear type according to the classification by Anderson (1972). They display, among other properties, steep risers generally lacking in vegetation; short, irregular treads normally not parallel over their entire length; and a sickle-moon shape. The average morphological properties of the terracettes, described in detail by Verster et al. (1985) are shown in Table 2.

Both areas, due to the high leaching potential, are composed of dystrophic soils. However, because of parent material differences (Table 1), the soil differs in respect of those properties normally used in terracette studies (Verster et al., 1985). On average the soils at Mike's Pass have higher clay and silt contents and corresponding higher liquid limit, plasticity index and linear shrinkage values (Table 3). A detailed exposition of the methods used to determine these properties is given by Verster et al.(1985).

3. METHODS

At each site a soil pit was dug down to bedrock. (From these pits soil samples were also taken for laboratory analyses, the results of which appear in Table 3). Following the technique and concept reflecting in the Young Pit (Statham, 1981), a steel rod was driven into the rock close to one side of the pit, protruding about 200 mm. By means of a plummet, nails were inserted perpendicularly into the pit side about 100 mm apart until about 20 mm projected. The distances between the nails, as well as that between the bottom nail and the steel rod, were carefully recorded, using a caliper. The pit was infilled and restored as closely as possible to its natural state. This project commenced in September 1975 for the Long Tom's site and in January 1980 for the Mike's Pass site. Initially soil movement was measured annually for three years and eventually after respectively eight and six years. To measure soil movement, the pits were excavated and the deviations of the nails from the perpendicular and the relative distances between the nails were carefully recorded. The rate and direction of movements were calculated by means of trigonometrical techniques. A common linear regression model was used to evaluate the relationship between soil movement and the site of the slope angle.

4. RESULTS AND DISCUSSION

Soil movement, as represented by movement of nails over the periods measured is shown in Table 4 for the 0-10 and 10-20 cm soil layers while Fig. 1 reflects movement for the whole profile. Total movement varies from a maximum of 4.50 (LTI) to a minimum of 0.59 mm/yr (MP2) for the upper 10 cm soil layer. With the exception of MP2, there is a substantial decrease in total movement from the 0-10 cm to the 10-20 cm soil layer: the latter varies from a maximum of 1.70 (MP3) to a minimum of 0.29 mm/yr (LT2) (Table 4). Lower down the profile soil movement obviously decreases further (Fig. 1).

The most striking result of the study, however, is the direction of movement. In the first place, soil transport invariably displays an inward component; secondly, both downslope and upslope movement were measured. Figure 2 illustrates the breakdown of total movement into linear and inward components as well as the angle of direction in relation to vertical. Of the fourteen measurements recorded in Table 4, net downslope movement was encountered in nine instances, with a maximum amplitude of 3.25 mm/yr downslope (MP3) and 0.57 mm/yr upslope (MP2). It is not the first time net upslope transport has been reported: according to Finlayson (1985), he (in 1976) and others (notably Kirkby in 1967) recorded frequent and substantial upslope movements in Young Pits. He offered no explanation but postulated that much of the movement in soils seems to be recoverable and not simply directed. Therefore "...a serious question arises as to when in a monitoring programme it may be assumed that the net non-recoverable soil creep has occurred. Much longer series of messurement will be necessary before this issue can be resolved..." (Finlayson, 1985). Young (1978), on the other hand, using a similar technique, recorded only net downslope transport over a twelve-year period. In view of the experience reported in the international literature, the present authors

314

Table 4. Soil movement at Long Tom Pass (LT), Sabie and Mikes Pass (MP), Cathedral Peak over periods of 8 and 6 years respectively.

Site	Land surface slope (o)	Depth cm	Soil Movement			
			Total mm/yr	Direction from vertical (o)	Linear mm/yr	Inward mm/yr
LT1	8.5	0-10	4.50	22	2.25	3.88
		10-20	0.38	41	0.23	0.20
LT2	11.0	0-10	0.75	27	0.44	0.50
		10-20	0.29	-31*	-0.10+	0.25
LT3	14.5	0-10	1.88	14	0.94	1.63
		10-20	0.58	62	0.53	0.15
LT4	21.0	0-10	3.38	-5*	1.00	3.00
		10-20	0.44	-21*	-0.03+	0.39
MP1	26.0	0-10	1.50	41	1.33	0.50
		10-20	0.37	-70*	-0.30+	0.23
MP2	15.0	0-10	0.59	-27*	-0.13+	0.50
		10-20	0.93	-49*	-0.57+	0.77
MP3	20.0	0-10	3.33	59	3.25	0.58
		10-20	1.70	82	1.55	0.35#

* Negative values indicate directions upslope of vertical

+ Negative values indicate upslope linear movements

Value indicates outward movement.

agree that upslope rebound could take place especially if solution is a major process (see below), but the fact remains that the majority of cases indicates a net downslope transport. Indeed, terracette formation is further proof of downslope movement.

With regard to the large inward component, Young (1978) ascribes it to loss of weathered material in solution. It is evident from the results that the solution process has played a major role in the rearrangement and settling of particles and hence in downslope transport. If it is correct to assume that the linear and inward components are attributable to soil creep *sensu stricto* and solution respectively (according to the views of Young, 1978), then solution is the dominant process in the Long Tom area and soil creep in the Mike's Pass area (see Table 4). These con-

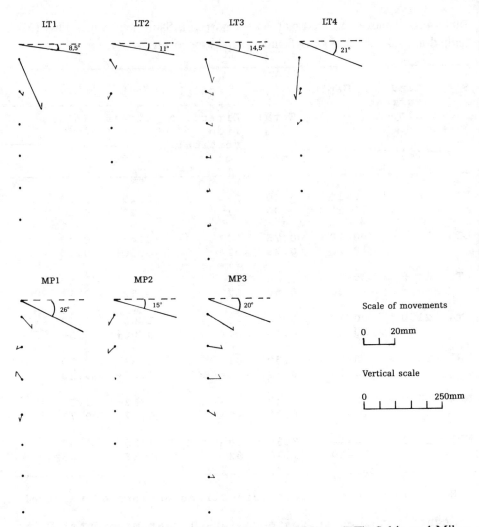

Fig. 1. Movements of nails in soil pits at Long Tom Pass (LT), Sabie and Mikes Pass (MP), Cathedral Peak over periods of 8 and 6 years respectively.

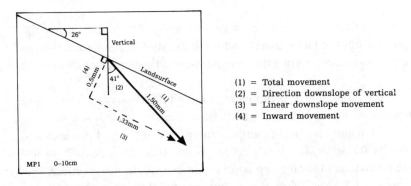

(1) = Total movement
(2) = Direction downslope of vertical
(3) = Linear downslope movement
(4) = Inward movement

Fig. 2. Breakdown of total soil movement into different components.

316

clusions are in accordance with the higher rainfall and hence higher weathering potential in the Long Tom Pass area. Furthermore, solution seems to be a more important cause of soil loss from soils derived from siliceous rock (Saunders and Young, 1983), as in the case of Long Tom. When comparing soil creep and solution as process involved in slope retreat and denudation, the general opinion among researchers (notably Carson and Kirkby, 1972; Young, 1972; Young, 1978; and Saunders and Young, 1983) is that solution is the dominant process in humid warm climates. The results of this study, however, indicate that both processes are active in triggering soil movement and that they are at least partly dependent on environmental factors. Concerning rates of slope retreat, for example, the inward component of LTI of 3,88 mm/yr (Table 4) represent slope retreat of 3880 Bubnoff units (1 Bubnoff unit = 1 mm/1 000 yr retreat; Saunders and Young, 1983), which compares favourably with rates referred to in international literature.

Comparing these soil transport data (average of seven sites = 1.47 mm/yr) with published data, especially those reported by Young (1978) over a twelve-year period (viz. 0.40 mm/yr), the faster rate of the former could probably be ascribed to active sites in the landscape due to terracette formation. By contrast, a possibly comparable study by Black and Hamilton in 1971 (referred to by Saunders and Young, 1983) for soil movement in a terracette situation, yielded rates of 5.0 mm/yr. Generally speaking, comparison of this nature are rarely possible due to spatial variability in environmental factors. Even the Long Tom and Mike's Pass results cannot be compared because of differences in the climatic and pedological components.

The temporal variation in soil movement is reflected in Table 5. With the exception of LTI and MPI, the general trend seems to be an initial relatively large transport and a subsequent decrease with increasing number of years. It is assumed that the disturbance of the site has caused this anomaly and that movement will stabilize over time. Long-term measurements will clearly be necessary to characterize the general trend.

Kirkby (1967) has suggested that soil creep rates should be related to slope angle. A linear regression model of soil movement (broken down into the different components as illustrated in Fig. 2) was run on the sine of the slope angle for the data contained in Table 4. The results indicate variability in statistical significance. Only the regression with direction of movement from vertical and inward movement appears to show some significance. Some of the highest values, although not statistically significant at the 95 % confidence level, are as follows: For the Long Tom sites the direction of soil movement from vertical for the 0-10 cm soil layer is $F = 0.06$; and for the Mike's Pass sites the inward movement for the 10-20 cm soil layer is $F = 7.38$ with 1 and 1 degrees of freedom; and $p = 0.22$. Furthermore, with regard to slope angle, the regression explains 88.4 % of the variance in the direction from vertical and 88.1 % of the inward component respectively whereas both show $r = -0.94$ values. However, most of the regression values, notably those of total soil movement, bear no significant relationship. In view of the limitations of these results (viz. number of replications and length of measurements), no definite

Table 5. Temporal variation in soil movement at 0-10 cm depth measured at Long Tom Pass (LT) and Mikes Pass (MP).

Site	Time Period	Soil Movement			
		Total	Direction from vertical	Linear	Inward
	yr	mm/yr	(o)	mm/yr	mm/yr
LT1	1	4.00	-	-	4.00
	2	2.00	16	2.25	4.55
	3	3.17	30	1.57	2.73
	8	4.50	22	2.25	3.88
LT2	1	3.70	79	3.50	0.50
	2	2.30	59	2.20	1.40
	3	1.37	53	1.13	0.43
	8	0.75	27	0.44	0.50
LT3	1	7.00	8	2.60	6.10
	2	4.10	5	1.40	3.70
	3	3.07	14	1.33	2.67
	8	1.88	14	0.94	1.63
LT4	1	15.20	8	7.00	13.20
	2	10.50	3	4.00	9.50
	3	9.07	3	3.50	8.13
	8	3.38	-5	1.00	3.00
MP1	1	1.60	36	1.30	0.10
	2	5.10	36	4.45	2.50
	3	3.40	42	3.23	1.27
	6	1.50	41	1.33	0.50
MP2	1	3.00	0	0	3.00
	2	1.35	-18	-0.15	1.35
	3	1.50	-23	-0.30	1.43
	6	0.59	-27	-0.13	0.50
MP3	1	9.50	42	8.50	4.00
	2	7.75	49	7.10	2.80
	3	5.13	44	3.60	3.50
	6	3.33	59	3.25	0.58

conclusions can be drawn except that total soil movement seems to be independent of slope angle. It is interesting to note that Finlayson (1981) came to the same conclusion and stated that movements in the soil occur in response to a variety of environmental factors which are largely independent of slope angle.

5. CONCLUSIONS

This study substantiated definite trends in downward soil movement, even though spatial and temporal variability do occur. The latter could possibly be regarded as random, short-term phenomena. It could also be assumed that more replications and long-term measurements would provide accurate estimates of rates of downslope movements. It seems evident that the measuring technique followed here has proved sensitive for detecting soil transport in any direction, including upslope transport. Although total soil movement appears to be independent of slope angle, there is some indication of a relationship between the downslope components, e.g. direction from vertical and inward, and slope angle. More research is necessary, however, to confirm this deduction.

The importance of both the solution and soil creep processes in shaping hillslopes in humid warm climates has also been established. These processes are likewise dependent on the environmental factors. Finally the study may well contribute substantially to the understanding of terracette formation.

6. SUMMARY

Two areas displaying visible soil movement in the form of tear terracettes were selected for study. Four sites at Long Tom Pass, Sabie and three at Mike's Pass, Cathedral Peak occurring on different slope angles varying from 8-26 degrees, were investigated. Nails inserted into soil pits in 1979 and 1980 were measured for soil movement, initially on a yearly basis and finally after an eight- and a six year period respectively. Average total soil movement of 1.47 mm/yr (n = 14) was recorded in the upper 20 cm of the soil layer, including net downslope (9 cases) and net upslope (5 cases) transport. It is suggested that these seemingly random directions in soil transport are short-term phenomena. Longer-term measurements would be necessary to produce accurate estimates of downslope transport rates. Furthermore, a significant movement inwards towards the saprolite was observed at most sites. The latter is interpreted as being caused by loss of weathered material in solution. It is therefore evident that a combination of processes, regulated by several factors, are responsible for soil movement on selected steep hillslopes in humid South Africa. Although total soil movement appears to be independent of slope angle, a linear regression model shows some indication of a relationship with the downslope components.

7. ACKNOWLEDGEMENTS

The financial support of the University of South Africa is gratefully acknowledged. A special word of thanks to Mr S.S. van Dyck who provided the statistical analysis,

Mrs W. Koch who typed the original manuscript and Ms. J.C. Oelofse who drew the illustrations.

8. REFERENCES

Anderson, E.W. 1972. Terracettes: a suggested classification, Area, 4, 17-20.

Carson, M.A. and Kirkby, M.J. 1972. Hillslope form and process, Cambridge University Press, Cambridge, 475p.

Culling, W.E.H. 1983. Rate process theory of the geomorphic soil creep. In: De Ploey, J. (ed.), Rainfall Simulation, Runoff and Erosion, Catena, Supplement 4, 191-214.

Finlayson, B.L. 1981. Field measurements of soil creep, Earth Surface Processes and Landforms, 6, 35-48.

Finlayson, B.L. 1985. Soil creep:a formidable fossil of microconception. In: Richards, K.S., Arnett, R.R. and Ellis, S. (eds.), Geomorphology and soils, Richards, George Allen and Unwin, 141-158.

Kirkby, M.J. 1967. Measurement and theory of soil creep, Journal of Geology, 75, 359-378.

Saunders, I. and Young, A. 1983. Rates of surface processes on slopes, slope retreat and denudation. Earth Surface Processes and Landforms, 8, 473-501.

Schulze, B.R. 1947. The climates of South Africa according to the classifications of Koppen and Thornthwaite, The South African Geographical Journal, 29, 32-42.

Statham, I. 1981. Slope processes: techniques for observing and measuring soil creep. In: Goudie, A.S. (ed.), Geomorphological Techniques, George Allen and Unwin, 170-174.

Verster, E., Van Rooyen, T.H. and Liebenberg, E.C. 1985. Relations of terracette morphology with soil properties and slope angle, The South African Geographer, 13, 113-120.

Young, A. 1972. Slopes, Oliver and Boyd, Edinburgh, 228 pp.

Young, A. 1978. A twelve year record of soil movement on a slope. Zeitschrift für Geomorphologie, Supplement Band 29, 104-110.

SPHEROIDAL WEATHERING IN THE LESOTHO FORMATION BASALTS

ROBERT SARRACINO
Department of Physics, National University of Lesotho
GISELA PRASAD
Institute of Southern African Studies, National University of Lesotho

Spheroidal weathering of Jurassic tholeiitic basalts from Lesotho is analyzed. A mathematical model which is then applied both to initial weathering from the corners and to weathering of central region. A double rate weathering model is shown to explain the peculiarities observed in initial weathering from a corner. The salient features of the geometry of the central region are explained by the model: ellipsoids weather initially into ellipsoids of greater or lesser eccentricity and finally into spheres. This is in agreement with the geometry of observed weathering patterns and is a feature which would be difficult to explain by Liesegang-type diffusion, with which it has until now been associated.

Many authors have described spheroidal weathering patterns, which are observed in a variety of rocks and sites (Augustithis et al., 1980; Augustithis, 1982; Eggleton et al., 1987; Sarracino et al., 1987). The basalts of our samples were extruded 187 Ma ago, when the rock was fractured, water seeped into the fractures, and intensive weathering began, probably in the early Tertiary (Prasad, 1983). Weathering begins at the corners of the polyhhedra formed by the fracturing, and proceeds until a closed surface, ellipsoidal in shape, is formed (Fig. 1). The closed surface weathers into an ellipsoid of greater or lesser eccentricity, and eventually into a sphere. Three types of spheroids are formed during weathering; (1) unweathered cores, (2) partially decomposed and leached shells, and (3) reprecipitated Fe-rich zones (Fig. 2). The geometry of the Fe-depleted shells parallels the geometry of the unweathered cores, leading to the hypothesis that the weathering rings recapitulate the time-history of the shrinking unweathered cores.

A geometric model of weathering has been developed which is based on a minimum number of general assumptions. The model produces a non-linear partial differential equation whose solution, a hypersurface in the four-dimensional space composed of the three spatial dimensions and time, gives a time-history of the

Geomorphological Studies in Southern Africa, G.F.Dardis & B.P.Moon (eds)
© *1988 Balkema, Rotterdam. ISBN 90 6191 831 6*

Fig. 1. Spheroidal weathering in the tholeiitic basalts of the Jurassic Lesotho Formation. The locality is on the Blue Mountain Road near St. Michael's, Lesotho, just
above the Clarens Formation/Lesotho Formation contact.

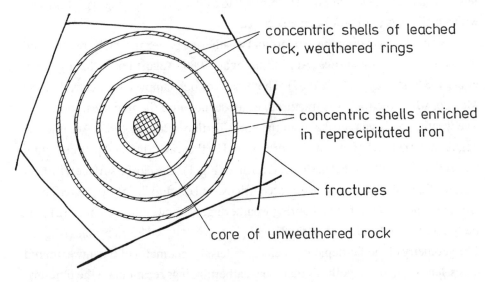

Fig. 2. Schematic section through a spheroidal weathering structure. Weathering
starts from a network of fractures along which water percolates and attacks the unweathered rock from all sides. Three types of spheroids are formed (see text). (After
Sarracino et. al. 1987).

shrinking unweathered core (Sarracino et al., 1987).

The differential equation is used to analyse both corner weathering and central weatherng. A double-rate weathering model is shown to explain the particular feature that until the first closed curve inscribing the initial polygon is formed the sides of the polygon weather very little, if at all, while considerable weathering takes place at the corner (Prasad and Sarracino, in prep.).

The equation is solved numerically for single-rate, central weathering from an initial ellipsoid. The numerical results show that, whereas it is possible initially for ellipsoids to become more eccentric through a given plane, eventually all ellipsoids, no matter how eccentric, will weather into sphere.

Twinning, an apparent anomaly in that its natural occurrence as part of the weathering process cannot be explained by the model, is discussed (Sarracino and Prasad, in prep.). Our observations show that twinning arises from further weathering during the period of intensive weathering.

REFERENCES

Augustithis, S.S. 1982. Atlas of Speroidal Textures and Structures and Their Genetic Significance. Theophrastus Publications, Athens, 329pp.

Augustithis, S.S. , Mposkos, E. and Vgenopoulos, A. 1980. Diffusion rings (spheroids) in bauxite. Chemical Geology, 30, 351-362.

Eggleton, R.A., Foudoulis, C. and Varkevisser, D. 1987. Weathering of basalt: Changes in rock chemistry and mineralogy. Clays and Clay Minerals, 35, 161-169.

Prasad, G. 1983. A review of the early Tertiary bauxite event in South America, Africa and India. Journal of African Earth Sciences, 1, 303-313.

Prasad, G. and Sarracino, R.S. (in preparation).Initial stages in speroidal weathering.

Sarracino, R.S. and Prasad, G. (in preparation). Investigation of spheroidal weathering and twinning.

Sarracino, R.S., Prasad, G. and Hoohlo, M. 1987. A mathematical model of spheroidal weathering. Mathematical Geology, 19, 269-289.

FREEZE-THAW WEATHERING: NEW APPROACHES, NEW ADVANCES AND OLD QUESTIONS

KEVIN J. HALL

Department of Geography, University of Natal, Pietermaritzburg

1. INTRODUCTION

In a 1973 review of notable gaps in our knowledge pertaining to arctic and alpine geomorphology, Ives (1973, p.1) identified four major areas of which one was "...The efficiency of freeze-thaw processes in the role of bedrock disintegration...". The hope of Ives was that by identifying key questions, it might stimulate research in that direction. However, in 1981, French (1981, p. 267) in a further review, reiterated the continued inadequacy of our knowledge regarding rock weathering under cold conditions. McGreevy (1981), in the same year, in the first of his series of key reviews on weathering processes (McGreevy and Whalley, 1984; Whalley and McGreevy, 1985) detailed the state of knowledge regarding freeze-thaw from the point of view of experimentation, mechanisms and field observations. Once more our lack of understanding is emphasised, but now it is clearly noted that it is the lack of base line field data, as the basis for experimentation or mechanism determination, that is the problem. At that time of writing the only studies that contained the essential field data in the English language (i.e. work may well exist in such as Russian, e.g. Konischev and Rogov, 1978), were those of Gardner (1969), Thorn (1979, 1980), Thorn and Hall (1980) and Hall (1975, 1980).

Subsequently, in two further key papers, McGreevy and Whalley (1982, 1985) identified and summarised the state of knowledge regarding the importance of rock temperature variation and rock moisture content respectively, in freeze-thaw weathering. For the former they concluded, amongst others, that there is a need for field data acquisition and that field and laboratory experiments need to be refined. With regard to rock moisture status, there is a call for data as none exist and that cognisance of this should be taken in simulations if they are to be of any meaning. To further stimulate 'thought' and mitigate against simplistic field judgements, Williams and Robinson (1981), McGreevy (1982) and Fahey (1985) all showed that the saline nature of the water which is subject to freezing indicates that both 'frost' and 'salt' weathering are operative, and not just the former. The actual relationship and inter-operation of 'frost' and 'salt' is, though, still unclear.

Geomorphological Studies in Southern Africa, G.F.Dardis & B.P.Moon (eds)
© 1988 Balkema, Rotterdam. ISBN 90 6191 831 6

Many questions have been asked, areas of data inadequacy specified, and new mechanisms hypothesised, but have any advances been attained? In short, yes, enormous advances have been made in the context of theoretical modelling, data acquisition, simulations and the application of new technology over the last five years. With respect to the investigation of weathering in South Africa little cognisance has been taken of these advances, and judgements still tend to be both subjective and qualitative (Hall, 1988). Presented here are some of the new techniques, approaches and findings regarding freeze-thaw, with particular emphasis on work that has been undertaken in South Africa as part of a joint investigation with British Antarctic Survey.

2. NEW APPROACHES

One new approach, which has grown in importance in recent years, is that of theoretical modelling. Hallet (1983), and later Walder and Hallet (1985, 1986), suggested a theoretical model, to explain the breakdown of rock due to freezing, using the well known theory of frost heaving in soils as an analogue. Hallet argued that the pressure exerted by ice growth is not primarily the result of volumetric expansion, concomitant upon the water to ice phase change, but rather that "... the induced pressure is assumed to arise thermodynamically because mineral surface effectively decrease the chemical potential of water in the close proximity...". Thus, there is unfrozen water existing during subzero conditions and this water flows towards the mineral surface and exerts a pressure; really a form of hydration shattering. Ultimately, this led to the formulation of a simple model based upon the stress intensity factor K_{IC} (the strength of the singularity in the cracktip stress field that tends to produce opening mode failure). Idealising rocks as isotropic linear elastic media, the suggested model is:

$$K_{IC} = \left(\frac{\pi \ell}{2}\right)^{\frac{1}{2}} (P + \sigma)$$

where ℓ is the length of a two-dimensional crack, P is the pressure applied inside the crack, and σ is the "applied" normal pressure perpendicular to the crack plane.

The model recognised a number of factors that exert an influence upon its operation, namely lithology, thermal regime, moisture content and moisture chemistry. However, it was predicted that the most rapid breakdown will occur in the temperature range -5° C (Hallet, 1983), with slow rates of cooling being the most conducive to destructive pressures (Walder and Hallet, 1985) but with less-than-saturated conditions decreasing the efficacy of the process (Walder and Hallet, 1986). Ultimately their findings led Walder and Hallet (1986) to suggest the general applicability of their model. However, despite its wholehearted adoption and further development by some workers (e.g. Tharp, 1987) there are still some problems. Whilst the model has introduced a major new approach, it is argued (Hall, 1986a) that its full application is inhibited by the lack of field data pertaining to the controls upon the model. For instance, as will be discussed in more detail below, the recent findings on rock moisture content (Trenhaile and Mercan, 1984; Hall, 1986b), rock moisture chemistry (Hall et al., 1986), the effects of rock anisotropy on freeze penetration (Hall, 1986c), the rate of fall of temperature in the outer, wetter part of the rock and the subsequent nature of the freeze (Hall, 1987a) do not all well agree with the generalised hypotheses regarding these assumed factors within the model. However, despite these reservations, this model of Hallet (1983) and its further development by Tharp (1987) introduces an exciting new approach which augurs well for the future.

Other innovative approaches are associated with the application of new technology to the study of freeze-thaw weathering. One recent approach that yields information direct relevance to the model of Hallet (1983) is the photoelastic study of ice pressure in rock cracks (Davidson and Nye, 1985). A new technique to measure the change in the stresses as water by means of photoelastic tchniques, utilising a photometric approach with digital processing of the resulting signals, was developed. The approach uses circularly-polarised light and a rotating analyser which allow for the fast and precise measurement of the lines of constant principal stress difference (isochromats) and the lines of constant principal stress direction (isoclinics). Upon freezing of water in a slot cut into a perspex block, the progression of the ice front could be monitored and the pressures exerted calculated. Two regimes were distinguished; (1) where an ice plug extrudes and so pressures are less, and (2) where the ice plug is fixed and grows *in situ*. The association of freezing front progression and the manner of cooling give very similar results to those found by Hall (1987a) using ultrasonic techniques.

Non-destructive ultrasonic testing is a technique that has been available for some time (e.g. Ide, 1937; Hornibrook, 1939; Timur, 1968; Fukada, 1971; New, 1976) but which has been employed in recent years, as was suggested by Aguirre-Puente (1978) and Fahey and Gowan (1979), to gain new insights into the breakdown of rock (e.g. Filonidov, 1982; Zykov, et al., 1984; Mak, 1985; Crook and Gillespie, 1986; Hall, 1987a). Variation of ultrasonic pulse propagation through rock or building material can be used to indicate changes in the quality of that medium through time. As the ultrasonic velocity in air is c.330 m/sec, in water it is c.1400 m/sec, and in ice c.3000-4000 m/sec (Press, 1966), it has been possible to discern a number of factors related to freeze-thaw by means of ultrasonic testing. For instance, data on the following have been obtained; the timing and progression of the water to ice and ice to water phasechanges (Hall, 1987a), calculation of moisture content and its variation resulting from freeze-thaw cycling (Matsuka, 1984), estimation of the amount of interstitial water that is frozen (Hall, 1987a), the effects of anistropy upon freezing (Zykov et al., 1984; Hall, 1986c), and the effects of water adsorption and desorption during the thaw phase (Hall, in press). Thus, this technique has offered a great deal of new information regarding freeze-thaw (see below).

A further new approach offers great potential for the determination of rock moisture content in the laboratory situation, this is time domain reflectometry (TDR). Originally a technique developed for the analysis of moisture content in soils (e.g. Patterson and Smith, 1981) it has now been applied to rocks (Hare, 1985). TDR is a type of pulse-reflection measurement with a broad band pulse travelling down the transmission line of the TDR from which, knowing the start and end points of the transmission line, the horizontal trace length can be measured. From the then known time of the pulse in the sample, it is possible to calculate the dielectric constant of the material. The dielectric property of rock or building material is a function of such factors as constituent materials, their structure and density, the presence of water and ice and salt content and temperature (Hare, 1985, p.89). The dielectric constant of most rock forming minerals is between 2 and 12 whilst that of water is between 80.1 and 87.7 (Hare, 1985). Thus, the measurement of the relative permittivity of the rock can provide a good indication of volumetric water content. Whilst suitable for use in soils, it is nevertheless difficult and time consuming when applied to rocks. Other techniques that can be utilised for determining rock content are such as neutron moderation (Bundey, 1982), differential thermal analysis (Mellor, 1970), suction-moisture content tests (Keune and Hoekstra, 1967) and dilatometry.

Dilatometry involves the measurement of volume change of material. Davison and Sereda (1978) developed a technique for monitoring the linear expansion of a brick due to freezing, whilst Pissart and Lautridou (1984) and Hames et al.(1987) undertook a similar approach to determine volumetric changes in materials due to the uptake of moisture. Yet another very new technological development which offers a whole new approach to the study of freeze-thaw weathering is that of optical fibre sensors (Hale, 1984). Somewhat akin to the use of strain gauges (Douglas et al., in press) the fibre optic crack detection system offers a whole new insight. Fibre optic crack detection gauges are bonded to the rock and infrared signal of known amount is fed through an optical fibre loop, the attenuation of which is continually monitored. If and when attenuation exceeds a preset amount a relay is triggered which can operate recording equipment. This technique is ideally suited to the monitoring of volumetric change upon water uptake (Hall, in press) and for use during freeze-thaw experiments (Hall, in prep a). One advantage of this system is that, upon failure the infrared light source can be uncoupled and a laser attached such that visible light is then emitted from the failure point(s). This allows exact determination of where failure occurred and whether it was at one or more places.

Finally, the last new advance that is readily available, and in use, is that of the application of microcomputers for the running of simulations and the simultaneous multiple channel monitoring of a variety of sensors. With the ever increasing power and accessibility plus the decreasing costs of microcomputers, they are ideal tools for the controlling, monitoring and data manipulation of weathering simulations. As part of a joint weathering project, in the Maritime Antarctic, between the University of Natal and British Antarctic Survey, a computer-controlled simulation cabinet was constructed. By means of purpose-made hardware, the microcomputer

328

may be programmed for temperature cycling of any length and complexity based upon data logged in the field. During running, the computer continually monitors the cabinet temperature, compares it against that which was programmed and initiates corrective action, if required. Data from six temperature sensors, a humidity sensor, ultrasonic transducers and the fibre optic crack detection system are read, stored, displayed and printed at intervals varying between 10 seconds and 99 minutes (together with the actual time) dependent upon what was chosen for that particular part of the simulation. Thus, a very sophisticated system is available that is able to undertake a variety of functions with great precision, for long periods of time, without the need of continous operator presence, and that can handle enormous amounts of data.

3. NEW ADVANCES

The new advances are largely as a result of the information derived from the application of the technology detailed above. However, one realm in which highly pertinent new progress has been made is that of fundamental field data acquisition. As was stated above, with regard to the constraints upon the use of hypothetical models, the lack of field data on such as temperatures, moisture content and chemistry, rock properties, and natural weathering rates, all serve to inhibit our understanding of freeze-thaw. Rock moisture content, as noted by McGreevey and Whalley (1985), was a largely unknown factor. However, the recent studies of Trenhaile and Mercan (1984), Hare (1985) and Hall (1986) have all begun to make available, albeit to a limited degree, data on field moisture content of rock. Three main points emerge from the available data. First, that, contrary to White's (1976) contention regarding rocks being greater than 50 % saturated and subject to freezing, a significant number of samples have been obtained (Hare, 1985, Table 4.2; Hall, 1986b, Table vi) that were in excess of 50% saturation. Secondly, that samples obtained from the faces of cliffs show very low moisture contents (Hall, 1986b). Thirdly, that the degree of saturation used in simulations is a poor representation of field conditions (Trenhaile and Mercan, 1984). All the data to date are, however, with respect to the 'averaged' moisture content of a sample and show no respect for moisture gradients within that sample (see below).

As was noted by Hallet (1983), and was a major inadequacy of the 'frost' and 'salt' weathering experiments of Williams and Robinson (1981), McGreevy (1982) and Fahey (1985), there is an almost complete absence of data pertaining to interstitial rock water chemistry. Prior to the development of utilisation of a new technique by Hall et al.(1986) there was only one analysis available, that of Kinniburgh and Miles (1983). However, the development of this new, relatively simple technique offers the potential for more data acquisition. Evidence available to date, from the Maritime Antarctic environment under investigation, indicates $NaCl$ molarities between 0.34 and 0.57 ($\bar{x}$ = 0.47), values very small to those suggested by Mc-Greevy (1982) to be the most efficacious for rock breakdown.

Another aspect of field data where new advances have been made, is with respect to the physical properties of rocks (Hall, 1987b). As part of the study of weathering in the Maritime Antarctic, the following rock properties were determined; compressive strength, indentor penetration, porosity, microporosity, absorption coefficient, saturation coefficient, rock mass strength and the size range of the weathering products. This gives fundamental background data for comparison with other studies, whilst the size range of the weathering products allows for direct collation with that from simulations. A further development of this is the long-term study of rock tablets in the field and the daily monitoring of a large tablet. In the former, a large number of small blocks whose properties had been determined, were placed in the field close to data loggers that monitored the climatic factors to which they were subject. A number of tablets are retrived each year and the properties reassessed to give some idea of weathering effects and rates (Hall, in prep). The large tablet, on the other hand, was weighted daily for a whole year and the climatic conditions noted. This gave data on daily changes in moisture content plus a standard regarding the amount of weathering against which simulation results could be compared (Hall, submitted). This is what is, perhaps, most significant in that, in addition to the providing of base-line data, the information is highly pertinent to the running of real world simulations.

The use of the computer-controlled cabinet together with the application of ultrasonics and fibre optics, has led to many new advances in our knowledge of freeze-thaw processes and their controls. The sort of new information derived from these simulations include such as the effects of rock anistropy on freeze penetration and freeze mechanism (Hall, 1986c), that the rate of fall of temperature appears to be less significant than the final temperature to which the freeze is going (Hall, 1987a), and that the rate of fall of temperature in the outer, wetter part of the rock is faster than that suggested to be most effective for breakdown by Walder and Hallet (1985, p. 342). Another important finding (Hall, 1987a) was that the cooling rate of the rock, which is largely controlled by the fixed final temperature, affects the nature of the phase change,with some being rapid and extensive whilst others are slower and progressive. Further, for low amplitude freezes (to c.-6° C) it was found (Hall, 1987a) that there was a need for temperatures to be maintained for a period of time before a phase change would take place. Finally, amongst the other specifics noted in Hall (1987a), it was also found that ice, during the thaw phase, returned to water between -0.7° and -1.9°C., and that during freezing c.80 % of the water that will freeze under natural conditions had done so by c.-6°C.

Other experiments have shown that small, omnidirectionally frozen samples of rock, with high moisture contents, do not replicate large, undirectionally frozen blocks (Hall, in prep b). This result therefore questions the applicability of the majority of earlier laboratory situations to many field situations. Yet other new evidence regarding changes in the elastic properties of rocks during water absorption and desorption, together with associated hysteresis effects, suggests that wetting and drying, like the saline solutions, is an intimate part of the freeze-thaw mechanism (Hall, in press).

The above comprise, albeit extremely briefly, but a part of the many new advances with respect to our understanding of freeze-thaw. Much new information on fracture mechanics is becoming available in engineering texts, and applied aspects are ever increasing. Details of all past and present information, together with future prospects regarding 'pure' and 'applied' aspects of weathering in cold climates will shortly be provided by Hall and Walton (in prep).

4. OLD QUESTIONS

In spite of the many new techniques and advances in our understanding, some of which have been cited above, the fundamental questions with respect to freeze-thaw remain essentially the same. In truth, the question of Ives (1971) given in the introduction still remains; we do not yet know the efficacy of freeze-thaw. Really what has happened is that rather than answering questions, we have been making a start with respect to filling-in the gaps in our knowledge that are required before the questions themselves can be addressed. Concomitant with this is the fact that our data base is still pitifully small. There are inadequate data, from a variety of environments and for any length of time, on such as the thermal regime the rocks are subject to, their moisture content and the chemistry of that moisture. Other problems such as the question of moisture gradients within rocks still remain to be solved.

At the same time , the number of questions have now increased due to the recognised interaction of saline solutions, wetting and drying and thermal effects within, or parallel with the freeze-thaw process itself. We have yet to fully understand these other mechanisms before their role within freeze-thaw needs can be assessed. Biological activity, particularly endolithic and chasmolithic bacteria and lichens, is yet another unknown that may well exert an influence on freeze-thaw.

Thus, whilst we have made major steps forward so our vistas have increased. Considering the depth and complexity of the topic, the number of active workers are few, the longevity of most research programmes too short and the cost of the technology becoming an inhibiting factor for many. What we are able to do, though, is to refrain from making the old, simplistic qualitative comments that have become so glib regarding 'freeze-thaw'. Just because, for instance, the Drakensburg are high mountains where it is cold for part of the year, then we can no longer justifiably say "freeze-thaw takes place". We must now rather ask the (same, old) questions regarding such factors as how cold does it become, is there any water in the rock when it freezes, is there sufficient moisture to effect damage, what other processes are operative and is it perhaps not these that actually cause the breakdown? The exciting thing is that the questions are still there to be answered and that many are being investigated within this country.

5. SUMMARY

A review is given of some of the advances that have been made with respect to freeze-thaw weathering, with particular emphasis on work undertaken in South Africa. Much of the recent findings result from the application of new technology. It is shown that 'freeze-thaw' is a far more complex process than may have been hitherto thought, and that despite our advances many of the original questions remain.

6. REFERENCES

Aguire-Puente, J. 1978. Present state of research on the freezing of rocks and construction materials. Proceedings of the Third International Conference on Permafrost, 1, 600-607.

Bungey, J.H. 1982. The Testing of Concrete in Structures. Chapman and Hall, New York, 207 pp.

Crook, R. and Gillespie, A.R. 1986. Weathering rates in granitic boulders measured by P-wave speeds, In: S.M. Colman and D.P. Dethier (eds), Rates of Chemical Weathering of Rocks and Minerals, Academic Press, Orlando, 395-417.

Davidson, G.P. and Nye, J.F. 1985. A photoelastic study of ice pressure in rock cracks. Cold Regions Science and Technology, 11, 141-153.

Davison, J.R. and Sereda, P.J. 1978. Measurement of linear expansion in bricks due to freezing. Journal of Testing and Evaluation, 6, 144-147.

Douglas, G.R., McGreevy, J.P. and Whalley, W.B. In press. The use of strain gauges for experimental frost weathering. Proceedings of the 1st International Geomorphological Symposium.

Fahey, B.D. 1985. Salt weathering as a mechanism of rock breakup in cold climates: an experimental approach. Zeitschrift fur Geomorphologie, 29, 99-111.

Fahey, B.D. and Gowan, R.J. 1979. Application of the sonic test to experimental freeze-thaw studies in geomorphic research. Arctic and Alpine Research, 11, 253-260.

Filonidov, A.M. 1982. Use of ultrasound for checking the frost resistance of concrete. Hydrotechnical Construction, 16, 278-283.

French, H.M. 1981. Periglacial geomorphology and Permafrost. Progress in Physical Geography, 5, 267-273.

Fukada, M. 1971. Freezing-thawing process of water in pore space of rocks. Low Temperature Science, Series A, 29, 225-229.

Gardner, J.S. 1969. Snowpatches: their influence on mountain wall temperatures and the geomorphic implications. Geografiska Annaler, 51A, 114-120.

Hall, K.J. 1975. Nivation processes at a late-lying, north-facing snowpatch site in Austre Okstindbredalen, Okstindan, northern Norway. University of Reading, unpublished M.Phil. thesis, 307 pp.

Hall, K.J. 1980. Freeze-thaw activity at a nivation site at northern Norway. Arctic and Alpine Research, 12, 183-194.

Hall, K.J. 1986a. The utilisation of the stress intensity factor (K_{IC}) in a model for rock fracture during freezing: an example from Signy Island, the Maritime Antarctic. British Anarctic Survey Bulletin, 72, 53-60.

Hall, K.J. 1986b. Rock moisture in the field and the laboratory and its relationship to mechanical weathering studies. Earth Surface Processes and Landforms, 11, 131-142.

Hall, K.J. 1986c. Freeze-thaw simulations on quartz-micaschist and their implications for weathering studies on Signy Island, Antarctica. British Antarctic Survey Bulletin, 73, 19-30.

Hall, K.J. 1987a. A laboratory simulation of rock breakdown due to freeze-thaw in a Maritime Antarctic environment. Earth Surface Processes and Landforms, 12.

Hall, K.J. 1987b. The physical properties of quartz-micaschistand their application to freeze-thaw weathering studies in the Maritime Antarctic. Earth Surface Processes and Landforms, 12, 137-149.

Hall, K.J. 1988. Weathering. In: B.P. Moon and G.F. Dardis (eds.), The Geomorphology of Southern Africa, Southern Book Co., Johannesburg (in press).

Hall,K.J. In press. The interconnection of wetting and drying with freeze-thaw: some new data. Zeitschrift für Geomorphologie.

Hall, K.J. submitted. The daily monitoring of a block of indegenous rock in the Maritime Antarctic: moisture and weathering data. British Antarctic Survey Bulletin.

Hall, K.J. In preparation a. The use of fibre optic crack detection gauges in freeze-thaw simulations. Earth Surface Processes and Landforms.

Hall, K.J. In preparation b. The utilisation of rock tablets for weathering studies in the Maritime Antarctic. British Antarctic Survey Bulletin.

Hall, K.J., Verbeek, A.A. and Meiklejohn, I.A. 1986. The extraction and analysis of solutes from rock samples and their implication for weathering studies: an example from the Maritime Antarctic. British Antarctic Survey Bulletin, 70, 79-84.

Hall, K.J. and Walton, D.W.H. In preparation. Rock Weathering in Cold Environments. Cambridge University Press, Cambridge.

Hallet, B. 1983. The breakdown of rock due to freezing: a theoretical model. Proceedings of the 4th International Conference on Permafrost, National Academy Press, Washington D.C., 433-438.

Hale, K.F. 1984. Optical fibre sensors for inspection monitoring. Physics Technology, 15, 129-135.

Hames, V., Lautridou, J.P., Jeannette, D., Ozer, A. and Pissart, A. 1987. Varietions dilatometriques de roches soumises a des cycle "humidifaction-sechage". Geomorpholo gie Physique et Quaternaire.

Hare, M.J. 1985. Conditions associated with frost action in rocks: a field and laboratory investigation. Carleton University, Unpublished M.A. thesis, 168 pp.

Hornibrook, F.B. 1939. Application of sonic method to freezing and thawing studies of concrete. American Society for the Testing of Materials Bulletin, 101, 5-8.

Ide, J.M. 1937. The velocity of sound in rocks and glasses as a function of temperature. Journal of Geology, 45, 689-716.

Ives, J.D. 1973. Arctic and Alpine geomorphology - a review of current outlook and notable gaps in knowledge, In: B.D. Fahey, and R.D. Thompson (eds.), Research in Polar and Alpine Geomorphology, Geo Abstracts, Norwich, 1-10.

Keune, R. and Hoekstra, P. 1967. Calculating the amount of unfrozen water in frozen ground from moisture characteristic curves. Cold Regions Research and Engineering Laboratory, Special Report 114.

Kinniburgh, D.G. and Miles, D.L. 1983. Extraction and chemical analysis of interstitial water from soils and rocks. Environmental Science and Technology, 17, 362-368.

Konischev, V.N. and Rogov, V.V. 1978. An experimental model of the cryogenic resistance of basic rock-forming minerals, in Problems of Cryolithology, 7, Moscow University Press, Moscow, 189-198 (in Russian).

McGreevy, J.P. 1981. Some perpectives on frost shattering. Progress in Physical Geography, 5, 56-75.

McGreevy, J.P. 1982. 'Frost and salt' weathering: further experimental results. Earth Surface Processes and Landforms, 7, 475-488.

McGreevy, J.P. and Whalley, W.B. 1982. The geomorphic significance of rock temperature variations in cold environments: a discussion. Arctic and Alpine Research, 14, 157-162.

McGreevy, J.P. and Whalley, W.B. 1984. Weathering. Progress in Physical Geography, 8, 543-569.

McGreevy, J.P. and Whalley, W.B. 1985. Rock moisture content and frost weathering under natural and experimental conditions: a comparative discussion. Arctic and Alpine Research, 17, 337-346.

Mak, D.K. 1985. Ultrasonic methods for measuring crack location, crack height and crack angle. Ultrasonics, 23, 223-226.

Matsuoka, N. 1984. Frost shattering of bedrock in the periglacial regions of the Nepal Himalaya. Seppyo, 6, 19-25. (in Japanese).

Mellor, M. 1970. Phase composition of pore water in cold rocks. Cold Regions Research and Engineering Laboratory, Report 292, 61 pp.

New, B.M. 1976. Ultrasonic wave propogation in discontinuous rock. Transport and Road Research Laboratory, Report 720, 19 pp.

Patterson, D.E. and Smith, D.W. 1981. The measurement of unfrozen water content by time domain reflectometry: results from laboratory tests. Canadian Geotechnical Journal, 1, 131-144.

Pissart, A. and Lautridou, J.P. 1984. Variations de longueur de cylindres de pierre de Caen (calcaire bathonien) sous l'effect de sethage et d'humidification. Zeitschrift fur Geomorphologie, 49, 111-116.

Press, F. 1966. Seismic Velocities. Geological Society of America Memoir, 97, 195-218.

Tharp, T.M. 1987. Conditions for crack propogation by frost wedging. Geological Society of American Bulletin, 99, 94-102.

Thorn, C.E. 1979. Bedrock freeze-thaw weathering regime in our alpine environment, Colorado Front Range. Earth Surface Processes, 4, 211-228.

Thorn, C.E. 1980. Bedrock microclimatology and the freeze-thaw cycle: a brief illustration. Annals of the Association of American Geographers.

Thorn, C.E. and Hall, K.J. 1980. Nivation: an arctic-alpine comparison and reappraisal. Journal of Glaciology, 25, 109-124.

Timur, A. 1968. Velocity of compressional waves in porous media at permafrost temperatures. Geophysics, 33, 584-595.

Trenhaile, A.S. and Mercan, D.W. 1984. Frost weathering and the saturation of coastal rocks. Earth Surface Processes and Landforms,9, 321-331.

Whalley, W.B. and McGreevy, J.P. 1985. A theoretical model of the fracture of rock during freezing. Geological Society of America Bulletin, 96, 336-346.

Walder, J.S. and Hallet, B. 1986. The physical basis of frost weathering: toward a more fundamental and unified perspective. Arctic Alpine Research, 18, 27-32.

Williams, R.B.G. and Robinson, D.A. 1981. Weathering of sandstone by the combined action of frost and salt. Earth Surface Processes and Landforms, 6, 1-9.

Zykov, Y.D., Lyakhovitskiy, F.M. and Chervinskaya, O.P., 1984. Experimental investigation of transverse isotropy in ice/clay thin-layered periodic models. Geophysical Journal of the Royal Astronomical Society, 76, 269-272.

LATE QUATERNARY ENVIRONMENTAL CHANGES IN THE KAROO, SOUTH AFRICA

M.E. MEADOWS, J.M. SUGDEN
Department of Environmental and Geographical Science,
University of Cape Town

1. INTRODUCTION

The Karoo (Fig. 1) occupies extensive semi-arid areas in southern Africa and encompasses a great diversity of climates, landform, soils and vegetation. The biome contributes considerably to the gross domestic product of South Africa through wool, mohair, goat meat and mutton production and there are important economic motives for a greater comprehension of how Karoo ecosystems operate (Cowling, 1986). Part of this understanding concerns the search for information on how the semi-arid areas have responded to the climatic and associated environmental changes which have taken place during the recent geological past. The Quaternary period, in particularly the late Pleistocene and Holocene, has been a time of great climatic dynamism in southern Africa generally (Tyson, 1986) and it is vital that a perspective is gained on the frequency, magnitude and environmental effects of this dynamism on the Karoo, so that the contemporary ecosystem may be managed more effectively. Acocks (1953) has alluded to the spread of Karoo vegetation eastwards, purportedly under the influence of inappropriate land management, but to what extent is this one of many such advances in the geological past, many of which may have occurred under the influence of mainly natural, as opposed to anthropogenic, causes?

The aim of this paper is to review the current knowledge on late Quaternary environments in the Karoo, laying particular emphasis on geomorphology, with a view to elucidating broad patterns of environmental change. Examples of multidisciplinary studies from the Nuweveldberg and Cederberg are used to illustrate the dynamism of Karoo environments and the importance of human activities in determining ecosystem characteristics. It is concluded that both sedimentological and palaeoecological studies are necessary in reconstructing late Quaternary environmental change in the Karoo.

Geomorphological Studies in Southern Africa, G.F.Dardis & B.P.Moon (eds)
© 1988 Balkema, Rotterdam. ISBN 90 6191 831 6

337

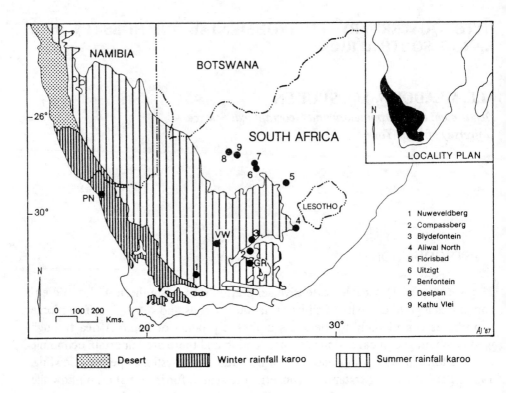

Fig. 1. The Karoo (after Cowling, 1986). PN = Port Nolloth, VW = Victoria West, GR = Graaff Reinet. Numbers 1-9 indicate sites of organic deposition.

2. EVIDENCE FOR LATE QUATERNARY ENVIRONMENTAL CHANGE IN THE KAROO

Much has been written over the past few years on environmental changes in southern Africa, culminating in the publication of two important edited volumes (Klein, 1984; Vogel, 1984) and the extensive review by Tyson (1986). For a variety of reasons, however, the Karoo region remains relatively poorly documented with respect to palaeoenvironmental studies, a situation in conflict with, firstly, the enormous geographical extent of the biome in relation to southern Africa (Rutherford and Westfall, 1986) and, secondly, the apparent sensitivity of Karoo environments to climatic changes (Meadows, 1988a).

Perhaps the most obvious problem in conducting palaeoenvironmental investigations in semi-arid environments is the scarcity of organic deposits suitable for radiocarbon dating. The result is that, while sediments containing palaeoenvironmental information may well exist throughout the Karoo, it is frequently impossible to fix these chronologically and reliance on relative dating is a somewhat unsatisfactory compromise. Furthermore, the standard palaeoecological techniques, especially of pollen analysis, are often difficult to apply where depositional environments are subject to lengthy periods of dessication and where oxidation of potentially fossil-forming organic material occurs. For this reason, rather few palynological studies have been successful in the Karoo region. For geomorphological evidence, the situation seems a little more promising; the semi-arid environment is one in which the "imprint" of past climates often remains visible for relatively long time-spans since, following climatic change, geomorphological processes often act so slowly as to preserve previous landforms and sediments. This is not always the case, however, for geomorphological events of low frequency but high magnitude occur in such environments, and may occasionally wipe clean the geomorphological "memory" in depositional situations.

2.1 Contemporary geomorphology and hydrology of the Karoo:a word of caution

Climatically and geologically the Karoo is complex, and this renders contemporary geomorphological conditions highly variable. Nonetheless, several pertinent generalisations can be made in order to capture the essential characteristics of today's Karoo environments. Climate is notable for extremes in both temperature and variability in amount and timing of rainfall. The climatic parameter most significant in terms of geomorphology is undoubtedly the low precipitation values, amounts which are often highly seasonal in distribution. As little as 61 mm falls at Port Nolloth in the winter rainfall Karoo in the west and increases eastwards through 254 mm at Victoria West and 346 mm at Graaff Reinet in the summer rainfall area (Cowling, 1986). Physiographically, lower-lying areas lie to the south of the Great Escarpment and include the Great Karoo, Little Karoo and Robertson Karoo. North of the escarpment, which incorporates the Roggeveldberg, Nuweveldberg and Sneeuberg Mountains, the extensive peneplain of the Upper Karoo is found with its vast undulating plains interrupted by occasional residual mountain ranges. Much of the geology of the biome is dominated by sedimentary rocks of the Karoo sequence, although Kalahari sands occur in the north and Cape Supergroup parent materials outcrop in the south (Cowling, 1986).

The combinations of complex climates and geology "... dictate a rather tenuous hydrological equilibrium in Karoo catchments..." (Gorgens and Hughes, 1986, p.53), although as with other semi-arid systems, the runoff component is relatively low in relation to rainfall. Gorgens and Hughes (1982) cite the proportion of runoff as a percentage of rainfall input in the Karoo to be as low as 4.75 %, which compares with the humid east of southern Africa where values may be as high as 30 % (Braune and Wessels, 1980). These mean runoff values, however, conceal a marked variability and the coefficient of variability should also be considered.

Gorgens and Hughes (1982) place this at 1.14, about double that of humid South Africa. Rainfall only in part accounts for this high coefficient of runoff variability and it must, therefore, be assumed that other factors, principally land use, play important roles in runoff generation. Sediment loads in the catchment systems are also somewhat capricious. Periods of drought encourage greater periods of chemical and mechanical weathering and reduced vegetation cover which, if followed by a flood event, produces high sediment loads. Alternatively, periods of above average precipitation with consequent improved vegetation cover retards sediment removal (Gorgens and Hughes, 1986).

It may be concluded that the geomorphology and hydrology of the Karoo is dominated by a variable rainfall regime acting on a catchment surface whose vegetation cover is determined both by variable rainfall and by land-use practices. This leads to complex shifts in the erosional-depositional balance in the catchments and makes the interpretation of potential palaeoenvironmental information embodied in accumulated sediments and soils extremely hazardous. Notwithstanding this word of caution, an attempt is made below to interpret the dated geomorphological evidence for the Karoo biome as a means of interpreting late Quaternary environments.

2.2 Late Quaternary geomorphological change

Much of the evidence for geomorphological changes in the Karoo during the late Pleistocene and Holocene is in the form of fluvial, alluvial and colluvial sediments; the problems of interpreting such evidence are manifold, and are reviewed by Meadows (1988a). Some information has also been gleaned from dunes, pans and lacustrine sediments, geoarchaeological material including cave deposits, palaeosols and duricrusts (Meadows, 1988a). As mentioned above, the overwhelming problem lies in the difficulty of chronological accuracy owing to the scarcity of carbon-rich remains in sediments. However, Butzer (1984a, 1984b) has made enormous advances in the interpretation of fluvial, colluvial and alluvial sediments, and many of his sites are within the Karoo.

As expected, the indicated geomorphological changes are frequent and complex, a function of the overall sensitivity of Karoo environments to climatic inputs, although a broadly consistent pattern does emerge on closer inspection. Perhaps the most striking aspect of the environmental history of the Karoo is the rapidity of change, especially in terms of effective moisture conditions and this has led Butzer (1984b, p.259) to comment on the "... sometimes small shifts in temporal definition, sometimes difference in amplitude, and sometimes radical differences in the direction of change...".

Little well-dated evidence exists for the period prior to about 16000 years B.P., an unfortunate situation since the period around 18000 years B.P. represents the last glacial maximum, at least in the northern hemisphere, and environmental conditions probably reached an extreme at this time. Such evidence as exists reflects markedly cooler temperatures and somewhat less available moisture (Tyson, 1986). The period subsequent to this is better documented and indications are that, while temperatures may have shown little amelioration, precipitation increased, and indeed produced a substantive lake in the pan of Alexandersfontein by 16000 years B.P. and improved spring discharge in other areas of the Karoo (Butzer, 1984a). Indeed, at Alexandersfontein (Serfontein and Uitzigt) mean annual precipitation between 16000 years B.P. and 13600 years B.P. may have been more than double the 400 mm which is received there today (Butzer, 1984a). Environments from about 14000 years B.P. into the early Holocene are characterized by major oscillations in available moisture, although temperatures seem to have remained relatively cool. A strong moisture signal re-emerges in the latest part of the Pleistocene, and the lake becomes re-established at Alexandersfontein around 11500 years B.P. (Butzer, 1984a). Elsewhere, the later Pleistocene (which may in fact encompass the last temperate glacial maximum in poorly-dated sites) seems to have been a period of cooler and more xeric climates, certainly at Kathu vlei and at Rose Cottage and Wonderwerk Cave sites (Butzer, 1984a) and at Blydefontein (Bousman et al, 1988). In general glacial climates seem to have been more xeric than non-glacial, as appears to be the case for much of the rest of Africa (Street, 1981), while Butzer (1984a, p.63) continues to be sceptical of attempts to define a glacial mode of climate anywhere in Africa.

The last 10 000 years of environmental history in the Karoo produces slightly less controversy, although the picture is by no means unequivocal. Almost everywhere in southern Africa, the Holocene is characterised by a fairly rapid rise in mean annual temperature, the hypsithermal occurring some time towards the mid-Holocene. Precipitation values across the interior do appear to show signs of increases early in the period; at Alexandersfontein, for example, a lake occupied the pan until around 6000 years B.P. (Butzer, 1984a) and at Voightspos, lake waters occupied a pan between 6900 years B.P. and 6300 years B.P. (Butzer, 1984a). But the strongest environmental signal emerging is that the latter part of the Holocene was even moister than the earlier. The archaeological invisibility of Later Stone Age human populations across much of the interior from 8000 years B.P. to 4000 years B.P., as evidenced by an occupational hiatus in many cave sites may, perhaps, offer supporting evidence for a somewhat drier first half to the Holocene, although the geomorphological and geoarchaeological evidence is free-standing in this respect.

In general, the last 5000 years in the Karoo was characterised by conditions both warmer and moister than the previous 5000 years. Lake sedimentation commenced shortly before 5000 years B.P. at both Voightspos and Deelpan and conditions at Kathu Vlei became more humid after 7500 years B.P., remaining so well into the late Holocene (Butzer, 1984a). At Alexanderfonteins, the lake resumed occupation of the pan from about 4500 years B.P. onwards (Butzer, 1984a). At Blydefontein, Bousman et al. (1988) document greater moisute availability during the later Holocene and between about 5000 years B.P. and 1000 years B.P. conditions are viewed as being wetter than today.

2.3 Radiocarbon Chronologies of Organic Sediments in Vlei, Spring and Pan Sites in the Karoo

Meadows (1988b) has tabulated the radiocarbon dates on organic sediments for the late Quaternary of southern Africa. On the general principle that organic sediments either commence accumulation, or at least accelerate in rate of deposition, when there is a shift to moister climate, he argues for the recognition of at least two major periods of greater effective moisture during, firstly, the later Pleistocene and, secondly, the latter half of the Holocene. When the data which relate specifically to the Karoo are separated (Table 1) and plotted as a frequency histogram of dates per 1000 year time period since 15000 years B.P. (Fig. 2), this pattern appears to be upheld. From the radiocarbon chronologies of nine sites within and immediately adjacent to the Karoo, there are strong indications of increased moisture from around 5000 years ago onwards. This is in broad agreement with the environmental interpretation as sketched out for the arid interior of southern Africa by Butzer (1984a), in which the second half of the Holocene is viewed as being somewhat wetter than the first and which is supported by archaeological evidence (Deacon, 1974; Deacon and Thackeray, 1984).

The radiocarbon data on organic facies for the later Pleistocene in the Karoo is relatively scarce by comparison with the rest of southern Africa (see Meadows, 1988b). This may reflect one of two contrasting scenarios, viz: either conditions were insufficiently moist to promote organic accumulation in the Karoo, or such sediments as were initiated were subsequently dessicated and eroded during the more xeric early Holocene which followed. A third possibility may be mooted: that precipitation conditions were fluctuating very markedly during the late Pleistocene and organic sedimentation was only sporadically encouraged. Whichever of these hypotheses is accepted will depend on the availability of further dates on sites of organic accumulation, although the balance of the evidence thus far favours the idea of marked fluctuations in precipitation occurring over relatively short time spans.

2.4 Late Quaternary environments of the Karoo: Patterns of change

Taken as a whole, the evidence for late Quaternary environmental change in the Karoo is beginning to suggest a broad pattern of change. Notwithstanding the difficulty of interpreting so few sedimentary sequences based on a variety of geomorphological situations, the following tentative climatic reconstruction, illustrated in Fig. 3, is suggested.

342

Table 1. Radiocarbon dates on organic sediments in the Karoo (Source: Meadows, 1988b).

SITE	RADIOCARBON DATE	LABORATORY NUMBER
1. Nuweveldberg	760 +- 50 BP	Pta-4351
2. Compassberg	3590 +- 70 BP	Pta-4342
3. Blydefontein	7790 +- 90 BP	Pta-4461
	5080 +- 70 BP	Pta-4273
	4430 +- 70 BP	Pta-4273
	4260 +- 60 BP	Pta-4458
	4010 +- 60 BP	Pta-4392
	3290 +- 60 BP	Pta-4390
	2000 +- 60 BP	Pta-4465
	1360 +- 100 BP	Pta-4417
	290 +- 40 BP	Pta-4259
4. Aliwal North	12600 +- 110 BP	GrN-4011
	4320 +- 110 BP	I-2108
5. Florisbad	5530 +- 110 BP	Pta-1128
6. Uitzigt	4075 +- 300 BP	SI-2232
	3450 +- 60 BP	SI-2049
	2220 +- 265 BP	SI-2231
	1880 +- 110 BP	SI-2228
	1555 +- 240 BP	SI-2233
7. Benfontein	4900 +- 125 BP	SI-2548
	1 3950 +- 110 BP	SI-2583
	3450 +- 60 BP	SI-2049
8. Deelpan	3890 +- 90 BP	Pta-3868
	2950 +- 80 BP	Pta-4183
9. Kathu vlei	7350 +- 90 BP	Pta-3073

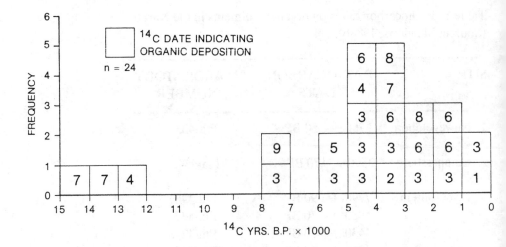

Fig. 2. Frequency histogram of radiocarbon dates on phases of organic deposition in and around the Karoo.

2.4.1 Last Glacial Maximum (Isotope Stage 2), c.18000 years B.P.

What little dated evidence as exists indicates very much cooler conditions about this time and, although much confusion seems to have surrounded the quantities of precipitation, the body of opinion appears to rest with a drier last glacial maximum in the Karoo. Even at sites outside the contemporary Karoo as defined by Rutherford and Westfall (1986), such as Kathu Vlei (27°40'S;23° 01'E), Wonderwork Cave (27°50'45"S, 23°33'19"E) and Rose Cottage Cave (29°12'5"S, 27°28"19"E) there are consistent indications of aridity (Butzer, 1984a). The late Pleistocene at Kathu Vlei was relatively dry, there was greater seasonality of rainfall and reduced vegetation cover in the vicinity of Wonderwork during this period, and at Rose Cottage Cave there was no spring influx during the last glacial maximum (Butzer, 1984a).

2.4.2 The Late Glacial (Isotope Stage 2/1), 15000 to 10000 years B.P.

For the earlier part of this period, the evidence for the Karoo unequivocally suggests moister conditions than the present day, with the lake at Alexandersfontein at its deepest and most extensive. For the later Pleistocene, however, some indications of fluctuating precipitation values are encountered and, as temperatures rose widely in the sub-continent, oscillations occurred in Alexandersfontein lake. Perhaps climatic conditions at this time were exceptionally dynamic, as globally the

344

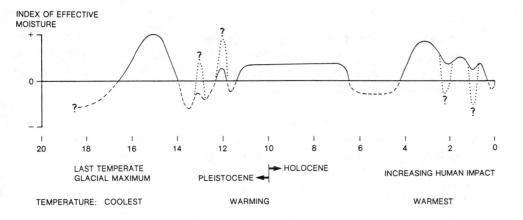

Fig. 3. Climatic reconstruction of the last 20000 years in the Karoo. The time scale is given in radiocarbon years B.P. x 1000.

transition was under way from a generally glacial to inter-glacial mode of climate. The onset of organic accumulation in vleis and springs both within and around the Karoo (Meadows, 1988b) certainly supports the idea of greater effective moisture, albeit somewhat intermittently.

2.4.3 The Holocene (Isotope Stage 1), 10000 years B.P. to the Present Day

Temperature amelioration is indicated throughout the sub-continent and the Karoo is no exception. The picture concerning moisture conditions is still uncertain for the earlier Holocene but rapidly becoming coherent for the most recent 5000 years of environmental history. Few sites of organic deposition demonstrate the onset of sedimentation for the first half of the Holocene and the archaeological hiatus in the interior around this time (Deacon, 1974) perhaps supports the concept of a drier period. The lake at Alexandersfontein was, however, maintained throughout this period and, pitched against a backdrop of rising temperatures, it becomes difficult to escape the conclusion that conditions, until around 6500 years B.P. at least, were a little moister than in the Karoo of today. During the second half of the Holocene, however, the evidence everywhere in the Karoo supports the idea of increased available moisture and a more effective vegetation cover. Towards the present day, marked oscillations again occur in the level of the lake at Alexandersfontein and, coupled with the cyclical sedimentation patterns recorded at Blydefontein, it must be concluded that these generally moister conditions were sporadically interrupted by periods of greater erosion and more marked climatic seasonality. The possibility that increasing human activity was at least in part responsible for this, especially during the past few hundred years, is also worth con-

345

sidering.

Having outlined the apparent larger-scale changes in the Karoo environment, the following sections examine detailed changes in the characteristics of ecosystems in, firstly the Sneeuberg Mountains north of Graaff Reinett and, secondly, the Nuweveldberg north of Beaufort West.

3. MID-LATE HOLOCENE ENVIRONMENTS AT COMPASSBERG, SNEEUBERG MOUNTAINS, EASTERN CAPE

3.1 Environment

Bousman et al.(1988) have described valley fills in the Blydefontein Basin near Noupoort in the east-central Cape. Here, a valley-in-valley configuration has been produced by incision into a series of valley fills, in which sediments up to 8 m thick have accumulated during the later Quaternary. The situation some 75 km to the southwest, in the vicinity of the highest peak in the Sneeuberg range, Compassberg (2502 m), appears to be identical. In the headwaters of the Klein-Seekoerivier, a vlei some 300 m wide has been incised by a donga, thereby exposing some 3 m of alluvial sediments. The surrounding vegetation is Karroid Merxmuellera mountain veld (Acocks, 1953) and the mean annual precipitation around 600 mm with a summer maximum.

3.2 Sediments and relation of Blydefontein palaeosols

The sequence of alluvial sediments is most interesting for the fact that, between about 30 cm and 85 cm below the surface, an organic soil is evident. The basal material, containing around 10-15 % of finely divided organic detritus, rests on a less organic sandy alluvium and grades upwards into a fibrous, more carbon-rich horizon. Similar palaeosols are described by Bousman et al. (1988), although at Blydefontein a more detailed geomorphological survey of sections has revealed a series of such palaeosols which are formed in their "younger" fills which are of later Holocene age. The age of the Compassberg palaeosol is fixed by a radiocarbon date on the basal organic material (at 90 cm) of 3590 years B.P., remarkably akin to the age of one of the palaeosols at Blydefontein, which is accorded an age of 3290 + - 60 years B.P. Bousman et al. (1988) argue that palaeosol formation in this environmental context would represent of the order of a few hundred years, and although no date for the upper horizons is available here, it may be assumed that the vegetation history revealed by the palaeosol represents a few hundred years post-3500 years B.P.

Absolute pollen counts were made on samples extracted from several depths both within the palaeosol and below it and prepared according to standard procedures (Faegri and Iversen, 1974). From the pollen diagram (Fig. 4) it may be seen pollen concentrations are much lower in the sandy alluvium, around 2500 grains per cc, and are dominated by the high percentages of both grass and sedge pollen. Conditions for pollen preservation are not ideal and the possibility of differential preservation cannot be overlooked, but the relatively low values of Asteraceae pollen below 200 cm and the peak in Cyperaceae pollen at 205 cm are strongly indicative of conditions somewhat moister than today in the catchment area. The onset of organic sedimentation at 3590 years B.P. is paralleled by a sharp increase in pollen concentrations to over 7000 grains per cc at 80 cm. Poaceae pollen remain high and, although sedges contribute a consistently significant proportion of the spectrum, higher counts of pollen from the asteraceous shrubs are indicative of conditions not dissimilar to the Karroid *Merxumellera* mountain veld found at Compassberg today. Some interesting changes occur within the period of palaeosol formation, and the shallower horizons are notable for decreases in grass pollen and a further increase in the pollen of karroid shrubs, especially *Elytropappus*. The general impression gained is that of a consistent decrease in effective moisture from the base of the palaeosol, from 3590 years B.P. onwards.

This interpretation is entirely consistent with the pollen analysis of identical deposits at Blydefontein (Bousman et al., 1988) in which gradual drying of a floodplain pool is indicated upwards through the organic horizons.

The geomorphological model put forward by Bousman et al. (1988) to explain the sequence of events is as follows: accelerated erosion under enhanced stream competence initiates the cycle, followed by the deposition of alluvium, pond deposits and, finally, pedogenesis (representing drier climate) before the commencement of the second cycle and return to moister conditions. An alternative interpretation is possible, however, especially if the accelerated erosion commencing each cycle is seen as representing drier, not moister, conditions with greater seasonality of precipitation and a more unstable hydrological regime under the influence of reduced vegetation cover in the catchment. Meadows (1988c) has argued that a shift to drier conditions in semi-arid climates may lead to the removal of sediment from vleis (i.e. a 'cut' rather than 'fill' phase). In such a scenario, the organic horizons of the palaeosol would represent the terminal phase of a wetter period immediately prior to downcutting brought about by a return to more xeric conditions: the progressive dessication evident through the organic layers at Compassberg is consistent with such an interpretation. Quite clearly, more of these deposits require investigation for, however they may be explained, they definitely represent valuable palaeoenvironmental tools. As Bousman et al. (1988) have noted, the local sedimentary system is sensitive and, especially in headwater vleis, may respond to quite subtle shifts in the hydrological characteristics of the basin in

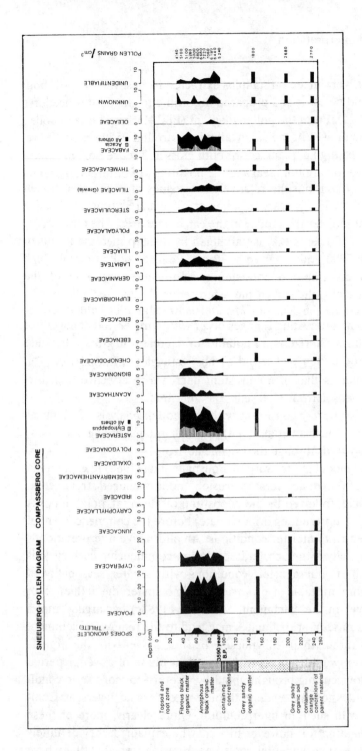

Fig. 4. Pollen diagram for Compassberg.

terms of precipitation inputs, vegetation cover and runoff. Unravelling these relationships should form a major goal of future investigations in the region.

4. RECENT ENVIRONMENTAL CHANGES IN THE NUWEVELDBERG, CENTRAL CAPE

4.1 Environment

At Bokkraal vlei high up (1820 m) on the upper plateau surface of the Nuweveldberg, a sequence of some 130 cm of organic deposits has accumulated which spans approximately the past 750 years. Again Maxmuellera mountainveld, replaced by Karoo (Acocks, 1953), dominates the landscape, although the karroid and more xeric communities of the middle and lower plateau may be expected to contribute some pollen to the fall-out over the vlei and this makes Bokkraal an ideal site to investigate vegetation change. In contrast to the complex stratigraphy of the vlei at Compassberg and of other similarly-situated vleis in southern Africa (see Meadows and Meadows, 1988) the accumulation of sediments here seems to have been relatively simple during the period involved. Radiocarbon dates are still awaited from various depths in the sequence, and for the present, it can only be assumed that accumulation rates have remained approximately constant throughout the core. At any rate, there are no signs of marked geomorphological and hydrological changes within the sedimentary record other than the onset of organic deposition some 750 years ago.

4.2 Pollen diagram and interpretation

Methods of collection and preparation are as for Compassberg and the pollen diagram is given as Fig. 5. The overall frequency of taxa within the local environment does not show much change, which supports the stratigraphical conclusions of a relatively moist vlei environment throughout deposition. Within the surrounding vegetation, however, some interesting changes are evident and a moister phase soon after the onset of sedimentation is indicated by higher quantities of the grasses, Poaceae, the tree *Cliffortia* and Caryophyllaceae. Somewhat drier conditions prevail after this and there are higher Asteraceae, Bignonaceae and Chenopodiaceae frequencies, possibly related in part to the arrival of Khoi pastoralists in the area (Sugden and Meadows, in prep.). Towards the surface, some very unusual community changes occur, including the simultaneous rise in Poaceae, *Stoebe*, *Elytropappus* and Chanopodiaceae. The rise in grass pollen at this time seems somewhat paradoxical, especially as conventional wisdom purports a decline in veld quality (associated, presumably, with a decrease in grassiness). However, application of an advanced statistical technique in the form of multiple dis-

349

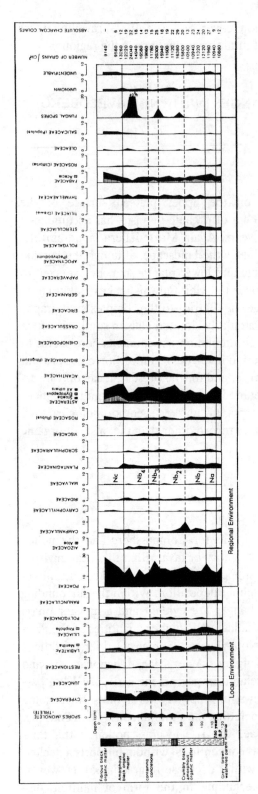

Fig. 5. Pollen diagram for Nuweveldberg.

350

criminant analysis (Sugden and Meadows, in prep.) coupled with a detailed contemporary pollen rain analysis, reveals that the recent pollen spectra are indicative of a very disturbed environment and that the rise in grass pollen may be attributed to increased flowering frequency of *Merxmuellera* spp., which are non-palatable to domestic stock and have probably flourished following local extinction of wild herbivores. The arrival of Trekboers and the introduction of sedentary farming activities may well correlate with the period of disturbance as documented in zone Na of the pollen diagram.

5. CONCLUSIONS

Both Compassberg and Nuweveldberg analyses show detailed conditions during time slices of the later Holocene. In the case of Compassberg, conditions for the few hundred years following 3500 years B.P. remain broadly constant and, although probably moister than the contemporary Sneeuberg, show some signs of being more xeric towards the end of organic deposition. The geomorphological thresholds were presumably then exceeded and at least one new cycle of erosion and sedimentation proceeded. On the Nuweveldberg we witness the latest phase of organic sedimentation in the vlei, during which time there is once again evidence for relative stability in the vlei environment. Here, however, there are strong signs of human activity and disturbance, firstly by the Khoi herders and secondly by Trekboer farmers. In both cases the disturbance was sufficient to affect vegetation, but in neither instance were the geomorphological and hydrological thresholds exceeded.

Late Quaternary environmental changes in the Karoo have been complex and the tentative pattern which seems to be emerging will require modification as further evidence accumulates. The last glacial maximum was definitely cooler and probably drier across much of the semi-arid interior, while the late glacial shows a number of major fluctuations in precipitation. The early Holocene was more xeric than the last 5000 years, although even then the geomorphological equilibrium has been disturbed a number of times, resulting in several sequences of erosion and deposition. Human activity appears not, on the basis of the evidence presented, to have effected significant geomorphological change.

Analysis of the changes in Karoo environments requires careful consideration of the relationships between rainfall, runoff, vegetation cover, sediment yield and anthropogenic activities. The apparent sensitivity of the Karoo to land-use management, which led Acocks to speculate about the relentless eastwards march of desert and Karoo must be viewed, therefore, against a backdrop of repeated oscillations in the environment, brought about by natural forces. Human activity is but one influence in the complex and inter-related dynamic system which is the contemporary Karoo of southern Africa.

6. ACKNOWLEDGEMENTS

This research was carried out with the aid of a CSIR/FRD Karoo Biome Research Grant. Cartography is by A. Vinnicombe.

7. REFERENCES

Acocks, J.P.H. 1953. Veld types of South Africa. Memoir of the Botanical Survey of South Africa, 28.

Bousman, C.B., Partridge, T.C., Scott, L., Metcalfe, S.E., Vogel, J.C., Seaman, M. and Brink, J.S. 1988. Palaeoenvironmental implications of late Pleistocene and Holocene valley fills in Blydefontein Basin, Noupoort, C.P., South Africa. Palaeocology of Africa, 19 (in press).

Braune, e. and Wessels, H.P.P. 1980. Effects of land use on runoff from catchment yield of present and future shortage. Paper delivered at the Workshop on the Effects of Rural Land Use and Catchment Management on Water Resources, CSIR, Pretoria.

Butzer, K.W. 1984a. Archaeology and Quaternary environments in the interior of southern Africa. In: Klein, R.G. (ed.), Southern African Prehistory and Palaeoenvironments. Balkema, Rotterdam, 1-64.

Butzer, K.W. 1984b. Late Quaternary environments in South Africa. In: Vogel, J.C. (ed.), Late Cainozoic Palaeoclimates of the Southern Hemisphere, Balkema, Rotterdam, 235-263.

Cowling, R.M. 1986. A description of the Karoo Biome Project. South African national Scientific Progress Report, 122, 43pp.

Deacon, H.J. and Thackeray, J.F. 1984. Late Pleistocene environmental changes and implications for the archaeological record in southern Africa. In: Vogel, J.C. (ed.), Late Cainozoic Palaeoclimates of the Southern Hemisphere, Balkema, Rotterdam, 375-390.

Deacon, J. 1974. Patterning in the radiocarbon dates for the Wilton/Smithfield complex in southern Africa. South African Archaeology Bulletin, 29, 3-18.

Faegri, K. and Iversen, J. 1975. Textbook of Pollen Analysis, 3rd Ed., Blackwell, Oxford.

Gorgens, A.H.M. and Hughes, D.A. 1986. Hydrology. In: Cowling, R.M., Rowe, P.W. and Pieterse, A.J.H. (eds.), The Karoo Biome: a preliminary synthesis. Part 1 - physical environment. South African National Scientific Programmes Report, 124, 53-83.

Gorgens, A.H.M. and Hughes, D.A. 1982. A synthesis of streamflow information relating to the semi-arid Karoo biome of South Africa. South African Journal of Science, 78, 58-68.

Klein, R.G. (ed.). 1984. Southern African Prehistory and Palaeoenvironments. Balkema, Rotterdam, 404pp.

Meadows, M.E. 1988a. Landforms and Quaternary climatic change. In: Moon, B.P. and Dardis, G.F. (eds.), The Geomorphology of Southern Africa, Southern Book Co. Johannesburg (in press).

Meadows, M.E. 1988b. Late Quaternary peat accumulation in southern Africa. Catena (in press).

Meadows, M.E. 1988c. Vlei sediments and sedimentology: a tool in the reconstruction of palaeoenvironments in southern Africa. Palaeoecology of Africa, 19 (in press).

Meadows, M.E. and Meadows, K.F. 1988. Late Quaternary vegetation history of the Winterberg, E. Cape. South African Journal of Science, (in press).

Rutherford, M.C. and Westfall, R. 1986. Southern African biomes. Memoirs of the Botanical Survey of South Africa.

Street, F.A. 1981. Tropical palaeoenvironments. Progress in Physical Geography, 5, 157-185.

Sugden, J.M. and Meadows, M.E. in prep. The use of multiple discriminant analysis in reconstructing vegetation changes on the Nuweveldberg, South Africa.

Tyson, P.D. 1986. Climatic Change and Variability in Southern Africa. Cape Town, Oxford University Press, 220pp.

Vogel, J.C. (ed.). 1984. Late Cainozoic Palaeoclimates of the Southern Hemisphere. Balkema, Rotterdam.

THE SEDIMENTOLOGY OF VLEIS OCCURRING IN THE WINTERBERG RANGE, CAPE PROVINCE, SOUTH AFRICA

FELICITY J. DEWEY

South African Sugar Association Experiment Station, Mount Edgecombe

1. INTRODUCTION

The fact that major environmental fluctuations occurred throughout southern Africa during the Quaternary period is no longer in dispute, and it is accepted that the physical environment has been significantly affected by climatic oscillations. Although the nature and extent of Late Quaternary and Holocene environmental change remains unclear for most of the subcontinent, a great deal of evidence has gradually been accumulated from a variety of palaeoecological sources. This evidence, although complicated and sometimes conflicting, is presented with the intention of producing a clear pattern of past climatic phases and improving the understanding of contemporary ecosystems.

The evidence for environmental change examined in this paper is derived from the radiocarbon-dated stratigraphy of accumulated organic and clastic sediments which occur in geomorphological features such as vleis. Their stratigraphy is taken to reflect changes in the environment affecting vlei development. Accumulation of such sediments appears to have commenced in the Late Pleistocene, spanning the whole of the Holocene (from c.12000 years B.P.), and could thus be available in the identification of major climatic fluctuations within a region, at least.

2. CONTEMPORARY VLEI ENVIRONMENTS FOR SEDIMENT ACCUMULATION

Vlei sediments are generally peat-like in appearance and for their formation require cool, moist climates where decomposition processes are limited by waterlogging, acidity and a relatively narrow range of climatic and hydrological conditions (Millington, 1985; Moore amd Webb, 1978). For this reason, suitable sites for palynological and sedimentological studies in South Africa are somewhat limited by the generally dry climate (Meadows, in press; Scott, 1982).

Geomorphological Studies in Southern Africa, G.F.Dardis & B.P.Moon (eds)
© 1988 Balkema, Rotterdam. ISBN 90 6191 831 6

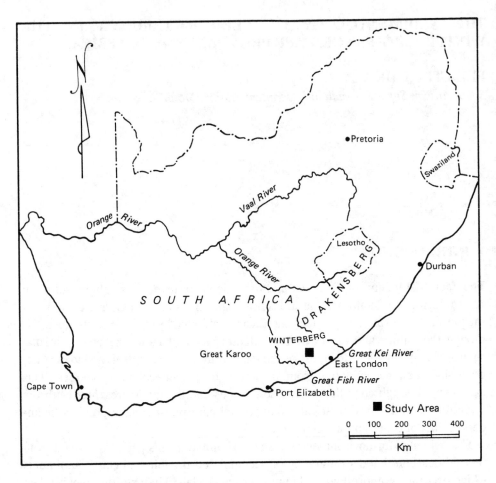

Fig. 1. Study site in Winterberg range.

Sites in southern Africa where suitably deep organic sediments exist, some having been accurately dated, are the dambos of Malawi (Meadows,1985), Zimbabwe (Tomlinson, 1973; Whitlow, 1985), Zambia (Mackel, 1974), and South Africa (Meadows et al., 1987; Meadows and Sugden, 1988). The type site for these linear, waterlogged, treeless depressions is thought to be southern Central Africa, where the term 'dambo' is most commonly used (Mackel, 1974; Meadows, 1985). Within this study, the term 'vlei' will be substituted and considered equivalent to 'dambo' when discussing the above geomorphological features in a South African context, although the term 'vlei' has a much less restrictive usage locally.

3. VLEI ENVIRONMENTS OF THE WINTERBERG

Vleis on the Winterberg plateau in the eastern Cape (Fig. 1) occur at high altitudes (over 1500 metres above sea-level) on gently rolling topography, where temperatures are cooler but precipitation is higher (the mean annual precipitation [MAP] exceeds 900 mm) than that of lower-lying slopes. Although gentle rains occur, frequent convectional storms provide the greatest sediment and fluvial input to vlei systems where impervious and geological outcrops impede drainage and promote waterlogging. The Winterberg vleis, although of lower spatial density, are more typical of the high-lying dambos of Malawi (Meadows, 1985) and Zimbabwe (Whitlow, 1985). They are linear and club-headed in shape and contain sediments several metres deep. They are surrounded by macchia vegetation and occupy fairly steep-sided valleys. Lower-lying vleis are found under a considerably lower rainfall regime in Zambia (Mackel, 1974) and the Karoo regions of South Africa (Meadows and Sugden, 1988) but differ markedly in morphology.

Continuous cores of sediment can be extracted from vleis, analysed for stratigraphic and palynological changes, and accurately radiocarbon dated to provide a chronological history of the period of their deposition.

4. WINTERBERG VEGETATION PATTERNS

A clear transition can be distinguished in vegetation on the slopes of the Winterberg, the drier foothills bearing false thornveld changing to temperate evergreen forests on higher slopes and protected valley heads, and finally giving way to macchia/grassland vegetation on colder exposed peaks. Within this hardy vegetation type, the Winterberg vleis occur along natural drainage lines. These exposed plateau areas offer no protection and soils are too shallow for woodland vegetation, thus larger species are not found in the vicinity of vleis. As vegetation boundaries are dynamic and in flux with climatic conditions, the significance of these potentially changing boundaries and vegetation types is important in determining the availability of sediment.

On the vlei sites existing on the Elandsberg plateau of the Winterberg range, a site on Dunedin farm was selected as the principal site because of its extensive organic deposits, accessibility, and relatively undisturbed state. This vlei remains waterlogged all year round due to an impeding dolerite dyke, and gives rise to mainly grasses and sedge which become more dense towards the centre, following the increasing moisture gradient. These hydromorphic species support a highly organic surface layer, deposition being more rapid in moist, warm periods, and slower in periods of increased aridity and lower temperatures. As such, organic sediment layers are possibly indicative of relatively moist environmental conditions. *Aristida* spp. and *Monocymbium* spp. occupy the sandier, better drained margins, while the central zone has various *Scleria* spp., *Scirpus* spp. and *Carex* spp. interspersed with patches of taller *Phragmites* spp. These dense hydromorphic stands play a vital role in sedimentation processes by filtering sediment-laden water and in providing increments of organic matter which are important for pollen preservation and radiocarbon dating.

5. SEDIMENT SOURCES AND DEPOSITIONAL PROCESSES

Winterberg vlei sediments are derived from weathered doleritic parent material entrained by overland flow, this process being greatly accelerated by the removal of vegetative cover. In depositional processes, larger particles require greater energy to be entrained because of their mass and so too do very fine particles due to their inherent cohesiveness (Morisawa, 1968). Once entrained, however, fine particles can be transported at lower velocities, and are thus often carried further into the vlei system.

The entrained particles are deposited when the current is no longer able to carry them. This loss of transporting ability is caused by several factors including decreased gradient and velocity when water flows into a wide basin filled with dense vegetation. Particles are deposited selectively and this sorting process depends primarily on grain size and density as described by Stokes' Law and Hjulstrom's deposition curves (Morgan, 1980). Definite textural zones exist in dambos as described by various researchers (Mackel, 1974; Whitlow, 1985). Changes in textural classes of sediment occur with changes in depth and away from dambo margins, coarser particles being most common along the margins and in headwater zones, while finer particles are found more often towards the centre of the water-body. These trends can be explained by the mechanics of streamflow (Morisawa, 1968; Weaver, 1978).

As inorganic sediments are derived from both outside (eroded slopes, stream input) and within vleis (decomposing parent material), they are both autochthonous and allochthonous in nature. Most organic deposits are, however, limited to in situ formation, organic matter growing and accumulating under suitable conditions. Accumulation ceases only when conditions are no longer suitable for plant growth, reflected in the stratigraphy by a reduction in the number of organic bands (Barber, 1981).

6. SEDIMENT STRATIGRAPHY AT DUNEDIN

6.1. Procedure

Fifteen undisturbed cores, each 3 metres in length, were extracted by means of a

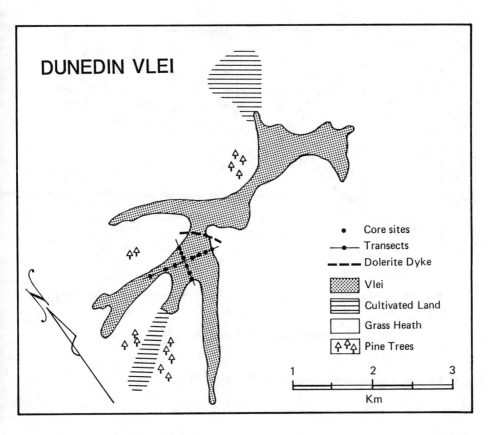

Fig. 2. Dunedin Vlei morphology.

Gouge auger used at selected intervals (core sites) along a series of transects across the vlei (Fig. 2). Each core was then subsampled and submitted for a series of chemical and physical laboratory analyses.

6.2. Results and discussion

Due to waterlogging throughout the year, Dunedin sediments are hydromorphic gleyed clays which are high in organic matter. A longitudinal transect (Fig. 3) indicated a clear textural pattern within the vlei, changing from a sandy loam at the edges and headwaters to a peat-like deposit of silt and clay in the vlei centre. Here clays were accumulated and mixed with organic debris. This sediment transition is related to the mechanics of sediment transport by water moving into the vlei (Mackel, 1974). Organic matter increased towards the centre due to a denser plant population occurring there, and there was a corresponding decrease in pH. The acidic environment was caused by high levels of exchangeable hydrogen released

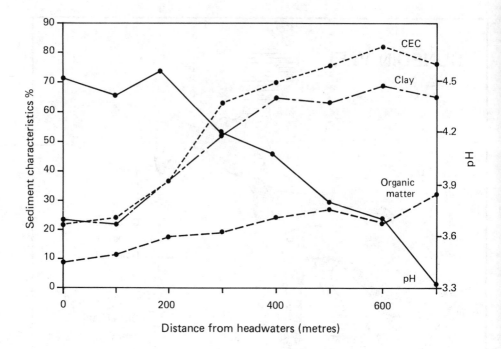

Fig. 3. Variation of surface sediments of Dunedin Vlei.

during plant decomposition and low base saturation. The variation of sediments away from the headwater zone and grain size distribution throughout the transect was indicative not only of the depositional environment, but also of erosion processes at the site.

Marked changes in stratigraphy occurred at different depths at the core sites (Fig. 4). A fibrous organic layer (0 to 50 cm), highly acidic (pH = 4) and of low bulk density occurs on the surface overlying a layer of amorphous black loam (50 to 100 cm) of lower organic content. This layer consists mainly of medium and fine grained sands with a lower clay content. Beneath this sandy layer is a black, highly organic peat-like layer (150 to 220 cm) with a very high clay content which rests on beds of inert grey sands (220 to 330 cm).

The peat-like layers, being high in clay and organic matter contents, have high cation exchange capacities (mean CEC = 80 meq/100 g). The CEC was, however, not as high as those of true peats, and these sediments are therefore classified as organic deposits. The positive linear relationship between organic matter and CEC

(r = 0.79) appeared to be well-correlated with depth, as was the relationship between organic matter and clay content (r = 0.88).

Subsequently, the organic layer at 150 cm was radiocarbon dated at around 8000 years BP, and from the characteristics of this layer, conditions at this time were warmer and more moist than at present, with sediment inputs largely restricted to

360

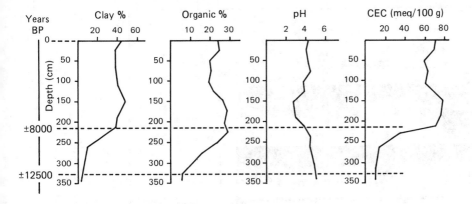

Fig. 4. Changes in sediment stratigraphy at Dunedin Vlei.

organic increments from a very varied dense hydromorphic population. Vegetation cover on adjacent slopes was probably of the dense montane type, providing protection for soil surfaces and reducing the erosive effects of overland flow. The chemical deposition of parent materials and leaching drainage waters resulted in fine silts and clays being moved into the vlei. The lowest beds, dated at 12 500 +- 160 years BP and of a sandier texture, suggested cooler, drier conditions with reduced organic inputs due to somewhat sparser vlei vegetation and probably poorly covered slopes. These conditions would have resulted in the retraction of the montane forest to protected valley heads, leaving a cover of grass and shrubs on exposed peaks. As fine sands are more easily eroded than clays, large increments to the vlei system would have been provided by runoff resulting from rainfall if the vegetation cover was poor.

Variations in sediment characteristics, in particular organic matter and particle size, are particularly indicative of changes in the environment leading to the availability of sediment.

6.3. Palynological evidence at Dunedin vlei

The reconstruction of environmental change at Dunedin has been attempted from a pollen diagram (Meadows et al., 1987) where comparisons with a pollen reference collection provided some indication of fluctuations in the patterns of vegetation over time. The arid/moist and warm/cool phases indicated in the pollen diagram correspond well to the pattern of environmental change suggested by the sediment characteristics (compare Figs. 4 and 5). The pollen diagram (Fig. 5) clearly indicates harsh cool/dry conditions at 330 cm (dated at 12 500 +- 160 years BP) when vegetation was dominated by grass/heathland with low shrub species of

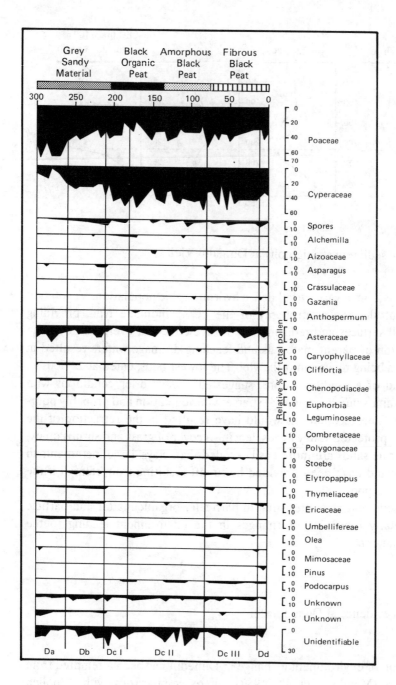

Fig. 5. Pollen diagram for Dunedin.

Ericaceae, Asteraceae and Chenopodiaceae. Organic sediments appear to increase from around 8 000 BP, indicating a change to montane forest species including *Podocarpus* and *Olea* spp. Xerophytic species also decrease, white Cyperaceae show a dramatic increase. These changes suggest that the environment was more moist and possibly warmer, allowing the forest species to flourish in areas larger than those found at present. The more recent past shows a return of adverse conditions, the retraction of montane forest to protect valley heads, and a concomitant increase in sclerophyllous shrub species (e.g. Ericaceae, Thymeliaceae and Crassulaceae). Climatic fluctuations throughout this recent period, although frequent, were not of great significance, and the vegetation could well have been influenced by human interference. Contemporary conditions do not appear to be as severe as those at 12500 years BP. This is shown by the more varied shrub/macchia occurring there now, rather than open grass/heathland which, it is suggested, tends to grow denser in harsher conditions (Story, 1952).

7. CONCLUSION

The sediments found preserved in vlei sites in South Africa are valuable as indicators of Holocene climatic change and compare well with similar sites throughout southern Africa. The complicated stratigraphy of Dunedin vlei has been well corroborated by additional palynological evidence, which has produced a pattern of past environments with reasonable clarity. This evidence can assist the management and conservation of these delicately balanced geomorphological phenomena, in addition to fitting one more piece to the jigsaw puzzle which constitutes the record of environmental change in southern Africa.

8. SUMMARY

Paleoenvironmental reconstruction of the Holocene period for central and southern Africa has been hampered by the erratic distribution of suitable sites and the limited amount of published data. One geomorphological feature which has supplied valuable evidence for interpreting fluctuations in past environmental conditions is the "vlei" or "dambo". Being a product of their environment, the clastic and organic sediments contained within these waterlogged features are affected by, and therefore reflect to some degree, the conditions of the environment as it existed during their development. With the scarcity of available literature, the history of environmental change in the eastern Cape region is less well-understood than documented sites in Malawi, Zambia and Zimbabwe. A site in the Winterberg Range (Cape Province) enabled a study to be made of a vlei system to establish an accurate chronology for the entire catchment area. Radiocarbon dates from the organic strata indicated that the age of these sediments spans the whole of the

Holocene period, from approximately 12000 years BP. The sediments of the vlei were analysed in terms of morphology, particle size and distribution, organic matter, and other physical and chemical characteristics. These analyses facilitated the construction of detailed stratigraphic diagrams and a chronological summary of sediment accumulation, from which the period and governing processes of vlei development under changing environments are inferred.

9. REFERENCES

Barber, K.E. 1981. Peat stratigraphy and climatic change. A.A. Balkema, Rotterdam.

Mackel, R. 1974. Dambos: a study in morphodynamic activity on the plateau regions of Zambia. Catena, 1, 327-365.

Meadows, M.E. 1985. Dambos and environmental change in Malawi, Central Africa, Zeitschrift für Geomorphologie, Supplement Band, 52, 147-169.

Meadows, M.E. (in press). Late Quaternary peat accumulation in southern Africa. Catena.

Meadows, M.E., Meadows, K.F. and Sugden, J.M. 1987. The development of vegetation on the Winterberg escarpment. The Naturalist, 31, 26-32.

Meadows, M.E. and Sugden, J.M. 1988. Late Quaternary environmental changes in the Karoo, South Africa. (this volume).

Millington, A.C., Helmisch, F. and Rhebergen, G.J. 1985. Inland valley swamps and bolis in Sierra Leone. Hydrological and pedological considerations for agricultural development. Zeitschrift für Geomorphologie, Supplement Band, 52, 201-222.

Morisawa, M. 1968. Streams. Their dynamics and morphology. McGraw-Hill.

Moore, P.D. and Webb, J.A. 1978. An illustrated guide to pollen analysis. Hodder and Stoughton, London.

Morgan, R.P.C. 1980. Soil erosion. Longman, London.

Scott, L. 1982. A Late Quaternary pollen record from the Transvaal bushveld (S.A.) Quaternary Research, 17, 339-370.

Story, R. 1952. A botanical survey of the Keiskammahoek district. Botanical Survey of South Africa Memoir, 27. Government Printer, Pretoria.

Tomlinson, R.W. 1973. The inyanga area: an essay in regional biogeography. University of Rhodesia Occasional Paper 1.

Weaver, A. van B. 1978. The bottom sediments of the Howieson's Poort reservoir. Unpublished M.A. proposal, Rhodes University, Grahamstown.

Whitlow, R. 1985. Dambos in Zimbabwe: a review. Zeitschrift für Geomorphologie, Supplement Band, 52, 115-146.

SEDIMENTOLOGY OF DEBRIS SLOPE ACCUMULATIONS AT RHODES, EASTERN CAPE DRAKENSBERG, SOUTH AFRICA

COLIN A. LEWIS
Department of Geography, University of Zululand
PATRICIA M. HANVEY
Department of Geography, University of Transkei

1. INTRODUCTION

This paper describes the sedimentology and genesis of stratified deposits which occur at Rhodes in the eastern Cape Province, South Africa. These deposits display many characteristics of periglacial stratified slope deposits (cf Dylik, 1960; Watson, 1969; Embleton and King, 1975; French, 1976). This, together with increasing evidence of periglacial conditions having prevailed throughout much of Southern Africa during the Pleistocene period (Alexandre, 1962; Butzer, 1973; Dyer and Marker, 1979; Harper, 1969; Hastenrath and Wilkinson, 1973; Lewis, in press; Linton, 1969; Sparrow, 1964,1967a,1967b,1973; Van Zinderen Bakker, 1965; Verhoef, 1969), and more specifically in the immediate vicinity of the present study (Lewis and Dardis, 1985; Hanvey et al., 1986), suggests that these deposits may have formed as a result of former periglacial activity.

2. REGIONAL SETTING

The site of the present investigation is located 1 km west of Rhodes in the Eastern Cape Province (Fig. 1) at an altitude of 1800 m. Morphologically the deposits form a concave slope which extends outwards from a steep, back-walled hollow (Fig. 2), similar to those described in the Orange Free State by Nicol (1973). The hollow is carved within Clarens Formation sandstone, capped by the Drakensberg Basalt Formation (South Africa Committee for Stratigraphy, 1980). The hollow has a maximum width of c.2500 m. with a back wall height of c.200 m., and is one of a series of such hollows which occur along the south-east facing flank of a ridge within the Ben MacDhui upland area of the Drakensberg.

Geomorphological Studies in Southern Africa, G.F.Dardis & B.P.Moon (eds)
© 1988 Balkema, Rotterdam. ISBN 90 6191 831 6

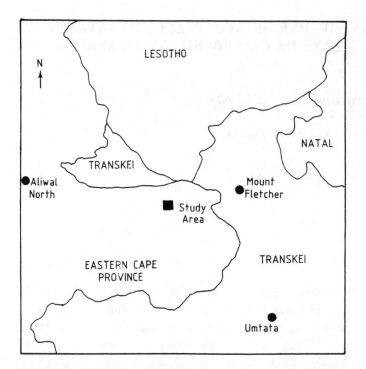

Fig. 1. Location of the study site.

Fig. 2. Debris lobe emanating from a cirque-like hollow.

3. DESCRIPTION OF THE SEDIMENTS

Two exposures (Fig. 3) within the deposits reveal the stratified nature of the component sediments. A detailed sedimentological analysis of the sediments was un-

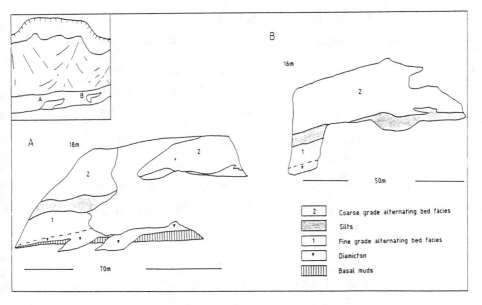

Fig. 3. Location and sedimentology of the two sections exposed within the debris lobe.

dertaken, using the lithofacies coding scheme used by Miall (1977) and modified for diamicton lithofaciés (Eyles et al., 1983). Five major lithostratigraphic units were identified (Table 1), all of which are continuous within both sections, except for the basal muds which are present only in section A (Fig. 3). The upper portion of the overall stratigraphic sequence is dominated by basalt lithologies while the lower portion consists predominantly of clasts derived from the Clarens Sandstone. Clasts throughout the sequence tend to be sub-angular or angular in shape.

3.1. Unit 1: Basal Mud Facies

This unit has a maximum thickness of 1.65 m and consists predominantly of massive silts and clay (Fm) (Table 2) interbedded with thin pebble (PGmm) layers (Fig. 4). Silt and clay layers range in thickness up to 21 cm and frequently display normal grading internally. Junctions within this sequence are sharp and often remobilised. The PGmm layers range up to 15 cm in thickness and are generally irregular and discontinuous. They consist predominantly of fine grade pebbles, sub-angular in shape and held within a silty matrix. A 12 cm thick diamictic mud (Fmd) layer and a 20 cm thick diamicton (Dmm) layer are also present within the sequence. The former consists of tiny pebbles dispersed within a clayey matrix while the latter consists of sub-angular clasts which average 10 cm in length held within a silty matrix. Unit 1 is only present at the base of section A (Fig. 3).

Table 1. General stratigraphy, Rhodes section.

Lithostratigraphic Unit	Description	Facies Composition	Environmental Interpretation
5	Coarse grade alternating bed facies	Dcm Fmd	Periglacial
4	Massive silts Pebble lags	Fm	Interstadial
3	Fine grade alternating bed facies	Dcm Fmd	Periglacial
2	Diamicton Dm(s)	Dmm	Periglacial
1	Basal muds	Fm, Fmd,PGmm Dmm	Interstadial/ Periglacial

3.2. Unit 2: Diamicton

This consists of a poorly sorted diamicton (Fig. 5) which is generally massive (Dmm) but in places shows internal crude stratification (Dm[s]). The lithofacies has a maximum thickness of 2 m and is characterised by angular/sub-angular clasts up to 10 cm in length held within a silty matrix. Within section B (Fig. 3) the lithofacies is transitional from being massive towards the base to being crudely stratified up-sequence and is gradational into an overlying alternating bed lithofacies (Unit 3). Within section A (Fig. 3) two distinctive lobes of the diamicton lithofacies truncate the underlying basal muds (Fig. 6). These lobes have a maximum thickness of 1.1 m, are crudely stratified and clasts are orientated downslope. The lobes also display normal to inverse grading.

Table 2. Sedimentary facies types, Rhodes section.

Code	Facies Type	Description
Dmm	Diamicton	Poorly sorted admixture of fines and clasts. Massive; matrix supported. Silt rich matrix. Clasts angular/ subangular, up to 10 cm in length.
Dm(s)	Diamicton	Similar composition to above facies, but shows internal crude stratification.
Dcm	Diamicton	Clast supported. Matrix deficient. Angular clasts. Massive.
PGmm	Gravels	Massive; matrix supported; silt rich matrix. Fine grade clasts averaging < 3 cm in length. Clasts subangular.
Fmd	Diamictic Fines	Matrix dominant with occasional dispersed clasts. Silt rich or clayey matrix. Clasts generally < 1 cm; angular/subangular.
Fm	Fines	Massive silts or clays. Frequently show internal inverse or normal grading.

3.3. Unit 3: Fine Alternating Bed Facies

Within both sections the diamicton lithofacies is overlain by a fine grade alternating bed lithofacies (Fig. 7). This comprises relatively continuous clast rich layers (Dcm) (Fig. 8) which alternate with clast deficient layers (Fmd). The former average up to 30 cm thick and consist of angular clasts which average up to 5 cm in length with occasional clasts up to 20 cm. The Fmd layers are matrix dominant (silt/fine sand) with dispersed angular clasts which average less than 2 cm in length, with occasional clasts up to 15 cm. The overall sequence of alternating beds attains a maximum thickness of 3 m and individual beds dip gently downslope. Sandstone lithologies are dominant within this unit.

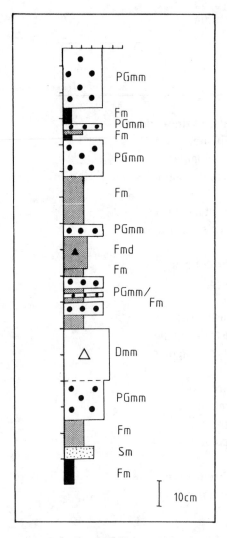

Fig. 4. Sedimentary log showing the detailed sedimentology of the basal muds.

3.4. Unit 4: Silts

This unit overlies Unit 3 and is continuous within both sections (Fig. 3). It consists predominantly of massive silts (Fm) with occasional discontinuous pebble lags (up to 1 m in length) of sub-angular/angular clasts up to 10 cm in length. Occasional clasts which average less than 1 cm are dispersed within the silts. The upper boundary of this unit is sharp but irregular with distinct solemarks (up to 40 cm in width and 50 cm in depth (Figs. 9 and 10) penetrating downwards from the overlying lithofacies.

370

Fig. 5. Poorly sorted massive diamicton facies.

Fig. 6. Lobe of stratified diamicton truncating the basal muds within section A.

Fig. 7. Fine grade alternating bed facies (unit 3) with overlying silt facies (unit 4).

Fig. 8. Close-up of a clast-rich layer (Dcm) within the fine grade alternating bed facies.

Fig. 9. Massive silts (unit 4) with occasional lags and an irregular upper boundary with the overlying facies.

Fig. 10. Close-up of a solemark penetrating downwards into the silts (Unit 4) from the overlying facies (unit 5).

3.5. Unit 5: Coarse Alternating Bed Facies

This is a coarser grade alternating bed lithofacies with individual beds being less distinctive and more discontinuous than within Unit 3 (Fig. 11) especially up-sequence. The facies is also more coarse and disorganised in section A (i.e. towards the central axis of the sediment lobe emerging from the hollow) than in section B

Fig. 11.Coarse grade alternating bed facies (unit 5) overlying the silt facies.

(which is closer to the periphery of the lobe). Clast-rich layers (Dcm) are up to 0.8 m thick and consist of coarse sub-angular clasts which average 30 cm in length. These layers are interbedded with silty layers with occasional dispersed clasts (Fmd) which average up to 0.3 m in thickness. Towards the end of section A an 'overfold' type structure is present within Unit 5 (Fig. 3). This is approximately 3 m in height and extends in a downslope direction. Unit 5 has an exposed thickness of 8 m and is composed largely of basalt lithologies.

4. INTERPRETATION

The restricted occurrence of the basal muds suggests that they represent localized deposition within a small ponded water body. The dominance of silts and clays indicates that much of the deposition has occurred from suspension sedimentation associated with low energy flows into the water body. This has periodically been interrupted by higher energy flows as evidenced by the thin pebble layers, and cohesive debris flows represented by the diamictic mud and diamicton beds.

The internal characteristics of the diamictons (e.g. cohesive matrix, evidence of grading) which replace the mud up-sequence, are suggestive of cohesive debris flows (cf. Johnson, 1965; Lawson, 1982; Innes, 1984). The angularity of component clasts suggests that they may be derived through frost shattering processes, have undergone minimum transport, and have not been subjected to extensive water activity. However, some water is believed to have contributed to the formation of this lithofacies, as debris flows, by definition (Innes, 1984), involve the downslope movement of debris mixed with minor yet significant amounts of water. A flow type origin of the diamictons is strongly supported by the two distinctive diamicton lobes present in section A which truncate the underlying muds, show internal stratification and contain clasts which are aligned downslope. It is apparent that the upper portion of the diamicton lithofacies has undergone additional winnowing by water as reflected by the presence of thin sand laminae in the vicinity of the upper boundary.

The fine grade alternating beds of unit 3 are similar to grezes litees or rythmically stratified slope deposits (Dylik, 1960), which are believed to result from intense frost action together with periglacial slopewash (French, 1976) possibly associated with meltwater from snow patches (cf. Watson, 1965). The alternation of clast-rich layers comprising angular particles, with layers of fines is generally thought to reflect freeze-thaw oscillations. The sorting and bedding, and the eluviation of the finer particles and their deposition downslope, is usually attributed to slopewash operating on vegetation-free slopes below snow patches, and to the melting of pore ice (French, 1976). However, it is not known whether such stratified slope deposits are indicative of diurnal or seasonal temperature oscillations.

The intermediate silt layer is believed to reflect some form of hiatus when freeze-thaw conditions were diminished or moderated. The lithofacies may represent aeolian deposition during a relatively cold phase with low frequency periodic sheet-wash events as reflected by the occasional pebble layers. An aeolian origin is supported by the stratigraphic position of the silts as they appear to form a thick drape over the underlying facies and do not appear to conform to any topographical depression which would be essential fer their formation within a ponded water body. They are also similar in character to massive silts found adjacent to glaciated areas in Alaska (Pewe, 1955) considered to be aeolian in origin and derived from glacial outwash. Alternatively, the full topographical disposition of the silts may be disguised, and thus in the absence of detailed grain-size and SEM analysis, which might differentiate between wind-blown or water-lain sediments, a localised lacustrine environment (i.e. a pond) of deposition must be considered. Thus the silts may alternatively represent a phase of colluvial sedimentation or low energy fluvial sedimentation into a water body during a climatic period of increased precipitation. Since no detailed analysis of particle shape within these silt deposits has yet been undertaken to determine whether they are of fluvial or aeolian origin their palaeoclimatic significance is uncertain. Nevertheless, they reflect a significant change in depositional processes which must in turn indicate a change in local climatic conditions at the time of deposition.

The upper coarse grade alternating bed lithofacies probably again reflects some form of freeze-thaw activity but this appears to have been more catastrophic than that responsible for unit 3, as reflected by the calibre of clasts, a greater degree of disorganisation, truncation of the underlying silts and the distinctive downward penetrating sole marks. The greater energy may be derived from annual phases of vigorous melting of a local snowpatch present within the hollow. Much of the disorganisation of the facies, especially towards the top of the profile, may have resulted from the final dissolution of this snowpatch. The indistinctive fold structure within the portion of section A has resulted from mobilization of the debris downslope.

5. DISCUSSION

The association of sediments at this site is believed to reflect phases of freeze-thaw activity within the rock depression, with the sediment derived through frost shattering of the confining rock walls, and mobilized during intervening periods of thaw. This would account for the dominance of Clarens Formation sandstones in the lower sedimentary units as these would be derived from the lower parts of the rock wall during the early activity. Subsequently as the sediment accumulation increases freeze-thaw processes would then operate on the higher basalts and this would account for the large proportion of these lithologies in the upper sedimentary units.

The basal mud unit represents ponded water and need not have any palaeoclimatic significance since ponded water can occur in many situations so long as a topographic hollow, or suitable obstruction to the free drainage of water is present. However, the occurrence of several pebble layers and debris flows into such a water body indicates periodic mobilization of surrounding debris and this may be related to thaw cycles within a freeze-thaw climatic regime.

The angularity of clasts within the diamicton facies (unit 2) suggests that they are derived from frost shattering and the internal characteristics of the overall facies indicates emplacement by debris flows. Once derived from frost shattering, the incorporation into, and transport of the material as debris flows requires the availability of water. Nicol (1973) has indicated that talus (shattered rock fragments) can behave as a viscous liquid and move downslope carrying large boulders in matrix, dumping unsorted, unstratified debris further downslope. Water for such flows can be derived from two sources; either from continuous rain for at least 24 hours (Rapp, 1960), or from melting snow (Nicol, 1973). The amount of water related to the proposed debris flows appears to have varied locally with the stratified diamictons within the lobes evidence of larger amounts of water and hence greater energy and mobility. Nevertheless, the overall massive character of the diamictons at Rhodes may reflect the ablation of a temporary snow patch (Fig. 12A) with concomitant production of meltwater, at an early stage of development as more regular freeze-thaw cycles or oscillations will likely give rise to distinctly stratified alternating fine and coarse sediments, such as the alternating bed facies of unit 3.

True cyclic sedimentation related to diurnal or seasonal temperature fluctuations are first evident within unit 3. This may be related to seasonal ablation of a snow patch or freeze-thaw processes operating at the periphery of a snow patch, since the occurrence of freeze-thaw action associated with snow patches is firmly established (Embleton and King, 1975). This is supported by the shallow steep-sided

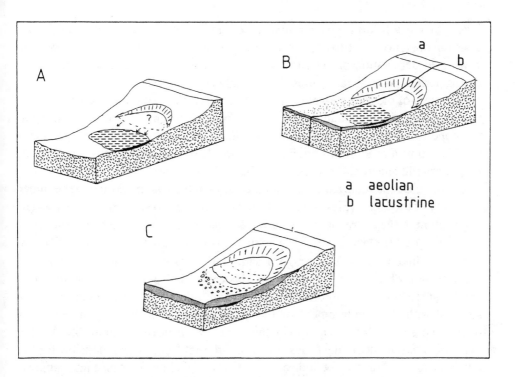

a   aeolian
b   lacustrine

Fig. 12. Schematic diagram illustrating the sequence of events at Rhodes as reflected in the sedimentary sequence. (A) Localized sediment flows into a ponded body of water. Flows are generated upslope and possibly related to the peripheral melting of a temporary snow bed, or prevailing freeze/thaw periglacial slope processes. (B) Intermediate period of no periglacial activity when the silts were deposited either within an aeolian or lacustrine environment. (C) Growth and decay of a snow patch within the hollow during a second phase of periglacial activity, resulting in an extensive apron of protalus deposits.

morphology of the hollow since nivation will be most intensive at the sides of such a snow patch and inhibited at the base, due to an insulation factor (cf. Thorn and Hall, 1980) thus preventing overdeeping of the base. Thorn and Hall (1980) have indicated that snow patch insulation precludes high freeze-thaw amplitudes at the base of a snowbed. Thus the importance of the snow itself is largely the provision of meltwater during periods of thaw (Embleton and King, 1975), which assists in the removal of debris downslope. Alternatively, the facies may simply reflect nivation within the hollow without the development of a snow patch if thaw periods within a freeze-thaw regime are sufficiently intense to prevent snow accumulation from one season to another. As previously pointed out the precise origin of stratified slope deposits is uncertain but the nature of component clasts indicates that they may have accumulated under freeze-thaw conditions. However, it is apparent that sheet flow is also an important component (Dylik, 1969) and this suggests the presence of a frozen subsurface at the time of their formation, preventing downward percolation of nival meltwater and causing concentrated surface flow during periods of seasonal melting (Lewis and Dardis, 1985).

The origin of the intermediate silts (unit 4) is at yet uncertain with the possibility of them being either aeolian or water-lain (Fig. 12B) during a period when periglacial conditions did not prevail. Similar aeolian deposits described by Pewe (1955) would support a wind-blown origin. As previously stated, internal clasts may result from periodic sheetwash events, or mobilization of clasts from local slopes followed by subsequent burial. If this is the case, then the sole marks along the upper boundary of the facies (Figs. 9 and 10) can be attributed to scouring and downward penetration of the overlying facies during emplacement of the latter. Alternatively, if the silts are water-lain, pebble layers may again represent some form of fluvial input, or sinking of clasts through a saturated mass of silts subsequent to the emplacement of the overlying facies. In this case the upper sole marks structures can be attributed to soft-sediment deformation, related to a density gradient associated with the emplacement of the coarse material above. Whatever their origin, the silts reflect a change in the nature of processes reponsible for the deposition of the underlying facies, and also different from those reponsible for the overlying facies. This may be indicative of substantial climatic change and supports the viewpoint of Harper (1960) and Lewis and Dardis (1985) that two separate periglacial phases occurred during the Quaternary in southern Africa, separated by a period of climatic amelioration. However, further detailed analysis of the silts is necessary before their precise origin is determined.

The upper lithofacies again shows alternating beds of clast-rich and clast-deficient layers which become disorganised and more coarse up-sequence and downslope. It is suggested that this represents the possible rejuvenation (Fig. 12C) and final disintegration of a snow patch within the hollow, following the proposed period of climatic amelioration. As discussed above, the more definite alternating beds towards the base may represent gelifluction terraces (cf. Embleton and King, 1975) which reflect oscillating freeze-thaw conditions at the periphery of the snow patch. The upper portion of the sequence reflects more vigorous melting and greater flow energy as indicated by the large clasts, less prominant and truncated fine layers and the overfold structure visible within the facies. It is suggested that much of this coarse debris may be derived as a result of rock fall material (cf. Thorn and Hall, 1980; Harris, 1986) sliding over the snow surface to the base of the snow patch (Richmond, 1962; Blagbrough and Breed, 1967; Washburn, 1979) to form protalus deposits (Harris, 1986; Ballantyne, 1987) around the lower limits of the snow patch. Although protalus deposits are often described as consisting of rock fragments with a lack of fines, Harris (1982; 1986) has described such features composed of a significantly higher content of fines than what is commonly reported. Therefore it is possible that the upper facies represents a series of protalus deposits related to a receding boundary of a contracting snow patch, as Harris (1986) has shown that more recent deposits in such a situation can merge

378

with older deposits downslope (developed when the snow patch was larger) to form a wide apron of debris. Alternatively, part of the upper facies may represent rock fall material which initially collected on the surface of the snow patch and was subsequently distributed downslope when support was lost through the final melting of the snow.

6. CONCLUSIONS

The deposits at Rhodes are believed to be predominantly periglacial in origin,resulting from nivation, gelifluction, and slope wash processes operating within a nivation hollow and possibly related to snow patch development within that hollow. Two distinct periods of periglacial activity are identified, separated by an interstadial in which the silt unit was deposited.

7. SUMMARY

The sedimentology of a debris lobe emanating from a nivation hollow at Rhodes, Eastern Cape Province is described. This reveals a complex range of stratified slope deposits. Clasts are angular or sub-angular throughout and are believed to be derived through freeze-thaw shattering accompanied by subsequent mobilization of sediment downslope. The sediments reflect two separate periods of periglacial activity.

8. ACKNOWLEDGEMENTS

The authors acknowledge financial assistance from the Research
Committee of the University of Transkei, and photographic assistance from Mr. B. Lewis.

9. REFERENCES

Alexandre, J. 1962. Phenomenes periglacaires dans le Basutoland et le Drankensberg du Natal. Biuletyn Peryglacjalny, 11, 11-13.

Ballantyne, C.K. 1987. Some observations on the morphology and sedimentology of two active protalus ramparts, Lyngen, northern Norway. Arctic and Alpine Research. 19, 167-174.

Blagbrough, J.W. and Breed, W.J. 1967. Protalus ramparts on Navajo Mountain, Southern Utah. American Journal of Science, 256, 759-722.

Butzer, K.W. 1973. Pleistocene 'periglacial' phenomena in southern Africa. Boreas, 2, 1-11.

Dyer, T.G. and Marker, M.E. 1979. On some aspects of Lesotho hollows. Zeitschrift fur Geomorphologie, 23, 256-270.

Dylik, J. 1960. Rhythmically stratified slope waste deposits. Biuletyn Peryglacjalny 8, 31-41.

Dylik, J. 1969. Slope development under periglacial conditions in the Lodz region. Biuletyn Peryglacjalny 18, 381-440.

Embleton, C. and King, C.A.M. 1975. Periglacial Geomorphology, Edward Arnold, London. 203pp.

Eyles, N., Eyles, C.H. and Miall, A.D. 1983. Lithofacies types and vertical profile models: an alternative approach to the description and environmental interpretation of glacial diamict and diamictite sequences. Sedimentology, 30, 393-410.

French, H.M. 1976. The Periglacial Environment. Longmans, London.

Hanvey, P.M., Lewis, C.A. and Lewis, G.E. 1986. Periglacial slope deposits in Carlisle's Hoek, near Rhodes, Eastern Cape Province. South African Geographical Journal, 68, 164-174.

Harper, G. 1969. Periglacial evidence in southern Africa during the Pleistocene epoch. Paleoecology of Africa, 4, 71-101.

Harris, C. 1982. The distribution and altitudinal zonation of periglacial landforms, Okstindan, Norway. Zeitschrift fur Geomorphologie, 26, 283-304.

Harris, C. 1986. Some observations concerning the morphology and sedimentology of a protalus rampart, Okstindan, Norway. Earth Surface Processes and Landforms, 11, 673-676.

Hastenrath, S. and Wilkinson, M.J. 1973. A contribution to the periglacial morphology of Lesotho. Biuletyn Peryglacjalny, 22, 156-167.

Innes,J.L. 1984. Debris flows. Progress in Physical Geography, 7, 469 - 501.

Johnson, A.M. 1965. A model for debris flow. unpublished Ph.D. thesis, Harvard University.

Lewis, C.A. 1987 in press. Periglacial features in southern Africa: an assessment. Palaeoecology of Africa, 19.

Lewis, C.A. and Dardis, G.F. 1985. Periglacial ice-wedge casts and head deposits at Dynevor Park, Barkly Pass area, north-eastern Cape Province, South African Journal of Science, 81, 673-677.

Linton, D. L. 1969. Evidence of Pleistocene cryonival phenomena in southern Africa. Paleoecology of Africa. 5, 71-78.

Miall, A.D. 1977. A review of the braided river depositional environment. Earth-Science Reviews, 14, 315-359.

Nicol, I.G. 1973. Land forms in the Little Caledon Valley, Orange Free State. South African Geographical Journal, 55, 56-68.

Pewe, T. L. 1955. Origin of the upland silt near Fairbanks, Alaska. Bulletin Geological Society of America, 67, 699-724.

Rapp, A. 1960. Recent development of mountain slopes in Karkevagge and surroundings, N. Scandinavia. Geografiska Annaler, 42, 165-200.

Richmond, G.M. 1962. Quaternary stratigraphy of the La Sal Mountains, Utah. U.S. Geological Survey Professional Paper. 454D, 41pp.

Sparrow, G.W.A. 1964. Pleistocene periglacial landforms in the southern hemisphere. South African Journal Science. 60, 143-147.

Sparrow, G.W.A. 1967(a). Southern African cirques and aretes. Journal of Geography (Stellenbosch), 2, 9-11.

Sparrow, G.W.A. 1967(b). Pleistocene periglacial topography in southern Africa. Journal of Glaciology, 6, 551-559.

Thorn,C.E. and Hall, K.J 1980. Nivation: An Arctic-Alpine comparison and reappraisal. Journal of Glaciology, 25, 109-124.

van Zinderen Bakker, E.M. 1965. Uber moorvegetation und den aufbau der moore in sud und ostafrika. Botanische Jahrbucher, 84, 215-231.

Verhoef, P. 1969. On the occurrence of Quaternary slope deposits. South African Geographical Journal. 51, 88-98.

Washburn,A.L. 1979. Geocryology. Edward Arnold, London.

Watson, E. 1965. Grezes litees ou eboulis ordonnes tardiglaciaires dans la region d'-Aberystwyth, au centre du Pays de Galles. Bulletin de l'Association de Geographes Francais, 38-9, 16-25.

Watson, E. 1969. The slope deposits in the Nant Iago Valley, near Cader Idris, Wales. Biuletyn Peryglacjalny. 18.

RECENT PERIGLACIAL MORPHODYNAMICS AND PLEISTOCENE GLACIATION OF THE WESTERN CAPE FOLDED BELT, SOUTH AFRICA

H. SÄNGER

Institute for Geography, University of Hamburg

1. INTRODUCTION

Although Sparrow (1971) reports on cirque findings from the Drakensberg range, some geographers, especially Butzer (1973), reject the idea of glacial and periglacial phenomena in the Western Cape mountain range: "...True "periglacial" forms and deposits of late (and middle?) Pleistocene age can be recognized in the Drakensberg and adjacent parts of the Cape Province ... Significant nivation in the Drakensberg is also indicated, but at higher elevations...Alleged "periglacial" phenomena in Rhodesia, the Transvaal, the Cape Folded Ranges and their coastal margin are not acceptable as such and include no evidence for cryonival or geliflual processes..." (Butzer, 1973; p. 8-10).

So Butzer wrote in 1973. Meanwhile further field work has revealed evidence for a Pleistocene glaciation of the Western Cape Folded Belt (February, 1979 - September, 1980 and from November, 1982 - February, 1983). An extended stay in the Hex River Mountains was necessary to obtain insight into the recent morphodynamic processes and climatic conditions.

The findings on the existence of a Pleistocene glaciation are based on and supported by an interpretation of the recent morphodynamics in relation to the non-recent morphologic phenomena in the summit area and the deposits in the piedmont zone and their morphogenetic and palaeoclimatic discussion and interpretation.

These geomorphologic features were discussed and compared in a critical discussion of the palaeoclimatic conditions, based on an interpretation of present climatic data.

The main aim of the work was to distinguish the morphologic features which depend on recent processes from those which are not connected to the present-day geomorphologic situation, but correlate more with a Pleistocene glaciation.

The complexity of the project demanded an investigation of the geomorphologic fea-

Geomorphological Studies in Southern Africa, G.F.Dardis & B.P.Moon (eds)
© 1988 Balkema, Rotterdam. ISBN 90 6191 831 6

tures and the morphodynamic/morphogenetic processes as well as climatic observations during a period of more than one year. Most important was the correlation of the results from the morpho-analysis with recent climatic data and the knowledge on the palaeoclimatic conditions in South Africa and Antarctica.

2. RECENT GEOMORPHOLOGIC PROCESSES

To find out what happens on the slopes long-term observations on experimental fields were necessary to identify processes connected to needle ice and its influence on denudation.

The processes of the present-day morphodynamics in the Western Cape Mountains justify an interpretation of the summit area as one of periglacial dynamics without the formation of patterned ground. The processes are those of congeli-solifluction within a subnival zone depending on extensive freeze and thaw processes with the frequent occurrence of needle ice, which produces a permanent surface creeping on the slopes.

Therefore, the summit area can be regarded as a subnival zone with periglacial dynamics.

3. PLEISTOCENE GEOMORPHIC PHENOMENA

The observations in the mountain range were supported by the interpretation of aerial photographs and flights over the mountains. The main observations were made in the area round Matroosberg (2249 m) and later extended to the surrounding mountains up to Stellenbosch.

The investigations in the field area justify the interpretation of a Pleistocene glaciation in the summit area of the Western Cape Mountains.

The most striking phenomena encountered are the wide-spread blockfields and glacial pavements in the higher parts of the mountains. The identification of these as products of frost-shattering is commonly accepted. This interpretation indicates cooler to periglacial climatic conditions in the past. The slope-covering blockfields of the summit area and the screes in the upper parts of the slopes are, however, inactive under the present climatic conditions. Today the subnival characteristics only allow processes of congeli-solifluction (cryonival and geliflual) which are not able to produce patterned ground).

The glacial pavements in the summit area suggest a Pleistocene glaciation in different parts of the mountains. In addition, special planation areas like small plateaux as well as cirques and ridges sharp enough to be called aretes can be observed in the higher parts of the mountains. The glaciation consisted of different plateau-glaciers and cirque-glaciers with some valley-glaciers; yet, real U-shaped valleys are rare. Therefore, the glaciation should not be misunderstood as having taking place on a grand scale over the whole mountain area.

Furthermore polished rock beds and rock faces point to a former glaciation. There are no striations, but small wave-like features, potholes and well-cut quartzitic pebbles embedded in the sandstone appear to be of glacial origin.

Some moraines and fluvioglacial as well as glacial deposits could be identified in the pediment zone of the mountains. Most of the so-called alluvial fans (Booysen, 1974) are interpreted here as fluvioglacial deposits.

The absence of striations in the bedrock can be explained by the material characteristics of the sandstone in which striations are not conserved under present climatic conditions. Therefore, observations taken over long time periods was necessary in the mountain range and its piedmont zone to determine the characteristics of the morphodynamic system which produced those geomorphologic features that cannot be explained by present-day climatic conditions.

If the geomorphologic features indicate a Pleistocene glaciation, however, there must also be palaeoclimatic indications, an investigation that has to start with the recent climate and lead to a discussion of palaeoclimate in connection to geomorphologic features.

4. PRESENT CLIMATE AND PALAEOCLIMATE

In discussing palaeoclimatic phenomena, present climatic conditions and their morphodynamic results have to be taken into account. At present, the mean annual temperature at Matroosberg (altitude: 1800 m; Matroosberg peak: 2249 m) is 3°C less than in Lesotho (altitude: 2375 m).

Snowfall is possible all through the rainy season from March/April to October/November (7-9 months). The snow does not cover the mountains the whole season, but for periods of days up to several weeks. Only a slight temperature decrease would be necessary to produce a snow-cover lasting for more than seven months.

Precipitation in the mountains is very high and intensive; the rainfall map 1:250000 (Worcester-sheet, 1966) shows an average rainfall of 3000 mm for the mountains. Yet, precipitation has never been measured and this estimate seems to be too high. In the pediment zone annual rainfall amounts to 400-500 mm. Extreme rates from Erfdeel farm (1200 m altitude, north from Matroosberg) range from 338 mm in 1960 to 964

Table 1. Pleistocene temperatures of four stations in the western Cape, assuming a temperature decrease of 10°C.

	Farm Erfdeel	Matroosberg Ski-Club Hut	Matroosberg Peak	Cape Town
	1200 m	1800 m	2249 m	17 m
Jan	6.7	2.9	- 1.3	11.2
Feb	7.6	2.6	- 0.4	11.5
Mar	5.0	1.2	- 3.0	10.3
Apr	2.6	-1.0	- 5.4	7.5
Ma	-0.5	-3.4	- 8.5	5.1
Jun	-2.9	-4.8	-10.5	3.4
Jul	-3.4	-7.0	-13.4	2.6
Aug	-2.4	-6.6	-12.4	3.2
Sep	-0.1	-5.9	- 8.1	4.5
Oct	0.6	-2.0	- 7.4	6.3
Nov	5.8	2.8	- 2.2	8.3
Dec	8.0	4.2	0.0	10.1

mm in 1977. Actual rainfall in the high-altitude zone has not been measured to date, but 2000 mm should be a reasonable estimate under present-day rainfall conditions.

The view, that a cooler and wetter climate may have prevailed during the Pleistocene age, with the formation of small cirque glaciers in the Drakensberg Mountains, in South Africa, has been put forward by some researchers (Sparrow, 1971). Yet, glaciers do not just form under very cold conditions, more important are the precipitation conditions.

Considering a very low lapse rate of 0.6°C/100 m and a temperature depression of 10°C during the Pleistocene (also confirmed by Butzer, 1973) an annual average Pleistocene temperature of c.-4°C for Matroosberg peak can be concluded; temperatures above zero were only possible during December (Table 1). The reasons for this are to be found in the oceanic location of the mountain area and the northward shift of the west-wind zone connected with the outspreading Antarctic zone during the Pleistocene. Consequently, comparing present climatic conditions with the knowledge about the climatic conditions during Pleistocene times, the result of the investigation point to a Pleistocene glaciation of the Western Cape Folded Belt.

5. CONCLUSIONS

Summarizing, the findings suggest that the higher parts of the Western Cape Folded Belt were cold and wet enough to have been glaciated during Pleistocene times. This cold phase was characterized by some plateau-, cirque-, and valley-glaciers. The main reason for the Pleistocene glaciation is not to be found in a strong temperature depression but rather depended on a northward shift of the west-wind and antarctic zone. This resulted in a cold and wet climate that persisted during the whole year in the Western Cape. In the light of known evidence it can be stated that some glaciers extended below the Pleistocene snow-line. To date it is impossible to give an exact altitude for the Pleistocene snow-line, but considering the discussed geomorphological phenomena and the present climatic data a local snow-line at 1500-1700 m for the Hex River Mountains and 1100- 1300 m in the Stellenbosch area can be assumed. The Western Cape cold phase lasted from about 30000-10000 years BP, contemporaneous with the Weichsel glaciation in Europe.

6. SUMMARY

The present day climatic conditions in the mountains of the Western Cape Folded Belt represent a periglacial morphodynamic in their higher parts with processes of congelisolifluction depending on freeze and thaw cycles within a subnival zone. During the Pleistocene age the climatic conditions have been cold and wet enough to produce different plateau-glaciers and Valley-glaciers in the high parts of the Cape Folded Belt. For the Hex River Mountains a Pleistocene snow-line at 1500-1700 m can be assumed.

7. ACKNOWLEDGEMENTS

The field work was made possible by financial support from the DAAD (Deutscher Akademischer Austauschdienst).

387

8. REFERENCES

Booysen, J.J. 1974. Alluviale waaiers in die Middelbree-rivierbecken: 'n morfometriese vergelyking. Die Suid-Afrikaanse Geograaf, 4, 390-394.

Borchert, G., Sanger, H. 1981. Research findings of a Pleistocene glaciation of the Cape mountain-ridge in South Africa. Zeitschrift für Geomorphologie, 25, 222-224.

Butzer, K.W. 1973. Pleistocene "periglacial" phenomena in Southern Africa. Boreas, 2, 1-11.

Sänger, H. (in press). Der geomorphologische Formenschatz der Kap-Ketten als Resultat der Wirkung von Frost und Eis. Eine Untersuchung zur periglazialen und pleistozän-glazigenen Morphodynamik.

Sparrow, G.W.A. 1971. Some Pleistocene Studies in Southern Africa. Tydskrif vir Aardrykskunde, 3, 809-814.

Van Zinderen Bakker Sr, E.M. 1976. The evolution of Late-Quaternary palaeoclimates of Southern Africa. Palaeoecology of Africa, 9, 160-202.

THE HOLOCENE EVOLUTION OF THE SUNDAYS ESTUARY AND ADJACENT COASTAL DUNEFIELDS, ALGOA BAY, SOUTH AFRICA

WERNER ILLENBERGER

Department of Geology, University of Port Elizabeth

1. INTRODUCTION

The Sundays River mouth is situated on the northern shore of Algoa Bay. The Alexandria coastal dunefield extends from the Sundays River mouth eastward for 50 km to Woody Cape, while the Schelmhoek coastal dunefield extends 8km westward from the Sundays River mouth (Fig. 1). Both dunefields have a maximum width of about 3km.

Deposits of estuarine and aeolian origin occur in the lower portions of the Sundays valley (Figs. 2 and 3). Foraminifera found in the estuarine sediments indicate a Holocene age (McMillan, pers. comm.). These Holocene sediments infill a valley cut into a sequence of Cretaceous clays and sandstones, which is overlain by the Tertiary Alexandria Formation.

The Alexandria and Schelmhoek dunefields have not been dated definitively. However, they are most likely Holocene in age, formed over the past 6500 years, since the sea had risen to the present level from its previous much lower levels (as much as 140m below present) during the last Glacial Maximum (Chappel, 1983). Similar coastal dunefields in New South Wales, Australia, have been assigned a Holocene age with radiocarbon dates of organic matter from the dunes and associated deposits (Thom et al., 1981). Portions of the dunefields are vegetated, in a manner which precludes the formation of the dunefields by a single continuous advance of sand.

In considering the morphologic development of the dunefields, it must be borne in mind that the net and major movement of sand within the dunefields is approximately towards ENE, the direction in which sand is blown by the dominant west-southwest wind (Fig. 1).

2. GEOMORPHOLOGY OF THE SUNDAYS ESTUARY AND ADJACENT AREAS

The lowermost Sundays estuary has a dog-leg shape, with the partly vegetated

Geomorphological Studies in Southern Africa, G.F.Dardis & B.P.Moon (eds)
© 1988 Balkema, Rotterdam. ISBN 90 6191 831 6

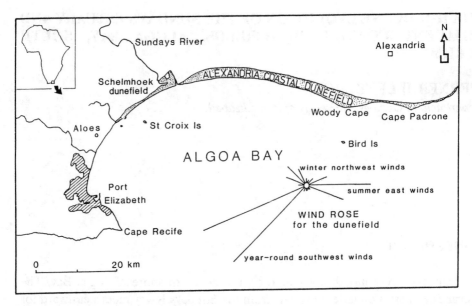

Fig. 1. Location map of the Sundays River mouth and adjacent coastal dunefields. The stylized wind rose shows combined data from Port Elizabeth, Aloes and Bird Island. The wind rose arms are plotted in the upwind direction. The urban area is Port Elizabeth is hatched.

Schelmhoek dunefield in the angle of the dog-leg (Fig. 2). Higher up the estuary is the presently active floodplain, largely underlain by Holocene estuarine sediments. The estuary channel follows a meandering path across the floodplain.

About half of the Schelmhoek dunefield is vegetated, in an irregular pattern. The vegetated dunes are mainly parabolic in form (Fig. 2). Parabolic dunes are also found in the much smaller proportion of vegetated dune patches in the Alexandria dunefield (Fig. 4). Apart from one vegetated patch at Woody Cape, these patches all lie along the landward edge of the dunefield (Fig. 4). The dominant dune type in the unvegetated dunefields is an akle pattern (Cooke and Warren, 1977), of which the transverse dune is the major element. Superimposed over the akle pattern are large mega-ridges of sand aligned roughly parallel to the shoreline (Figs. 4-6).

The source of sand for the dunefields is the sandy beach which bounds the dunefields on the seaward side. Sand is blown off the beach into the dunefields by the onshore-directed dominant wind (Illenberger and Rust, in press). The Alexandria coastal dunefield actively transgresses the hinterland: the landward edge advances over the hinterland at a rate of 0.25 m/yr, as determined by field

Fig. 2. The Sundays River mouth area. (V - vegetated dunes A - active dunes, characterised by an akle pattern of which the transverse component is dominant F - floodplain, underlain by Holocene sediments M - old meander channel of the estuary P - remnants of Pleistocene dunefields D - dolines in the Alexandria Formation.). Aerial photograph 802/17/2503 and 802/17/2504, 1978.

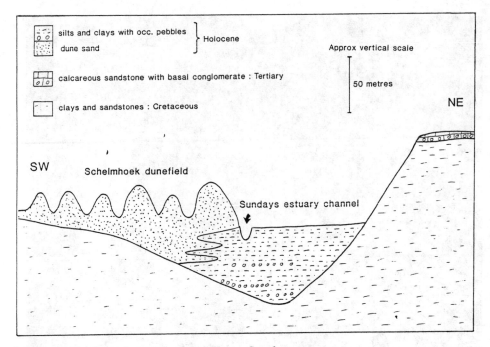

Fig. 3. Stylized section through the lower Sundays estuary.

mapping over a five year period (Illenberger, 1986). Sand from the Schelmhoek dunefield is for the most part blown into the Sundays estuary, to be periodically flushed out to sea by river floods (Reddering and Esterhysen, 1981). The landward boundary of the dunefield is a continuous slipface about 30 m high, extending along the whole 50 km length of the boundary. This is a "precipitation ridge" (Cooper, 1958); locally the term "main slipface" is used.

3. HOLOCENE DEVELOPMENT OF THE SUNDAYS ESTUARY

During the last Ice Age, when sea level was up to 140 m below the present level (Chappel, 1983), the Sundays River valley at the present mouth was cut broader and deeper than at present (Fromme, in Badenhorst, 1984). During the subsequent transgression this valley was inundated by the sea, eventually becoming a drowned river valley (c.6500 years ago)(Fig.7a).

The drowned valley then changed into a barrier island-lagoon estuary (Fig. 7b). The barrier island would have grown by accretion from the large amounts of sand transported by longshore drift in the Sundays River mouth area (Swart, 1986), sand blown off the barrier beach by the onshore-directed dominant wind (forming the embryonic Schelmhoek dunefield), and fluvial sediments from the Sundays River.

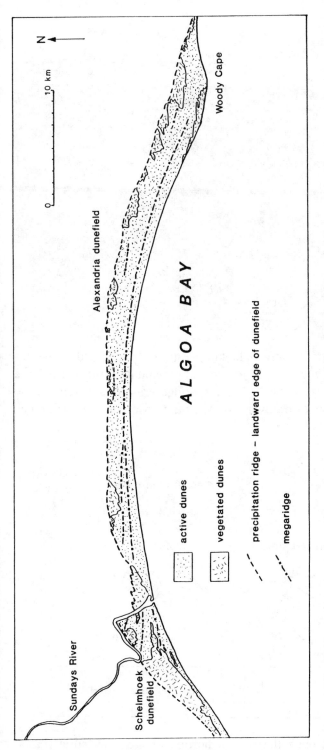

Fig. 4. The Alexandria and Schelmhoek dunefields, showing the distribution of vegetated areas and megaridges.

Fig. 5. The Alexandria coastal dunefield, looking westward from Woody Cape. A - megaridge crest B - trough between megaridge and precipitation ridge. The large slipfaces belong to transverse dunes. The megaridge crest is apparent as the highest portions of the crests of transverse dunes; the transverse dunes are much smaller in the trough.

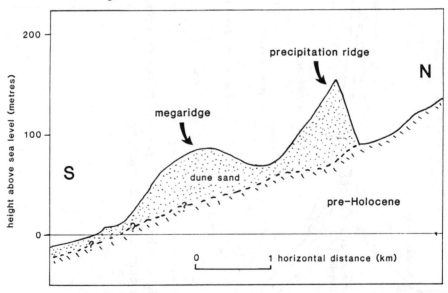

Fig. 6. Stylized section through the central portion of the Alexandria coastal dunefield (c. 20 km east of the Sundays River mouth), showing the precipitation ridge and a megaridge.

The net eastward movement of longshore drift and east-northeast movement of wind-transported sand steadily pushed the river mouth eastward, eventually pinning the river mouth against the relatively hard eastern edge of the valley, where it has remained to the present time, resulting in the dog-leg shape of the lower estuary.

The large back-barrier lagoon would have filled in with the relatively large amount of mainly muddy sediment that the Sundays River carries (Bremner and Du Plessis, 1980), with sand transported by flood tidal currents (Reddering and Esterhysen, 1981), and by encroachment of the incipient Schelmhoek dunefield into the palaeo-lagoon. Once the lagoon was filled completely, the upper estuary channel would have started to meander.

Chapman (pers. comm., in Perry, 1983) and Fromme (pers. comm., in Baden-horst, 1984) have described sequences of events similar to the above for the Holocene evolution of the Schelmhoek area.

4. HOLOCENE HISTORY OF THE DUNEFIELDS

The geomorphology of the dunefields can be interpreted as resulting mainly from a pulsing sand supply to the dunefields. Vegetation could spread over the dunefields between pulses, if the time interval between pulses was long enough. This vegetation would be swamped by the next pulse of sand.

The sand supply could fluctuate due to; (1) variations in the relative sea level during the Holocene. The sandy beach which supplies sand to the dunefield would be displaced landward or seaward, depending on the relative variation in the sea level, and thereby effect a pulsing sand supply to the area that is now dunefield; (2) large-scale variations in the total sand supply along the whole coastline; (3) climatic changes, e.g. variation in wind energy and/or variation in rainfall (Hesp, 1986) which would affect the rate at which sand is transported by wind, as well as whether pioneer vegetation could establish itself on shifting dunes; and (4) destruction of dune vegetation by human (Strandloper) activity (Tinley, 1985) or bush fires, which would release sand previously fixed by vegetation.

The following interpretation of the Holocene development of the dunefields appears to be realistic, assuming for simplicity that variations in relative sea level are responsible for sand supply fluctuations; this does not imply that other possible causes for variation in sand supply are not valid.

The precipitation ridge is here hypothesised to occupy approximately the present position of the first sand pulse which started advancing landward about 6500 years ago when the sea returned to the interglacial level after the last Ice Age (Chappel, 1983) (Fig. 7a). Sea level about this time was somewhat higher than present, as evidenced by   a precipitation ridge in the Schelmhoek area which is farther inland and at a higher altitude than the presently active precipitation ridge in the

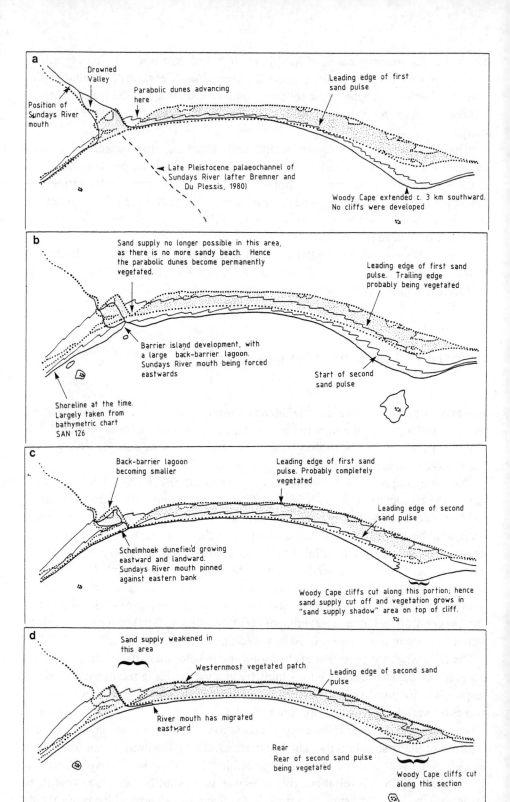

a

Drowned Valley

Position of Sundays River mouth

Parabolic dunes advancing here

Leading edge of first sand pulse

Late Pleistocene palaeochannel of Sundays River (after Bremner and Du Plessis, 1980)

Woody Cape extended c. 3 km southward. No cliffs were developed

b

Sand supply no longer possible in this area, as there is no more sandy beach. Hence the parabolic dunes become permanently vegetated.

Leading edge of first sand pulse. Trailing edge probably being vegetated

Barrier island development, with a large back-barrier lagoon. Sundays River mouth being forced eastwards

Start of second sand pulse

Shoreline at the time. Largely taken from bathymetric chart SAN 126

c

Back-barrier lagoon becoming smaller

Leading edge of first sand pulse. Probably completely vegetated

Leading edge of second sand pulse

Schelmhoek dunefield growing eastward and landward. Sundays River mouth pinned against eastern bank

Woody Cape cliffs cut along this portion; hence sand supply cut off and vegetation grows in "sand supply shadow" area on top of cliff.

d

Sand supply weakened in this area

Westernmost vegetated patch

Leading edge of second sand pulse

River mouth has migrated eastward

Rear
Rear of second sand pulse being vegetated

Woody Cape cliffs cut along this section

396

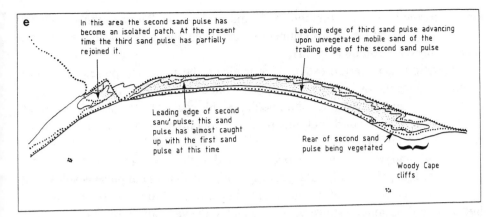

Fig. 7. The development of the Sundays estuary and adjacent dunefields over the past 6500 years, assuming that sea level changes were responsible for fluctuations in sand supply to the dunefields. The dotted lines show the present positions of the dunefields, coastline and river.

same area (Fig. 4). Estuarine shells (*Loripes Clausus*) found near the shore about 3 km east of the Sundays River mouth were deposited when the sea level was at least 1.5 m higher than present (J S V Reddering, pers. comm., 1987). The *Loripes* shells are about 6000 years old (according to a radiocarbon date Pta 4469; J C Vogel, pers. comm. 1986), confirming this higher mid-Holocene sea level. Flemming (1977), Yates (1986), Reddering (in press) and others also record a Holocene transgression for the southern African coastline 2-3 m higher than present sea level, about 6000 years ago. The fossil precipitation ridge in the Schelmhoek are is presumed to be the ultimate landward position reached by the first sand pulse before it was fixed by vegetation.

During this time (6500 years ago) parabolic dunes probably advanced on the area landward on the presently active precipitation ridge on the side of the estuary (Figs. 4, 7a). These parabolic dunes are at present vegetated and are more eroded than parabolic dunes in vegetated dune patches (Fig. 2), so these dunes were probably vegetated early in the life of the dunefield (about 5500 to 6000 years ago?), before the first sand pulse was finally vegetated. This implies that sand supply to these parabolic dunes was cut off prior to 5500 years ago, most likely because the sandy beach which used to supply the dunes with sand ceased to exist. This sandy beach would have disappeared as a result of the development of the barrier island which was growing across the Sundays estuary (Fig. 7b).

From a comparison of the 1942 and 1980 aerial photographs of the Woody Cape area, it is apparent that during these four decades the cliffs had been eroding at an average rate of about 0.3 m/year. Accepting this value as a long term average rate, the cliffs could have extended at least 2 km farther southward 6500 years ago. This is a minimum figure since the erosion rate was probably higher 6500 years ago because the concentration of wave energy at Woody Cape should have been higher since it formed a more prominent headland at the time. The coastline at Woody Cape 6500 years ago is thus presumed to have been about 3km farther southward than the present position (Fig. 7a).

Vegetation then started establishing itself on the first sand pulse. This establishment of vegetation could have resulted from a climatic change, or from a large reduction in sand supply to the dunefield, possibly caused by the shoreline retreating seaward about 2 km during this time, corresponding to a relative drop in sea level of about 15m (Fig. 7b).

A second sand pulse could have been initiated by the sea level rising to about the present level approximately 3500 years ago (Fig. 7c). This sand pulse is still visible at present at a few places at the dunefields, in the form of the megaridge. This sand pulse has at present almost everywhere engulfed the vegetation that has taken hold on the first sand pulse, the exceptions being the vegetated dune patches, which are remnants which have survived for the various reasons:

(1) at various times during the history of the dunefield, the Sundays River mouth migrated up to 3 km to the east of its present position, similar to the present behaviour of the Gamtoos River mouth. Evidence for this migration is the occurence of shells of the estuarine bivalve, *Loripes clausus*, and other estuarine fauna, e.g. *Mercierella*, in the living positions in sediments along the coastline about 3 km east of the present river mouth (Hesp, 1986). The remnants of a channel are also visible on 1939 aerial photographs (Dept of Surveys, Job 141/39). This occasional change of position of the river mouth would intermittently reduce sand supply in the western end of the Alexandria dunefield, and probably resulted in the survival of the westernmost vegetated patch (Fig. 7e).

(2) The second sand pulse has not yet caught up with the first sand pulse along the precipitation ridge in the Woody Cape area and at other vegetated patches along the precipitation ridge.

(3) The creation of the cliff at Woody Cape by wave erosion cut off sand supplies in that area, resulting in the vegetated patch in the "sand supply shadow" area above the cliffs (Figs. 7b-e).

The well-developed megaridge extending from Woody Cape almost to the Sundays River mouth (Fig. 4) represents a third pulse of sand, at the trough between this megaridge and the landward megaridge is interpreted to mean that the sand supply to the dunefield was again interrupted. This could have been caused by the shoreline retreating about 1 km during this time, probably about 2500 years ago, corresponding to a relative drop in sea level of about 5 m (Fig. 7d). Vegetation spread across part of the Schelmhoek dunefield during this lull in the sand supply (Figs. 2, 4, 7).

The third sand pulse could have been caused by the sea again rising to the present stand. From the dune movement rates (Illenberger and Rust, in press), the northward component of movement of a dune 30 m high is about 0.5 m/yr. The third sand pulse, which is 30 m high, would thus take about 1200 years to move to its present position 600 m from the shore, so one can infer that this pulse was initiated about 1200 years ago.

In the western end of the Alexandria dunefield the third sand pulse has developed a double crestline, and in the Schelmhoek dunefield has split into two megaridges (Fig. 4). This could have resulted from a retreat of the shoreline, corresponding to a 1 m drop in sea level about 600 years ago.

Figure 8 represents a sea level curve for Algoa Bay for the past 7000 years assuming that the sea level changes are responsible for fluctuations in the sand supply to the dunefield.

The precipitation ridge represents a sand pulse (or coalscent pulses) which has been steepened, slowed down and heightened by vegetation: plant growth on a dune will trap mobile sand, causing sand to pile up and steepen the dune slope, and slow down dune movement and heighten a dune because sand is accreting vertically rather than moving horizontally. The active, non-vegetated third sand pulse is at present moving landward at about 0.5 m/year, while the precipitation ridge is at present moving landward at about 0.25 m per year. The landward slope of an active, non-vegetated sand pulse within the dunefield is about 1 in 20, while the precipitation ridge slopes landward at about 1 in 2 (Fig. 6). In addition, the precipitation ridge usually forms the highest part of the dunefield (Fig. 6). This implies considerable slowing down, steepening and heightening of a sand pulse when it becomes or coalesces with the precipitation ridge.

Sand pulses should eventually coalesce in the precipitation ridge since they move faster than the precipitation ridge; at the present time the pecipitation ridge consists of the coalesced first and second sand pulses along the most of its length (Fig. 7).

If vegetation on the precipitation ridge is disturbed, a flood of sand which had been "dammed" by the vegetation should be released: this is indeed the case; disturbed precipitation ridge moves up to four times as fast as undisturbed precipitation ridge (McLachlan et al., 1982).

5. THE CAUSE OF THE FLUCTUATING SAND SUPPLY

The existence of an episodic sand supply to the dunefield seems to be well established; such fluctuations have also been observed in Holocene coastal dunefields at Cape Recife (Lord et al., 1985), Cape St Francis (inspection of aerial photographs) and other parts of the world (Thom et al., 1981). Four possible causes of fluctuating sand supply were suggested in the above section. Their relative merits will now be discussed.

5.1. Variations in the relative sea level in on-offshore displacement
    of the sandy beach

Fairly large, perhaps unrealistic, variations in sea level are required. Supporting evidence for such variations in relative sea level is at present lacking, apart from evidence for the +3m sea level 600 years ago. Otherwise this mechanism is feasible.

5.2. Large scale variations in the total sand supply

Fossil dunefields with ages which may range up to the Pliocene (McMillan, 1986) are found in the Algoa Bay area, suggesting that the sand supply in the area has been maintained for a very long time.

This does not preclude the possibility of fluctuations in the sand supply of the order of hundreds to thousands of years, but such fluctuations do not seem likely if one considers the supply of sand to the coastal system; from fluvial input, erosion of the coastline and in situ biogenic production. Coastal sand transport systems are very fast; erosion started downstream (with respect to longshore drift) of the Port Elizabeth harbour within a year of the construction of the breakwater, which formed a barricade to longshore transport. Further evidence for the efficacy of longshore drift is to be found in the beach erosion problem of Port Elizabeth beachfront. The Noordhoek dunefield which used to supply sand which was transported by longshore drift along the beachfront was artificially vegetated in 1970; 15 years later the beachfront had been almost completely stripped of sand by longshore drift (Lord et al., 1985).

Fluctuations in the order of hundreds to thousands of years would thus imply that sand supply from fluvial input, erosion of the coastline and in situ biogenic production ceased for the corresponding time periods. This is very unlikely. As long as there is a sand supply to the coastal sand system, there will be a sandy beach along the northern shore of Algoa Bay, according to the logarithmic spiral curve theory (Bremner, 1983), except in the unlikely event that the average wave regime in the area should change. This sandy beach would ensure a sand supply for the dunefield. Relatively minor fluctuations in the sand supply to the coastal sand transport system of the order of months to tens of years could result from the onshore movement of shallow marine sand bodies by storm waves, and episodic sediment input by river floods. Such variations are probably not significant in the magnitude or in the time scale relevant to the sand supply inferred in coastal dunefields. Thus it seems that this is not feasible as a cause of fluctuations in the sand supply.

5.3. Climatic changes

The presence of fossil dunefields in the Algoa Bay area in similar palaeotopographic localities along the northern shores of palaeocoastlines (Illenberger, 1986) as the present dunefields suggests that the average wind regime has been similar to the present sea level highstands over at least the past 4 million years. A calculation of sand budget for the Alexandria coastal dunefield during the past 6500 years showed that wind transport of sand over that time period was not very different from the wind regime over the period 1939 to 1980 (Illenberger and Rust, in press), again implying a fairly constant wind regime. Thom et al. (1981) have speculated that variation in wind and wave action resulted in sand pulses in coastal dunes along the east Australian coast. The evidence presented here refutes their speculation.

Variation in temperature and/or variation of rainfall at the scale of hundreds to thousands of years is, however, possible. An oxygen isotope temperature curve for the southern Cape (Talma and Vogel, in prep, in Tyson, 1986) is in phase with the inferred fluctuation of sand supply to the dunefields (compare Figs. 8 and 9).

The early Holocene (9000 to 4000 years ago) in southern Africa was probably slightly drier than the present climate, followed by a wetter period up to 1000 years ago (H.J. Deacon, pers. comm. 1986; Tyson, 1986). These rainfall variations are not in phase with the inferred fluctuation of the sand supply to the dunefields.

In a study of pollen records from the eastern Cape covering the last 12000 years, Meadows et al.(1987) found a marked change in vegetation 8000 years ago, to warmer and moister conditions. This change corresponds to the start of the present interglacial. From 8000 years again to the present, there was little change in the vegetation pattern, implying only slight climatic change.

Such slight climatic changes, partly out of phase with the sand pulses, are probably insufficient to cause the vegetation and devegetation of a transgressive coastal dunefield.

5.4. Destruction of dune vegetation

Destruction of dune vegetation by human activity or bush fire is possible, and could have caused the second sand pulse. The large number of middens in the dunefield, consisting mainly of shells of *Donnax serra*, with some pottery fragments, stone tools and vertebrate bones, is evidence that Strandlopers often inhabited the area. Tinley (1985) conjectured that their activities could have destroyed vegetation covering the dunefield.

However, the third sand pulse does not appear to have been caused by destruction of vegetation, making this mechanism less plausible. In addition, since fluctuations in sand supply occurred in other parts of the world (Thom et al., 1981), global destruction of dune vegetation by human activity is implied. This is possible, but not very likely.

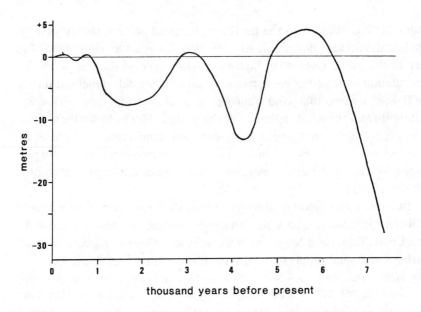

Fig. 8. Relative sea level changes in Algoa Bay for the past 7000 years, inferred by assuming that sea level changes are responsible for fluctuations in the sand supply to the dunefields.

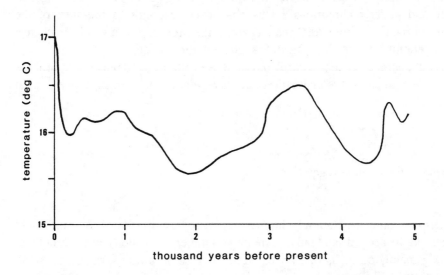

Fig. 9. Oxygen isotope temperature curve for the last 5000 yrs BP for the Southern Cape at Cango Cave (After Talma and Vogel, in Tyson, 1986).

6. CONCLUSIONS

The Sundays estuary developed from the infilling of a drowned river valley by the

402

rise in sea level after the last Glacial Maximum. A barrier island-lagoon estuary developed at first. The lagoon then filled in, and the estuary channel now follows a meandering path across the floodplain in the upper estuary.

The Schelmhoek dunefield grew out of the western barrier island, with sand transported off the beach and into the dunefield by the dominant west-southwest wind. As this dunefield grew, it pushed the lower estuary channel eastward, till the channel encountered the relatively hard eastern bank, preventing further movement. The channel is maintained by river floods. The distribution of vegetated dunes and the presence of large shore-parallel ridges of sand within the Schelmhoek and Alexandria coastal dunefields indicates that sand was supplied to the dunefields in pulses.

Three main pulses are discernible. These pulses could result from changing sea levels, climatic variation and from damage to vegetation resulting from Strandloper activity or natural bush fires. A combination of these causes is possible, with the first pulse resulting from an initial sea level high stand, and the second and third pulses resulting from vegetation changes due to Strandloper activity, aided by slight climatic changes. Large-scale variations in the total sand supply to the coastline, which might also cause the pulsing sand supply, are unlikely. The pulses of sand are analogous to swash running up a beach at an oblique angle (in the direction of net sand movement by wind, to the east-northeast). The present configuration of areas of vegetated dunes is elucidated by this analogy.

7. SUMMARY

The Sundays estuary developed from the infilling of a drowned river during the latter Holocene. The drowned valley was formed by the rise in the sea level after the last Ice Age. A barrier island-lagoon estuary developed at first. The lagoon then filled in, and the estuary channel now follows a meander path across the floodplain in the upper estuary. The Schelmhoek dunefield grew out of the western barrier island, with sand transported off the beach and into the dunefield by the dominant west-southwest wind. As this dunefield grew, it pushed the lower estuary channel eastward, till the channel encountered the relatively hard eastern bank, preventing further movement. The channel is maintained by river floods. The distribution of vegetated dunes and the presence of large shore-parallel ridges of sand within the Schelmhoek and Alexandria coastal dunefields indicates that sand was supplied to the dunefields in pulses. Three main pulses are discernible. These pulses could result from changing sea levels, climatic variation and from damage to vegetation resulting from Strandloper activity or natural bush fires. A combination of these causes is possible, with the first phase resulting from an initial sea level high stand, and the second and third pulses resulting from vegetation changes due to Strandloper activity, aided by slight climatic changes. Large-scale variations in the total sand supply to the coastline, which might result in a pulsing sand supply to the dune field are unlikely. The pulses of sand are analogous to swash running up a beach at an oblique angle (in the direction of net sand movement by wind, to the east-northeast). The present configuration of areas of vegetated dunes is elucidated by this analogy.

8. REFERENCES

Badenhorst, P. 1984. Proposed mining of dune sand Schelmhoek - Sundays River mouth. Council for Scientific and industrial Research Contract Report C/SEA 8416, Stellenbosch, South Afica, 22 pp.

Bremner, J.M. 1983. Properties of logarithmic spiral beaches with particular reference to Algoa Bay. In: McLachlan, A. and Erasmus, T. (eds.), Sandy beaches as ecosystems, W. Junk, The Hague, 97-113.

Bremner, J.M. and Du Plessis, A. 1980. Basement morphology and unconsolidated sediments in Algoa Bay. Geological Survey, Department of Mines, South Africa, Report no. 1980-0206, 7 pp.

Chappel, A. 1983. A revised sea level record for the last 300000 years from Papua New Guinea. Search 14, 99-101.

Cooke, R.U. and Warren, A. 1973. Geomorphology in Deserts.   University of California Press, California, 374 pp.

Cooper, W.S. 1958. Coastal sand dunes of Oregon and Washington.   Geological Society of America Memoir 72, 169 pp.

Flemming, B.W. 1979. Langebaan Lagoon; a mixed carbonate- siliclastics tidal environment in a semi-arid climate.   Sedimentary Geology, 18, 61-95.

Hesp, P.A. 1986. Notes on the Alexandria dunefield, Algoa Bay, Field guide: Workshop on structure and function of sand dune ecosystems, 20-22 January 1986; Institute for Coastal Research Report No 8, University of Port Elizabeth, South Africa, 95-103.

Illenberger, W.K. 1986. The Alexandria Coastal Dunefield: Morphology, Sand budget and History. Unpublished M.Sc thesis, University of Port Elizabeth, South Africa, 87 pp.

Illenberger, W.K. and Rust, I.C. in press. A sand budget for the Alexandria coastal dunefield. Sedimentology.

Lord, D.A., Illenberger, W.K. and Mclachlan, A. 1985. Beach erosion and sand budget for the Port Elizabeth beachfront.   Institute for Coastal Research Report no 1, University of Port Elizabeth, South Africa, 48 pp.

McLachlan, A., Sieben, P. and Ascaray C. 1982. Survey of a major coastline dunefield in the Eastern Cape. Zoology Department Report no. 10, University of Port Elizabeth, South Africa, 48 pp.

McMillan, I.K. 1986. Tertiary to Holocene foraminifera of the Algoa Bay area.

Proceedings of the Seminar on Tertiary to Recent Coastal Geology, 23-25 January 1986; Institute for Coastal Research, Department of Geology, University of Port Elizabeth, South Afica.

Meadows, M.E., Meadows, K.F. and Sugden, J.M. 1987. The development of vegetation on the Winterberg escarpment. The Naturalist, 31, 26-32.

Perry, J.E. 1983. The Sondags estuary. Hydrological/hydraulic study of Cape estuaries, National Research Institute for Oceanology Memorandum 8303, Council for Scientific and Industrial Research, Stellenbosch, South Africa, 6 pp.

Reddering, J.S.V. in press. Evidence for a middle Holocene transgression, Keurbooms estuary, South Africa. Palaeoecology of Africa, 19.

Reddering, J.S.V. and Esterhysen, K. 1981. Sedimentation in the Sundays estuary. Research on sedimentation in estuaries, Department of Geology, University of Port Elizabeth, South Africa, 47 pp.

Swart, H.E. 1986. Physical environmental interactions in the Sondags River/Schelmhoek area. Council for Scientific and Industrial Research, National Research Institute for Oceanology Report No 586, Stellenbosch, South Africa, 39 pp.

Tinley, K.L. 1985. Coastal Dunes of South Africa. South African National Scientific Programmes Report no. 109. Foundation for Scientific and Industrial Research, Pretoria, South Africa, 300 pp.

Thom, E.G., Bowman, G.M. and Roy, P.S. 1981. Late Quaternary evolution of coastal barriers, Port Stephens-Myall Lakes area, central New South Wales, Australia. Quaternary Research, 15, 345-364.

Tyson, P.D. 1986. Climatic Change and Variability in Southern Africa, Oxford University Press, Cape Town, 220 pp.

Yates, R.J., Miller, D.E., Halkett, D.J., Manhire, A.H., Parkington, J.E. and Vogel, J.C. 1986. A mid-Holocene high sea-level: a preliminary report on geoarchaeology at Elands Bay, western Cape Province, South Africa. South African Journal of Science, 82, 164-165.

STRATIGRAPHY AND GEOMORPHOLOGY OF THREE GENERATIONS OF REGRESSIVE SEQUENCES IN THE BREDASDORP GROUP, SOUTHERN CAPE PROVINCE, SOUTH AFRICA

J.ROGERS

Marine Geoscience Unit, University of Cape Town

1. INTRODUCTION

Stratigraphic and geomorphological criteria for mapping three generations of aeolianites within the Bredasdorp Group of the southern Cape are discussed in this paper, which expands on a summary by Rogers (1986). The geology and topography of the Pre-Cenozoic bedrock will first be discussed followed by a description of Cenozoic lithostratigraphy and regional geomorphology. The post-Palaeozoic geological history of the southern Cape coastal plain is then reconstructed.

2. BEDROCK GEOLOGY

Sandstones of the Table Mountain Group (TMG), crop out along the coast eastwards to Cape Vacca past the mouth of the Gouritz River and westwards past Ystervarkpunt to Bloukrans (Figs. 1,2). East of 22°E, Toerien (1979) has mapped anticlines of the TMG sandstone, overturned to the north, west of Cape St Blaize. Thus, although the sandstones on the coast dip southwards towards the Agulhas Bank they could be expected to be overturned to the north also. However, between Cape Vacca and Vleespunt, beside the shore of Visbaai (Fig. 2), is an intertidal outcrop of folded Bokkeveld sediments within a synclinorium, between the TMG sandstones which dip about 45° northwards at Cape Vacca and at the same angle, but southwards, at Vleespunt. Therefore, although Haughton et al. (1937) did not map the Bokkeveld outcrop beside Visbaai, their geological profile is probably correct in showing an anticline that is not overturned beneath Cenozoic cover, midway between Ystervarkpunt and the Gouritz River (Fig. 3). Detailed regional mapping of the Gouritsmond, Albertinia, Herbertsdale, Rietvlei and Stilbaai sheets has recently been completed by

Geomorphological Studies in Southern Africa, G.F.Dardis & B.P.Moon (eds)
© 1988 Balkema, Rotterdam. ISBN 90 6191 831 6

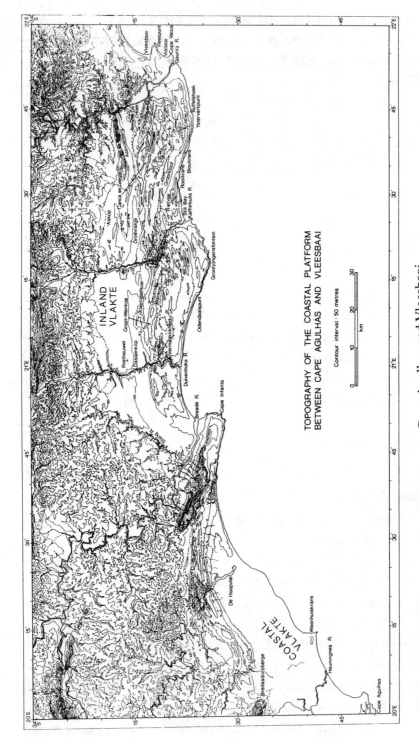

Fig. 1. Topography of the coastal platform between Cape Agulhas and Vleesbaai.

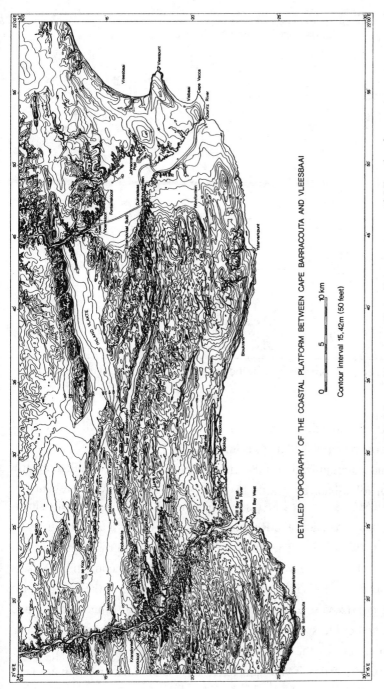

DETAILED TOPOGRAPHY OF THE COASTAL PLATFORM BETWEEN CAPE BARRACOUTA AND VLEESBAAI

Contour interval 15,42m (50 feet)

Fig. 2. Detailed topography of the coastal platform between Cape Barracouta and Vleesbaai.

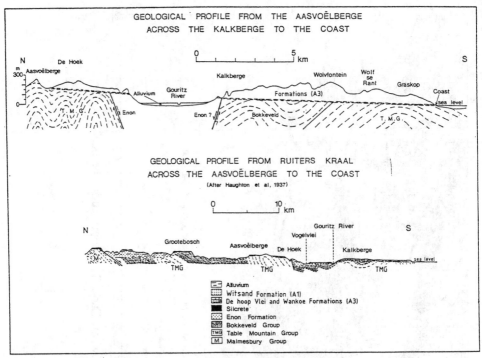

GEOLOGICAL PROFILE FROM THE AASVOËLBERGE
ACROSS THE KALKBERGE TO THE COAST

GEOLOGICAL PROFILE FROM RUITERS KRAAL
ACROSS THE AASVOËLBERGE TO THE COAST
(After Haughton et al, 1937)

Alluvium
Witsand Formation (A1)
De hoop Vlei and Wankoe Formations (A3)
Silcrete
Enon Formation
Bokkeveld Group
TMG Table Mountain Group
M Malmesbury Group

Fig. 3. Geological profile from the Aasvoelberge across the Kalkberge to the coast.

De V. Wickens of the Geological Survey (personal communication, 1984), so a complete revision of existing published maps is imminent.

Wickens (personal communication, 1984) has mapped a major E-W fault along the southern shore of Vleesbaai eastwards to Vleespunt (Fig. 2). At Vleesbaai the southerly-dipping TMG sandstones are coated on the northerly fault scarp with silica-cemented sedimentary breccia of the Robberg Formation (Rigassi and Dixon, 1972; Rogers, 1966; Rust, 1979, 1983). The clasts are very angular and range in size from pebbles to boulders. The lithology is restricted to TMG sandstone and the matrix is characteristically ferruginous. Within a few hundred metres north of the fault are outcrops of light gray quartzose sandstone containing lenses of well-rounded sandstone pebbles, cobbles and boulders and occasional lenses of angular sedimentary breccia. This is again reminiscent of the Robberg type-section at Plettenberg Bay, where sedimentary breccia overlies TMG quartzites and is in turn overlain by gray sandstones and well-developed lenses of well-rounded, silica-cemented conglomerates.

Iron-stained clayey gravel of the Enon Formation (Malan, 1987a; Malan and Theron, in press) is found in the Bredasdorp District (Figs. 1,4) where Whittingham (1969, p.3) reported: "...The clays and clay-gravels are seen in places to underlie the calcareous

410

Bredasdorp Formation in the "duine"... Boreholes drilled in the clay-gravel formation show that these rocks locally attain a thickness of over 300 feet (90 m)..." (Fig. 4). The best exposures according to Whittingham (1969, p.37) are in the gorges of the Kars River near Bredasdorp and of the Sout River north of De Hoop Vlei (Fig. 1). However the Inland Vlakte, especially between the Breede and Kafferkuils Rivers (Fig. 1), are strewn with iron-stained, often muffin-shaped, rounded cobbles of TMG sandstone up to 30 cm in diameter (cf. Wybergh, 1919, p.51).

The contact between TMG sandstones to the south and Bokkeveld shales and sandstones to the north runs from Cape Vacca in the east, across the Gouritz River south of Melkhoutfontein to Bloukrans, west of Ystervarkpunt (Fig. 2). It continues westwards across the inner continental shelf to Still Bay West, where the contact is obscured by Recent sand north of the harbour (De Bruin, 1971). The contact is again obscured beneath Cenozoic sediments north of Jongensfontein (near Cape Barracouta) (Fig. 2) and the TMG coastal outcrop ends to the west at Odendaalspunt (Fig. 1) (Whittingham, 1971). The Bokkeveld/TMG contact again crosses the inner shelf past the Duiwenhoks River to the mouth of the Breede River north of Cape Infanta (Fig. 1). Thus, from Cape Vacca to Cape Infanta the TMG/Bokkeveld contact has a consistent orientation of WSW, similar to the orientation of the calcarenite ridges.

The age of initiation of the faulting exposed at Vleesbaai is probably 162 Ma (i.e. mid-Jurassic [Bathonian]) if the faulting along the northern Algoa Basin (Hill, 1972) is contemporaneous. The cessation of such faulting is discussed by Dingle et al. (1983, p.99): "...The large, localized downward crustal movements that occurred at these times produced a rugged terrain in which sedimentation was locally very rapid, in tectonically unstable environments (taphrogenic phase of sedimentation)...These movements ceased in late Aptian-early Albian times in all areas, with the possible exception of southern Mozambique, and graben- infill sediments are overlain, often with a marked discordance, by thick Upper Cretaceous and Cenozoic strata which dip gently seawards and locally prograde over the continental margin into the deep ocean basins (epeirogenic phase of sedimentation)... sedimentation has probably been more or less continuous since at least Upper Jurassic times in the basins that underlie the continental margins, but inland of the present-day coast has been sporadic in time and space...in the intermontaine basins of the southern Cape, sedimentation was locally rapid (up to 4 km) for a short period (Middle Jurassic to Hauterivian), then apparently ceased altogether..."

Dingle et al. (1983, p.102) summarize their conclusions on the age of cessation of the major faulting as follows:"...Large-scale taphrogenesis had ceased by the end of

411

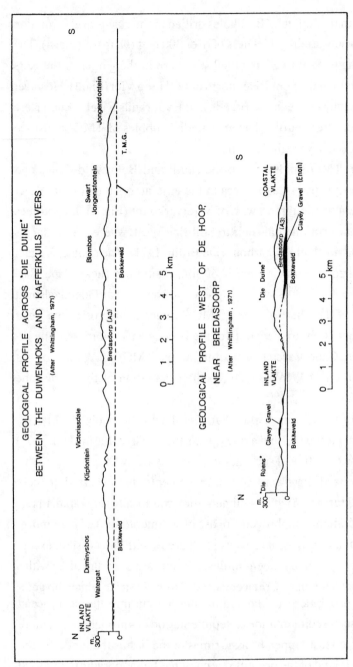

Fig. 4. Geological profile across "Die Dune" between the Duiwenhoks and Kafferkuils Rivers.

Aptian times around southern Africa, and Albian sediments spread over and beyond the previous boundary faults as a new, epeirogenic phase of continental subsidence began..."

Thus the Vleesbaai faulting was probably initiated in mid-Jurassic times (Bathonian) 162 Ma ago, but probably became dormant during Early Cretaceous times (end-Aptian) 109 Ma ago.

On the northern side of the broad Gouritz River floodplain (Fig. 2) is the well-exposed De Hoek fault, which truncates steeply dipping, white-weathering TMG sandstones of the Kouga Formation (Wickens, personal communication, 1984) (Fig. 3). Red-weathering, massive sedimentary breccia of the Enon Formation, containing pebble- to boulder-sized angular clasts, are well-exposed on the west bank of the river beside the Gouritsmond-Albertinia road. The fault, however, in this vicinity is not capped by younger formations. However, on the east bank a tributary near Vloerskloof (Fig. 2) has eroded eastwards along the fault and then northwards to expose the fault, where it is capped by tectonically undisturbed younger sediments (Fig. 3).

The road from Vogel Vlei to Cooper Siding runs northwards along the watershed east of this kloof and gives access to Fonteinskloof (Fig. 2) immediately east of the road. On the western side of Fonteinskloof is an unconformity between sedimentary breccia of the Enon Formation and marine gravelly sandstones of the De Hoop Vlei Formation, here forming the base of the Bredasdorp Group (Malan, 1987a). The Enon is characterised by distinctive angular pebble-size clasts of TMG-sandstone set in a ferruginous matrix. The De Hoop Vlei gravelly sandstones consist of well-rounded TMG pebbles set in a light yellowish-brown sandy matrix rich in well-preserved moulds of molluscs.

Haughton et al.(1937, Plate 1 and Fig. 1) present both a photograph and a geological profile to illustrate that the De Hoek fault is capped by younger sediments (Fig. 5) containing marine shells and glauconite. Haughton et al.(1937, p.28) do not offer any suggestions as to the age of the cessation of faulting, but are categorical as to its relative age: "...The faulting which affected the (Enon) conglomerates on Vogelvalley is, however, of the normal gravity type with downthrow to the south and is similar to that along the better-known Zuurberg, Oudtshoorn, and Worcester faults...Its age with respect to the supposed thrust is unknown...It is presumably later, but it definitely antedates the Tertiary peneplanation.... The Tertiary peneplanation affects all the rocks up to and including the Enon beds...Its remnants north of the main Cape road are covered with thin continental deposits; but on the seaward side of the Aasvogelberg and the east-west line continuing the ridge, the deposits are of marine origin and thicker..."

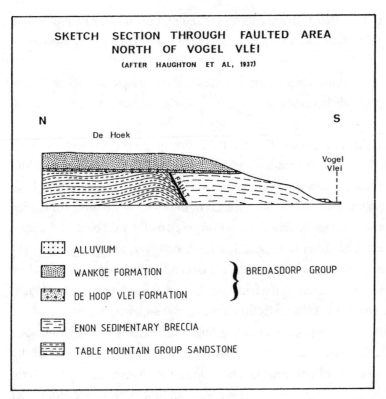

Fig. 5. Sketch section through the faulted area north of Vogel Vlei.

3. BEDROCK TOPOGRAPHY

Whittingham (1971) attempted a crude bedrock-topography map for the area between the Duiwenhoks and the Kafferkuils Rivers. He based his map on borehole data obtained during a 10-day field trip, mainly from farmers' recollections of the formations encountered when drilling for water. A general pattern has emerged that water was usually struck once sediments rich in coarse shell fragments were encountered just above bedrock. In other words, as Haughton et al. (1937), Whittingham (1971) and De Bruin (1971) have reported, rainwater percolates into the aeolianites of the Wankoe Formation (Malan, 1987a) but the aquifer lies just above the unconformity within the De Hoop Vlei Formation. Unfortunately, according to De Bruin (1971), only boreholes 48 and 49 near Eilandskop in Canca se Leegte are reported to have reached Bokkeveld bedrock, both at 65 m above sea level (Fig. 6). In Grootkloof, farther south, borehole 74 is nearest to the road but merely indicated that bedrock lies below +56 m. Similarly, borehole 55, between Grootkloof and Driefontein, proves

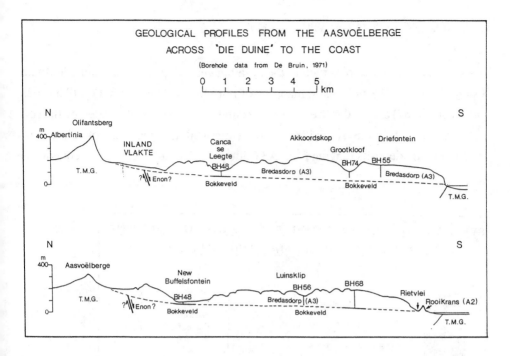

Fig. 6. Geological profiles from the Aasvoelberge across "Die Dune" to the coast.

that bedrock lies somewhere below + 60 m.

In general, previous workers (Wybergh, 1919; Whittingham, 1969, 1971; De Bruin, 1971) are of the opinion that the bedrock surface is a surf-cut plain that dips gently seawards by 6 m/km (33 ft/mile, Wybergh, 1919, p.47). The seaward slope of the Bokkeveld is best exposed in Still Bay East in road cuttings, just upstream of the bridge, which are best viewed from the west in the late afternoon. However echograms across the inner shelf, between Cape Hangklip and Danger Point, show that Bokkeveld rocks are easily planed to a relatively smooth surface, whereas Table Mountain Group sandstones often produce highly irregular topography (Rogers, 1985, Fig. 3). On the inner shelf west of the southern Cape Peninsula Simpson et al. (1970) also reported highly irregular TMG topography.

De Bruin (1971) reported that his borehole 80, west of Ystervarkpunt (Fig. 2) on Ystervarkfontein, failed to reach bedrock 32 m below sea level, whereas borehole 71, a few hundred metres inland, reached TMG sandstone 39 m above sea level. TMG sandstone crops out at sea level at the coast, south of borehole 80, implying that a narrow, but major bedrock depression underlies the coastal road west of Ystervarkpunt.

415

4. CENOZOIC LITHOSTRATIGRAPHY

The Cenozoic sequence observed in the area between Cape Agulhas and Vleesbaai is summarised in Figure 7. Field mapping of both the Knysna Formation (Thwaites and Jacobs, 1987) and of the Bredasdorp Group (Malan, 1987a,b) is aided by the soil classification of Schloms et al.(1983) who have postulated three sets of calcareous, chiefly aeolian, deposits ranging from A3 (oldest) to A1 (youngest). The sequence is

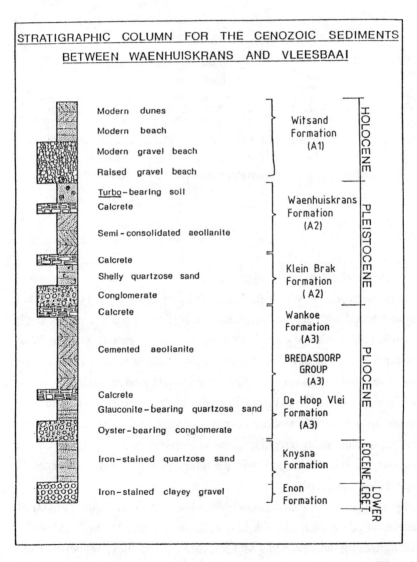

Fig. 7. Stratigraphic column for the Cenozoic sediments between Waenhuiskans and Vleesbaai.

as follows; (7) Witzand Formation (Unconsolidated coastal dunes- A1), (Contemporaneous beaches of sand and gravel), (6) Raised Holocene beach gravel, (5) Waenhuiskrans Formation (Regressional coastal dunes) (A2), (4) Klein Brak Formation (Transgressional beach sand and gravel) (A2), (3) Wankoe Formation (Regressional coastal dunes) (A3), (2)De Hoop Vlei Formation (Transgressional beach sand and gravel), (1) Knysna Formation (Iron-stained quartzose sand). Units 2 to 7 being part of the Bredasdorp Group (Malan, 1987a,b). Unit 1 is stratigraphically the oldest and unit 7 the youngest.

4.1. Knysna Formation (Iron-stained quartzose sand)

Wybergh (1919, p.51, Sections I to III) (Fig. 8) describes the iron-stained sands as follows: "...Above this pebble bed, along the northern edge of the formation, in the outliers and along the foot of the escarpment, there is a bed of soft reddish brown sandstone, which readily disintegrates, covering the surface with deep sand. The maximum thickness of this bed is probably, in its northern portions, about 100 feet, but, unlike the pebble bed, it thins out and completely disappears a couple of miles or so south of the escarpment. Only in one place, viz., at Vermakelykheid, are the red sands exposed in one of the little valleys which, within the area south of the escarpment, have dissected the formation down to bedrock, and this seems to mark its extreme southern limit. It appears to increase in thickness going westwards towards the Breede River, where for many miles it forms the face of the escarpment. In this locality (Vermaaklikheid) it exhibits marked false bedding, but nevertheless its general stratification approaches the horizontal and it forms a definite horizon overlain by the white shell limestone. This is particularly well seen in the outliers to the north. At Oude Muragie, the farthest north of these, about four miles north of the escarpment, the whole formation is thinner. The hill consists of a bed of sandstone about 25 feet thick, underlain by a thin sheet of reddish gravel, and overlain by a capping about 25 feet thick, which no doubt consists of the shell limestone, as its surface is covered by limestone tufa. In the wide valleys and the portion of the open plain which lies immediately north of the main escarpment the surface is often covered by a red sandy soil, which is no doubt the remains of this sandstone bed, the rest having been removed by denudation. No fossils were found in the sandstone, which is non-calcareous, often dark-coloured and highly ferruginous..."

Some of the outliers described by Wybergh were visited and sampled at Rietheuwel, near Oude Muragie, on the left (east) bank of the Duiwenhoks River, at Huis se Kop,

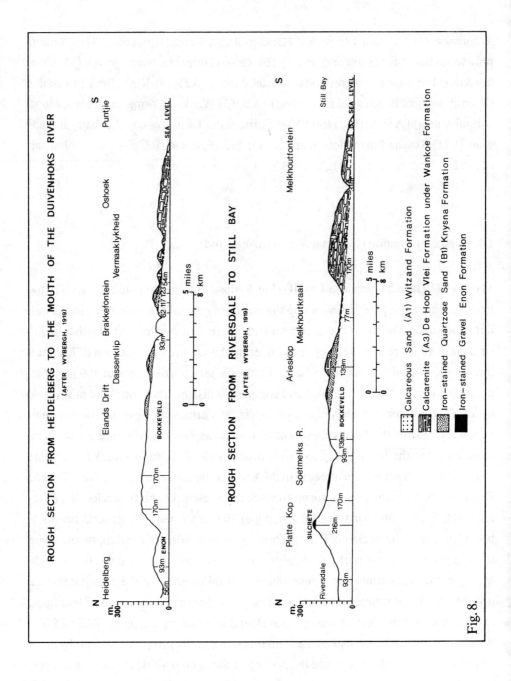

ROUGH SECTION FROM HEIDELBERG TO THE MOUTH OF THE DUIVENHOKS RIVER

(AFTER WYBERGH, 1919)

ROUGH SECTION FROM RIVERSDALE TO STILL BAY

(AFTER WYBERGH, 1919)

⬚⬚⬚ Calcareous Sand (A1) Witzand Formation

▨ Calcarenite (A3) De Hoop Vlei Formation under Wankoe Formation

░░ Iron-stained Quartzose Sand (B1) Knysna Formation

■ Iron-stained Gravel Enon Formation

Fig. 8.

418

on the eastern side of the Riversdale-Still Bay road, and at Askop, on the southern side of the national road, between Reisiesbaan and Dekriet sidings. The main escarpment was sampled at Kom se Kop, just north of Vermaaklikheid on the left bank of the Duiwenhoks River, and at Elbertskraal Suid on the right bank of the Gouritz River.

Whittingham (1971) describes "red sandy rock" from 2 boreholes (26 and 30), which failed to reach bedrock, at Blombos (Fig. 4) on the coastal road midway between Vermaaklikheid and Still Bay (Fig. 1). Three metres were penetrated in borehole 26, between +29 and +26 m, and 7 m were encountered in borehole 30, between +37 and +30 m. The powerful perennial spring at Jongensfontein (Fig. 2) west of Still Bay issues from a cliff of this red sandstone beneath aeolianite, whose heavily calcretized carapace is well exposed in the road cutting leading down to the resort. On the right bank of the Kafferkuils River at Grootfontein (Fig. 2), just upstream of the bridge, there is a prominent sand quarry in soft red sandstone, beneath calcarenite, and farther north near Kransfontein another outcrop was observed in a tributary.

4.2.De Hoop Vlei Formation (Transgressional beach sand and gravel)

The base of the calcrete-capped calcareous A3-sediments is a coarse sandy gravel zone, the De Hoop Vlei Formation (Malan, 1987a), which in many places lies directly on bedrock. A section was measured in the Still Bay East road cutting (Fig. 2) near the bridge, in an area where the overlying gray-weathered, calcrete-capped, aeolian calcarenite of the Wankoe Formation was about 63 m thick, and where the gravels overlie Bokkeveld bedrock:

0.6 m   White calcrete, grading upwards from powder to hardpan

0.6 m   Light green, glauconite-bearing, quartzose, massive, muddy fine sand

0.6 m   Light gray, glauconite-bearing, quartzose, rich in shell fragments, horizon tally laminated, slightly muddy fine sand with floating pebbles

0.6m   Orange, iron-stained, quart-rich, bearing shells (molluscs, oysters, barnacles), lithic (calcarenite minor, Bokkeveld subordinate, TMG dominant), coarse sandy gravel (well-rounded clasts 10 - 150 mm in dameter in a clast-supported fabric)

0.9m   Light yellow, bearing shell fragments, quartz-rich, lithic (well-rounded TMG clasts, 10 - 350 mm in diameter, in a clast-supported fabric) coarse sandy gravel.

419

This 3.3 m-thick sequence was photographed and sampled and then compared to similar sequences farther upstream (Fig. 2) at Kransfontein on the right bank and in a tributary of the Kafferkuils River on Riethuiskraal on the left bank, both over 10 km farther inland, where glauconite, so abundant on the Agulhas Bank (Birch, 1979), is associated with marine shell fragments. Rare sharks' teeth are a distinctive feature of this sequence (Siesser, 1972). One, obtained from the basal horizon in a borehole on Platbos in Still Bay West was donated to the writer by Dr O R Van Eeden and in turn donated to the S.A. Museum, where it was identified as *Odontaspis acutissima* by Mr J Pether. The same species was found in Skelmkloof, the valley east of Fonteinskloof (Fig. 2), in the same horizon, but at an elevation of about 100 m and about 15 km inland, on a field trip led by Mr M J Mountain (Galliers, personal communication, 1984). Unfortunately teeth of this species are not age-diagnostic, unlike those of the extinct giant shark, *Carcharodon megalodon* that was found at the base of the Pliocene Duynefontyn member of the Varswater Formation at Koeberg Power Station (Rogers, 1982). Dr Van Eeden also guided the writer to an excellent exposure in a tributary on Grootfontein, north of Still Bay West, where a good specimen of the extant, warm-water oyster, *Crassostrea margaritacea*, was found and donated to the S.A. Museum, where it was identified by Mr J Pether (personal communication, 1984).

4.3. Wankoe Formation (Regressional coastal dunes)

The regressional aeolianites of the Wankoe Formation (Malan, 1987a), volumetrically comprise the bulk of the Bredasdorp Group (Figs. 6,8). Schloms et al.(1983, p.77), refer to the *terra rossa* soil on the calcarenites as follows: "...The A3 subunit exhibits more advanced pedogenesis....". It forms distinctive, well-grassed, firm slopes and is locally the foundation of small agricultural communities, especially east and west of Still Bay (Figs. 2,6,8).

Exposures of Wankoe aeolianite are reasonable in parts of the more dissected northern slopes of Rietvlei, Grootkloof and Canca se Leegte east of the Kafferkuils River (Fig. 2). However exposures are excellent in the precipitous gorges incised into the northern slopes of Wankoe, west of the Kafferkuils, and in tributaries of the Duiwenhoks (Fig. 1). The aeolianites are distinguished by typical, large-scale, aeolian, tabular cross-bedding and although the fresh rock is a light yellowish brown dotted with specks of very dark green glauconite, the weathered surfaces are characteristically dark gray. This is in sharp contrast to the white, weathered exposures of the hardpan calcrete that encrusts the deeply dissected surface of the aeolianite. In general, (Fig. 8) the aeolianites thicken southwards, where some of the highest peaks occur above bedrock which is nearing sea level. The highest peak, Baken se Kop, occurs nearly 7 km NNW of Ystervarkpunt (Fig. 2) and reaches an elevation of 325 m. As the peak lies halfway between the coast, where TMG bedrock is at sea level, and Canca se Leegte (Fig. 2), where De Bruin (1971) reported Bokkeveld bedrock at + 65 m, it is probable that bedrock lies near + 30 m, so that the maximum thickness of the aeolianites may be an impressive 300 m. Immediately inland of Ystervarkpunt the aeolianite peaks reach elevations of 164 m, 210 m (Buffelskop), 237 m (Miskraalkop) and 278 m (Mierhoopbak) suggesting thicknesses of 150 to 250 m of aeolianites.

The aeolianite is calcite-cemented and does not disaggregate readily into its constituent parts of quartz, shell fragments and glauconite. The "*terra rossa*", in contrast, is unconsolidated, but its quartz grains are very distinctively pitted and iron-stained.

4.4. Klein Brak Formation

A younger set of sediments, termed A2 by Schloms et al.(1983), is exposed east of Still Bay on the coast at Rooikrans, near Rietvlei (Figs. 2,5). The vertical cliffs at Rooikrans are 60 m high and consist of a lower marine sequence capped by a non-pedogenic calcrete and overlain by an aeolianite sequence (the Waenhuiskrans Formation). The sediments are semi-consolidated and closely resemble the semi-consolidated regressive sequence exposed at Swartklip on the northern shore of False Bay (Barwis and Tankard, 1983). Assuming contemporaneity, the lower marine sequence, the Klein Brak Formation (Malan, 1987b) is correlated with the calcrete-capped marine Velddrif Shelly Sand Member of the Bredasdorp Group (Tankard, 1976) and a Late Pleistocene (Eemian) age of 125000 years is assigned to it. The Rooikrans deposit appears to have largely been eroded away farther east, although remnants were observed at Bloukrans (Fig. 2). Wickens (personal communication, 1984) reports that only the aeolianites are exposed along the coast west of Rooikrans, possibly indicating that the Eemian marine sediments were deposited along a more embayed coastline.

At Waenhuiskrans (Fig. 1), west of the Breede River, the exposure differs from that at Rooikrans in that beneath a bed of indurated, laminated, Late Pleistocene beach sand is a well-developed, imbricated, well-cemented, Eemian boulder beach. East of Ystervarkpunt, a calcrete-capped, unfossiliferous boulder beach up to 8m above sea level was excavated in a series of trenches and it is also correlated with the Klein Brak Formation.

4.5.Waenhuiskrans Formation

The well-exposed aeolianites at Waenhuiskrans (Malan, 1987a) and at the Rooikrans cliffs (Rogers, 1986) are correlated with the Langebaan Limestone Member (Visser and Schoch, 1963), which is attributed to increased aeolian activity during the Late Pleistocene Wurm IIb regression (Barwis and Tankard, 1983). The winds would have been strongest about 20000 years ago, when the coastline had retreated southwards to the present 130 m isobath on the Agulhas Bank (Dingle and Rogers, 1972). The Waenhuiskrans Formation's outcrop at Rooikrans is much narrower (300 to 1000 m) and thinner (up to 60 m) than that of the Wankoe Formation (12 km wide and up to 300 m thick) (Figs. 2,6).

4.6.Raised Gravel Beach

An unusually well-developed raised gravel beach extends westwards for about 12 km from the mouth of the Gouritz River to Ystervarkpunt (Fig. 2). The beach is variable in width (10 - 40 m) and has a plane surface of extensively weathered, angular to well-rounded, cobbles of TMG sandstone and subordinate calcarenite. The upper surface dips gently seawards to a berm above a foreshore slope leading to the modern rocky intertidal zone. Near Ystervarkpunt, the inland surface has been surveyed at 6.4 m above sea level, which is not too high for the storm beach of a 2-3 m post-Holocene sea-level high, 2000-3000 years ago (Flemming, 1977; Yates et al, 1986). Farther east, however, a soil rich in *Turbo*-shell Strandloper debris, extends downslope to reach the raised gravel beach about 2 m below its inland edge. This implies that the soil started forming during the Wurm regression and that the raised gravel beach represents the inland limit of the Flandrian transgression.

4.7.Witzand Formation (Unconsolidated coastal dunes)

Along the steep, seaward convex slope of the Wankoe Formation west of Ystervarkpunt, unconsolidated dune sand forms a scrub-covered veneer, which is mappable as A1 sediment (Schloms et al, 1983). In the Western Cape, similar sands have been tentatively designated as the Witzand Calcareous Sand Member (Rogers, 1982). Along the south coast extensive development of coastal dunes is found at Witzand, east of the Breede River mouth (Fig. 1) where the dunes are mobile and unvegetated. Malan (1987, personal communication) therefore favours the name Witzand Formation for such dunes.

Just east of Ystervarkpunt a modern boulder beach is contemporaneous with the modern beaches feeding the modern (A1) dunes. The clasts of the raised beach are similar in angularity and size to those of the modern boulder beach, but the modern clasts are not weathered and do not, apparently, contain calcarenite cobbles. Both the younger beaches contain cobbles of the closely jointed and sometimes brecciated TMG sandstone seen along the intertidal outcrop, whereas the Late Pleistocene calcrete-capped beach (Klein Brak Formation), exposed in trenches landward of the Holocene raised gravel beach, contains an overwhelming predominance of well-rounded cobbles of homogeneous, unbrecciated TMG sandstone.

5. GEOMORPHOLOGY

5.1 The Post African 1 surface

Partridge and Maud (1987) have labelled the marine peneplain beneath the Bredas-dorp Group as the Post African 1 surface. In the study area it terminates along the seaward slopes of the Aasvoelberge, having cut right across a major taphrogenic (rift-ing) fault, at an elevation of 200 m (Fig. 3). This is similar to the elevation of the coastal platform east of Plettenberg Bay (Partridge and Maud, 1987, Section 12), but it is rare that richly fossiliferous marine sediments, as extensive as those north of Vogel Vlei, overlie the platform.

In the Bredasdorp area, Whittingham (1969) draws a clear distinction between the Inland Vlakte, north of the main outcrop of the Wankoe Formation and the Coastal Vlakte between Bredasdorp and Waenhuiskrans (Figs. 1 and 4). The Inland Vlakte are seen at the foot of the Aasvoelberge (Fig. 2), a doubly plunging anticline of TMG sandstone (Fig. 3) and are underlain by the DIL subunit of Schloms et al.(1983). The 20 km wide Coastal Vlakte reach hills capped by the Wankoe Formation near the 30 m contour and must have been cut by a post-Wankoe transgression.

5.2. Geomorphology of the Wankoe aeolianites

The topography of the Wankoe aeolianites is illustrated using a 50 m-interval in Figure 1, between Cape Agulhas and Vleesbaai. A more detailed map of the area between the Kafferkuils River and Vleesbaai was drawn using a contour interval of 15,42 m (50 feet) (Fig. 2).

Near the coast, gently undulating, fairly level terrain (Figs. 4, 6) is coated with a well-developed "*terra rossa*" soil supporting both *Restionacea* and grass. This is the zone of active farming and is where the coastal road runs between the farmsteads at elevations of 150-250 m. North of Grootkloof (Figs. 2, 6) the terrain becomes steadily more incised, the northern sides of the depressions usually being more extensively incised than the southern flanks. In the more dissected terrain the "*terra rossa*" soil has often been eroded away to reveal the white hardpan calcrete beneath. In the youthful gorges, particularly those on the northern flanks, the grey-weathering aeolianite is exposed. Thus, according to the degree of dissection, three different tones appear on aerial photographs. East and west of the actively farmed plateau around Driefontein (Figs. 2, 6), west of Ystervarkpunt, the aeolianite surface is characterised by rounded, convex hills incised by youthful "valleys", the hills reaching a maximum height of 325 m on Baken se Kop.

Marker and Sweeting (1983) have analysed the karst landscape on the aeolianites of the Nanaga Formation which overlies the marine Alexandria Formation east of Port Elizabeth. These formations are correlated with the Wankoe and De Hoop Vlei Formations, respectively (Malan, 1987a). A detailed study of the karst landscape of the Wankoe Formation between Cape Agulhas and Cape Infanta has been reported by Marker (1981) and Russell (1987). Attention is focused here on karst landscapes in the intervening area, especially between the Kafferkuils and Gouritz Rivers.

The most striking features of the area are the elongate depressions of Canca se Leegte, Grootkloof and Rietvlei (Fig. 2). Marker (1978, p.185), although chiefly dealing with landforms of potentially soluble dolomite, rather than calcarenite rich in insoluble quartz, comments: "...Enclosed hollows are accepted as the chief characteristic of karst...". Buckle (1978, Plate 6.7: Figure 6.18) illustrates two elongate closed depressions, about 1 km wide and 3-4 km long, that have developed on Jurassic limestones and dolomites in the Middle Atlas Mountains of Morocco. Buckle (1978, p.126) describes them as "poljes", which he defines as " ....a very large, shallow, steep-sided depression with a generally flat floor...It may be many kilometres in size. The flat floor is often emphasised by the deposition of terra rossa, a red clay material, which may form an impermeable layer and lead to flooding after heavy rains...Small residual hills, known as hums, form significant features in some poljes, and depending on the depth of the water table a polje may hold a temporary or permanent lake...Many of these giant depressions coincide with structural basins formed by folding or faulting. The large-scale solution involved in their origin has thus been guided by fault lines or fold axes...." On the same page Buckle also defined "ponorstet" or small cenotes (Marker, 1976): "...(A ponor) is a deep hole with nearly vertical sides leading to an underground cave system...Some result mainly from surface solution along joints and some from a combination of solution and subsurface collapse...Many are located on the floors of dolines, uvalas and poljes...".

Having defined the karst terms, let us return to the Atlas analogue of the Wankoe Formation's landscape. Buckle (1978, p.128) describes: "... a karst landscape of numerous solution hollows has been formed in the folded and faulted Jurassic limestones and dolomites...Dolines...are widespread, while the influence of faulting has concentrated solution into many downfaulted sectors and led to the formation of poljes...However, few of the limestones are pure enough or thick enough to favour the development of any underground drainage...Also, much of the karst may be fossilized since present rainfall totals are not sufficient to encourage very active solution...Ksouatene (Polje)....contains three ponore, while at the east end of Afriroua (Polje) is an aguelman, or shallow lake which lies at the lowest part of the polje...".

The surface features of the South Cape depressions relate well to poljes in their overall dimensions and shape (Fig. 2), in being floored by terra rossa, in containing ponore (sinks) and sometimes a lake. The eastern end of Wankoe (Fig. 1), west of the Kafferkuils River, has well-developed residual hills or "hums". In summary, the landscape is often deeply dissected, but, like the Middle Atlas, it is today also relatively desiccated.

5.3. Incised rivers

The degree of incision of rivers such as the Duiwenhoks, the Kafferkuils and the Gouritz is well displayed in Figs. 1 and 2. During the Late Pleistocene Wurm regression to today's 130 m isobath on the Agulhas Bank (Dingle and Rogers, 1972), the degree of incision was naturally even greater and today the lower courses of these rivers have been drowned by the Holocene Flandrian transgression. The seaward extensions of these drowned rivers have been located during seismic surveys of the inner continental shelf, the best-developed infilled palaeovalley being that off the Breede River (Birch, 1980).

The intensity of erosion was probably even greater during the Wurm, because although a cooler world climate would have resulted in less evaporation from the sea to initiate the hydrological cycle, any rainfall would have been subject to less evaporation and thus would have been more effective in eroding the landscape. Figure 1 shows that the Post-African 1 surface was later incised by the rivers, apparently after Late Pliocene uplift (Du Toit, 1926, p.440) (Partridge and Maud, 1987). Wybergh (1919, p.51) argues that the presence of gravels up to 15 m thick below the edge of the Duiwenhoks river at Dassenklip (Figs. 1, 8) indicates that a pre-existing gravel-filled gorge was planed by the transgression and then later re-excavated.

6. GEOLOGICAL HISTORY

The marine peneplain, which extends from the coast inland to the 200 m-contour at the foot of the Aasvoelberge (Fig. 1), may be as old as Late Cretaceous (about 70 Ma ago), according to Siesser and Miles (1979), Siesser and Dingle (1979) and Dingle et al. (1983). This date is derived from microfossiliferous Campanian to Maastrichtian sediments above 300 m at Needs Camp, near East London (McGowran and Moore, 1971; Dingle, 1981) which may have been deposited by a transgression up to 460 m above present sea level (Dingle et al, 1983). The succeeding regression, during the Cretaceous-Tertiary transition (65 Ma), was probably accompanied by major incision of the local rivers. A subsequent Early to Middle Eocene transgression, which reached 360 m at Needs Camp near East London, 204 m at Birbury (Siesser and Miles, 1979), and 170 m at Pato's Kop (Maud et al., 1987) near Bathurst in the Eastern Cape, would again have reached the foot of the Aasvoelberge. A major regression is postulated during the Oligocene to explain both the dearth of Oligocene marine sediments and the erosion levels on seismic records from the Agulhas Bank between about 530 and 425 m below present sea level (Siesser and Dingle, 1981). This would have been the period of maximum incision of the rivers and may also have been the period during which the iron-stained quartzose sands were leached and ferruginized. Du Toit (1926) suggested a correlation between these sands and the Knysna Formation (Rogers, 1909), which contain ferruginous sands and pollen-rich lignites that Thiergart et al.(1963) and Thwaites and Jacobs (1987) interpret as Early Tertiary (probably post-Eocene) in age.

Siesser and Dingle (1981), Dingle et al.(1983) and Hendey (1983) postulate a Middle to Late Miocene transgression, which reached elevations of about + 100 m and was followed by an Early Pliocene regression. Any Miocene sediments would have been vulnerable to erosion by the succeeding Pliocene transgression to over 200 m, once again to the foot of the Aasvoelberge (Fig. 2), but for the last time. Spies et al.(1963) quote a Mio-Pliocene age for an assemblage of bivalves, gastropods and benthic foraminifera from the De Hoop Vlei Formation near Bredasdorp. It is probable, therefore, that the fossiliferous sediments overlying the De Hoek fault are Pliocene in age but that the calcarenite clasts in the De Hoop Vlei Formation are derived from older Tertiary deposits.

Dingle et al. (1983, p.305) conclude their discussion of sea levels as follows: "...Further regression with strong erosion is inferred for late Pliocene times, based on widespread regressive aeolian deposits onshore (e.g. (Nanaga) Formation aeolianites) and the destruction of offshore phosphorite beds...". Their overall sea-level curve is strongly supported by evidence of a similar sequence of events in Western Australia (Quilty, 1977).

Spies et al.(1963) and Whittingham (1969) have drawn attention to the development of the Coastal Vlakte east of Bredasdorp, which appear to have been eroded into Wankoe aeolianites by a transgression as high as + 30 m, on the outskirts of Bredasdorp, some 20 km inland (Figs. 1 and 7). The Wankoe aeolianites may be Pliocene in age (2-5 Ma), but the + 30 m transgression may be as old as Late Pliocene, because it is higher than the 8-9 m sea level that Hendey and Cooke (1985) associate with Early Pleistocene coastal deposits at Skurwerug near Saldanha Bay. Krige (1927) referred to the transgression responsible for the surf-cut platform across TMG sandstone to + 40 m at Hermanus as his "Major Emergence", and ascribed the development of a 7-8 m bevel to his "Minor Emergence").

A Late Pleistocene raised beach on the left bank of the Klein Brak River east of Mossel Bay was first described by Rogers (1906). A shelly deposit, 4-6 m above sea level, was found between the junction of the Moordkuils River and the westward bend of the river. Twelve gastropods and 13 bivalves were identified, all except two (*Cerithium*, sp. n.sp.? and *Calliostoma*, sp. n.sp.?) being extant. Haughton et al. (1937, p.23) later identified the extinct gastropods as *Calliostoma mosselens* and *Cerithium (Vertagus) kochi* and reported other exposures in a road cutting on the left bank. The sequence is calcrete-capped, beneath a *"terra rossa"* soil and is underlain by a brown, sandy clay containing articulated bivalves. The sequence clearly resembles the deposit immediately east of the Swartklip cliffs on the shore of False Bay. Pether (personal communication, 1984) reports that *Calliostoma mosselense* Tomlin (1926) was considered by Barnard (1963) to be a synonym of *Cantharidus fultoni*. Kilburn and Tankard (1975) subsequently identified an extant grazing bivalve, *Cantharidus suarezensis suarensis* Fischer (1878) from the warm waters between Durban Bay and East Africa. The extinct Late Pleistocene *Cantharidus suarezensis fultoni* Sowerby (1889) has a type locality in the Swartkops River Beds and a range from Klein Brak River to Port Elizabeth. It is found on terraces 7 m, 4-5 m and 3.3 m above sea level. Malan (1987b) has proposed that the name Klein Brak Formation be formally assigned to these richly fossiliferous beds that are correlatable with the calcrete-capped fossiliferous Velddrif Member erected by Tankard (1976) and the calcrete-capped but unfossiliferous boulder beach immediately east of Ystervarkpunt (Fig. 2).

During the Wurm regression to -130 m (Chappell, 1974, p.421) Strandlopers occupied the coastal plain and left numerous shells of the gastropod, *Turbo sarmaticus,* on the colluvial soil overlying the calcrete-capped gravels east of Ystervarkpunt. More effective drainage during the cooler Late-Pleistocene hypothermal led to erosion of youthful valleys in the Wankoe-aeolianite hinterland and the deposition of small alluvial fans on the coastal terrace at the mouths of the valleys. The numerous fountains emanating from the De Hoop Vlei Formation at the Bredasdorp/Bokkeveld unconformity flowed more strongly during the hypothermal, giving rise to extensive calcareous tufa deposits in areas like Kransfontein on the right bank of the incised Kafferkuils River (Fig. 2).

At the end of the Flandrian transgression the sea had risen to 2-3 m above present sea level (Flemming, 1977; Yates et al., 1986) and formed a Holocene raised gravel beach, up to 6.4 m above sea level. Below seepages this raised beach and its capping dunes form a natural dam and marshes develop behind them, particularly between Ystervarkpunt and the Gouritz River (Fig. 2).

Recent unconsolidated (A1) dunes occur locally and are best developed at Witzand east of the mouth of the Breede River (Fig. 1).

7. CONCLUSIONS

The De Hoek fault, near the Aasvoelberge, is capped by unfaulted marine, fossiliferous, gravelly sands of the De Hoop Vlei Formation that is possibly Pliocene in age. The faulting at De Hoek may have ceased as recently as 5 m.y. ago. However it probably ceased over 100 m.y. ago, in Early Cretaceous times at the close of the rifting phase during the breakup of Gondwanaland.

The field characteristics of three generations of calcareous, coastal sands have been determined. A distinctive coarse sandy basal conglomerate (the marine De Hoop Vlei Formation) forms the aquifer for countless perennial springs along the Bredasdorp/Bokkeveld unconformity. The overlying aeolian Wankoe Formation (A3) is calcite-cemented, gray-weathering and has a well-developed calcrete profile capped by a "terra rossa" soil in an advanced state of pedogenesis. The Wankoe aeolianites form a belt up to 12 km wide, reach elevations of up to 325 m and are characterised by well-developed elongate depressions or poljes, with more deeply incised northern flanks. The Late-Pleistocene, A2-sediments are semi-consolidated, are not calcite-cemented, but contain calcrete horizons above a lower marine succession (the Klein Brak Formation) and capping an upper aeolian succession, the Waenhuiskrans Formation. They are rarely more than 0.5 km wide, are up to 60 m high, and form vertical and unvegetated cliffs at Rooikrans, between Still Bay and Ystervarkpunt, the type section being at Waenhuiskrans east of Cape Agulhas (Fig. 1). A1-sediments are the modern beaches and unconsolidated dunes (the Witzand Formation), whether partially vegetated or mobile. The Witzand dunes east of the Breede River mouth are good examples of the A1 dune fields forming, often at river mouths, under modern conditions.

8. SUMMARY

Calcarenites and calcirudites of the Cenozoic Bredasdorp Group have been deposited upon a surf-planed basement of folded and faulted Palaeozoic sediments of the Cape Supergroup between Cape Agulhas and Mossel Bay. Near Ystervarkpunt, west of the Gouritz River, southerly dipping Table Mountain Group sandstones on the southern limb of an overturned syncline are succeeded northwards by Bokkeveld Group shales and then by Table Mountain Group sandstones in the anticline of the Aasvoelberge. The Palaeozoic sediments are downfaulted in a graben of mid-Jurassic to Early Cretaceous age (162-109 Ma), between two East-West faults, one exposed at Vleesbaai and the other at the southern end of Gouritz River Poort. Sediments of the Enon Formation fill the graben. A marine peneplane truncates both the Enon and the Cape Supergroup sediments. Basal conglomeratic sandstones of the De Hoop Vlei Formation overlying the peneplane contain oysters and sharks' teeth and at Fonteinskloof, east of Gouritz River Poort, there is a rich deposit of well-preserved bivalve moulds. The oldest (Pliocene) generation of calcarenites often overlies an older, possible Early Tertiary, sequence of slightly muddy, iron-stained, quartzose sand, correlated tentatively with the Knysna Formation. The outcrop of the oldest calcarenites, chiefly the aeolian Wankoe Formation, is up to 300 m thick and 12 km wide. The calcarenites are characteristically calcite-cemented and calcrete-capped. The outcrop of the Late Pleistocene marine Klein Brak Formation and the overlying dominant aeolian Waenhuiskrans Formation, which are semi-consolidated, is only 60 m thick and only about 0.5 km wide. The Recent Witzand Formation consists of isolated modern beaches and patches of unconsolidated dune sand. The major elongate, closed depressions of Canca se Leegte and Grootkloof in the Wankoe Formation appear to be karst landforms called poljes. In a wetter Pleistocene climate they were probably coast-parallel lakes, like Rietvlei east of Still Bay.

9. ACKNOWLEDGEMENTS

The Principal of the University of Cape Town and Dr R.V. Dingle are thanked for permitting and encouraging me to embark on this investigation after a memorable reconnaissance flight along the south coast with Mr M.J. Mountain. Guidance at critical stages of my investigation was provided by Dr Dingle and Mr Mountain. Generous assistance was also given by Dr Q.B. Hendey and Mr J. Pether of the Cenozoic Research Unit at the South African Museum. Critical exposures at Still Bay were generously shown to me by Dr O.R. Van Eeden, a former Director of the Geological Survey, and by Mr De V. Wickens then mapping the Riverdale Sheet for the Geological Survey. Mr M. Vandoolaeghe of the Geohydrological Division of the Department of Environmental Affairs was instrumental in obtaining copies of the three key unpublished geohydrological reports on the region. Mesdames S.M.L. Sayers and S.N. Smith are thanked for draughting the diagrams and Mr A. Vinnicombe for the photographic reductions. Mr M. Smith undertook the textural and carbonate analyses and Mr H. Atkinson assisted with processing the textural data. Mesdames V. Morris and E.G. Krummeck are thanked for their typing expertise.

10. REFERENCES

Barnard, K.H. 1963. Contributions to the knowledge of South African marine Mollusca. Part III. Gastropoda: Prosobranchiata: Taeinioglossa. Annals of the South African Museum, 7, 1-99.

Barwis, J.H. and Tankard, A.J. 1983. Pleistocene shoreline deposition and sea-level history at Swartklip, South Africa. Journal of Sedimentary Petrology, 53, 1282-1294.

Birch, G.F. 1979. The nature and origin of mixed glauconite/apatite pellets from the continental margin off South Africa. Marine Geology, 29, 313-334.

Birch, G.F. 1980. Nearshore Quaternary sedimentation off the south coast of South Africa (Cape Town to Port Elizabeth). Bulletin of the Geological Survey of South Africa, 67, 1-20.

Buckle, C. 1978. Landforms in Africa: An Introduction to Geomorphology. Longmans, London, 249pp.

Chappell, J. 1974. Late Quaternary glacio- and hydro-isostasy, on a layered Earth. Quaternary Research, 4, 405-428.

De Bruin, C.G. 1971. Boorgat- en fonteinopname, Kafferkuils-Gouritz-Riviergebied, Distrik Riversdal. Technical Report Department of Water Affairs South Africa

Geohydrological Division, GH 1498, 1-11.

Dingle, R.V. 1981. The Campanian and Maastrichtian ostracoda of South East Africa. Annals of the South African Museum, 85, 1-181.

Dingle, R.V. and Rogers, J. 1972. Pleistocene palaeogeography of the Agulhas Bank. Transactions of the Royal Society of South Africa, 40, 155-165.

Dingle, R.V., Siesser, W.G. and Newton, A.R. 1983. Mesozoic and Tertiary Geology of Southern Africa. A.A. Balkema, Rotterdam, 375pp.

Du Toit, A.L. 1926. The Geology of South Africa. Oliver and Boyd, Edinburgh, 463pp.

Flemming, B.W. 1977. Depositional processes in Saldanha Bay and Langebaan Lagoon. Bulletin Joint Geological Survey/University of Cape Town Marine Geoscience Unit, 8, 1-215.

Haughton, S.H., Frommurze, H.F. and Visser, D.J.L. 1937. The geology of the country around Mossel Bay, Cape Province. Explanation Sheet 201 (Mossel Bay) South African Department of Mines Geological Survey, 48pp.

Hendey, Q.B. 1983. Cenozoic geology and palaeogeography of the fynbos region. In: Deacon, H.J., Hendey, Q.B. and Lambrechts, J.J.H. (eds), Fynbos Palaeoecology: A Preliminary Synthesis, Report South African National Scientific Programme, 75, 1-216.

Hendey, Q.B. and Cooke, H.B.S. 1985. Kolpochoerus paiceae (Mammalia, Suidae) from Skurwerug, near Saldanha, South Africa and its palaeoenvironmental implications. Annals of the South African Museum, 97, 9-56.

Hill, R.S. 1972. The Geology of the Northern Algoa Basin, Port Elizabeth. Unpublished M.Sc. thesis, Geology Department, University of Stellenbosch, 68pp.

Kilburn, R.N. and Tankard, A.J. 1975. Pleistocene molluscs from the west and south coasts of the Cape Province, South Africa. Annals of the South African Museum, 67, 183-226.

Krige, A.V. 1927. An examination of the Tertiary and Quaternary changes of sea-level in South Africa, with special stress on the evidence in favour of a recent world-wide sinking of ocean-level. Annals of the University of Stellenbosch, 5(A1), 1-81.

Malan, J.A. 1987a. The Bredasdorp Group in the area between Gansbaai and Mossel Bay. South African Journal of Science, 83, 506-507.

Malan, J.A. 1987b. Cainozoic sea-level movements, Hermanus to Mossel Bay Southern Cape coastal plain. Proceedings Sixth National Oceanographic Symposium, C-66, Stellenbosch.

Malan, J.A. and Theron, J.N. In press. Notes on an Enon basin northeast of Bredasdorp, southern Cape Province. Annals of the Geological Survey of South Africa.

Marker, M.E. 1976. Cenotes: A class of enclosed karst hollows. Zeitschrift fur Geomorphologie, 26, 104-123.

Marker, M.E. 1981. Karst in the Bredasdorp area: A preliminary analysis. South African Geographer, 9, 25-29.

Marker, M.E. and Sweeting, M.M. 1983. Karst development on the Alexandria Coastal Limestone, eastern Cape Province. Zeitschrift fur Geomorphologie, 27, 21-38.

Maud, R.R., Partridge, T.C. and Siesser, W.G. 1987. An Early Tertiary marine deposit at Pato's Kop, Ciskei. South African Journal of Geology, 90, 231-238.

McGowran, B. and Moore, A.C. 1971. A reptilian tooth and Upper Cretaceous microfossils from the Lower Quarry at Needs Camp, South Africa. Transactions of the Geological Society of South Africa, 74, 103-105.

Partridge, T.C. and Maud, R.R. 1987. Geomorphic evolution of southern Africa since the Mesozoic. South African Journal of Geology, 90, 179-208.

Quilty, P.G. 1977. Cenozoic sedimentation cycles in western Australia. Geology, 5, 336-340.

Rigassi, D.A. and Dixon, G. 1972. Cretaceous of the Cape Province, Republic of South Africa. Proceedings Ibadan University Conference on African Geology, 1970, 513-527.

Rogers, A.W. 1906. A raised beach deposit near Klein Brak River. Annual Report Geological Commission of the Cape of Good Hope, 10, 293-296.

Rogers, A.W. 1909. Notes on a journey to Knysna. Annual Report Geological Commission of the Cape of Good Hope, 13, 130-134.

Rogers, J. 1966. The Geology of Robberg. Unpublished Honours Project, Geology Department, University of Cape Town, 87pp.

Rogers, J. 1982. Lithostratigraphy of Cenozoic sediments between Cape Town and Eland's Bay. Palaeoecology of Africa, 15, 121-137.

Rogers, J. 1985. Geomorphology, offshore bathymetry and Quaternary lithostratigraphy around the Bot River Estuary. Transactions of the Royal Society of South Africa, 45, 211-237.

Rogers, J. 1986. Cenozoic geology of the southern Cape coastal plain, particularly the area between the Kafferkuils and Gouritz estuaries. Palaeoecology of Africa, 17, 3-11.

Russell, L. 1987. Karst development: the application of a systems model. In: Gardiner, V. (ed.), Geomorphology 1986, Pt II. Wiley, New York.

Rust, I.C. 1979. Excursion Guidebook for Geokongres 79. Geological Society of South Africa, Johannesburg

Rust, I.C. 1979. Excursion Guidebook for Sedplett 83. Geological Society of South Africa, Johannesburg

432

Schloms, B.H.A., Ellis, F. and Lambrechts, J.J.N. 1983. Soils of the Cape Coastal Platform. In: Deacon, H.J., Hendey, Q.B. and Lambrechts, J.J.N. (eds), Fynbos Palaeoecology: A Preliminary synthesis. Report South African National Scientific Programme, 75, 70-86.

Siesser, W.G. 1972. Petrology of the Cainozoic Coastal Limestones of the Cape Province, South Africa. Transactions Geological Society of South Africa, 75, 177-185.

Siesser, W.G. and Dingle, R.V. 1981. Tertiary sea-level movements around southern Africa. Journal of Geology, 89, 83-96.

Siesser, W.G. and Miles, G.A. 1979. Calcareous nannofossils and planktic foraminifera in Tertiary limestones, Natal and Eastern Cape, South Africa. Annals South African Museum, 79, 139-158.

Simpson, E.S.W., Du Plessis and Forder, E. 1970. Bathymetric and magnetic traverse measurements in False Bay and west of the Cape Peninsula. Transactions Royal Society of South Africa, 39, 113-116.

Spies, J.J., Engelbrecht, L.N.J., Malherbe, S.J. and Viljoen, J.J. 1963. Die geologie van die gebied tussen Bredasdorp en Gansbaai. Explanation Sheets 3419C and 3419D (Gansbaai) and 3420C (Bredasdorp) Geological Survey of South Africa, 39pp.

Tankard, A.J. 1976. Pleistocene history and coastal morphology of the Ysterfontein Elands Bay area, Cape Province. Annals South African Museum, 69, 73-119.

Thiergart, F., Frantz, U. and Raukopf, K. 1963. Palynologische Untersuchungen von Tertiarkohlen und einer Oberflachenproben nahe Knysna, Sudafrika. Advancing Frontiers Plant Science, 4, 151-178.

Thwaites, R.N. and Jacobs, E.O. 1987. The Knysna lignites: a review of their position within the geomorphological development of the southern Cape Province, South Africa. South African Journal of Geology, 90, 137-146.

Toerien, D.K. 1979. The geology of the Oudtshoorn area. Explanation Sheet 3322 Geological Survey of South Africa, 13pp.

Visser, H.N. and Schoch, A.E. 1963. The geology and mineral resources of the Saldanha Bay area. Memoir Geological Survey of South Africa, 63, 1-150.

Whittingham, J.K. 1969. Geohydrological/geophysical survey for underground water supplies in the Bredasdorp and South Swellendam districts, Cape Province. February to August, 1969. Report Geological Survey of South Africa, GH 1507:1-15.

Whittingham, J.K. 1971. Geohydrological investigation of underground water supplies in the Duine between Duiwenhoks and Kafferkuils Rivers, Riversdale District. Technical Report Department Water Affairs South Africa Geohydrological Division, GH 1566, 1-15.

Wybergh, W. 1919. The Coastal Limestones of the Cape Province. Transactions

Geological Society of South Africa, 22, 46-47.

Yates, R.J., Miller, D.E., Halkett, D.J., Manhire, A.H. and Parkington, J.E. 1986. A late mid-Holocene high sea-level: a preliminary report on geoarchaeology at Elands Bay, western Cape Province, South Africa. South African Journal of Science, 82, 164-165.

MUDBALL GENESIS IN A SUB-TROPICAL ESTUARINE ENVIRONMENT, MBASHEE RIVER MOUTH, TRANSKEI, SOUTHERN AFRICA

GEORGE F.DARDIS
Department of Geography, University of Transkei
EMILE PLUMSTEAD
Department of Zoology, University of Transkei

1. INTRODUCTION

Modern environments of deposition of mud balls have been described in some detail (Hass,1927; Bell, 1940; Kugler and Saunders,1967). These demonstrate that even though mud balls may be considered a diagnostic feature of coarse sedimentation (Potter, 1967), they are also found in freshwater lacustrine (Dickas and Lunking, 1968), flash flood (McKee et al., 1967; Karcz, 1969; 1972), debris slope (Fritz and Harrison, 1984), beach (Stanley, 1969; Hall and FRitz, 1984), submarine channel (Stanley, 1964), glacial outwash (Leney and Leney, 1957) and urban (Ojakangas and Thompson, 1977) environments. This suggests that mud ball genesis results from specific processes which operate within a wide range of environments.

This paper describes mud balls formed in a sub-tropical perennial stream - tidal flat environment at the mouth of the Mbashee River, Transkei, Southern Africa (Fig.1) and outlines some of the principal processes involved in their genesis.

2. PHYSICAL SETTING

The Mbashee River is one of a number of high gradient, high sinuosity superimposed streams which flow eastward off the Drakensberg Escarpment into the Indian Ocean. Average run-off is estimated at 4000 1/s (Eksteen et al., 1979). Mean annual rainfall varies from 900 - 1100 mm, with 75 % falling in the summer months (SA Met Office, 1985). The Mbashee River is perennial, with high flow conditions occurring in summer months (October - March) and low flow conditions prevailing in the winter months (April - September). Other minor streams in this area are ephemeral and are commonly cut-off at the mouth in winter months. Periodic flooding occurs in early and late-summer months as a result of above average precipitation during these periods. The study area at the mouth of the Mbashee River (Fig. 1) occurs along the southern coastal belt of Transkei, where deeply dis-

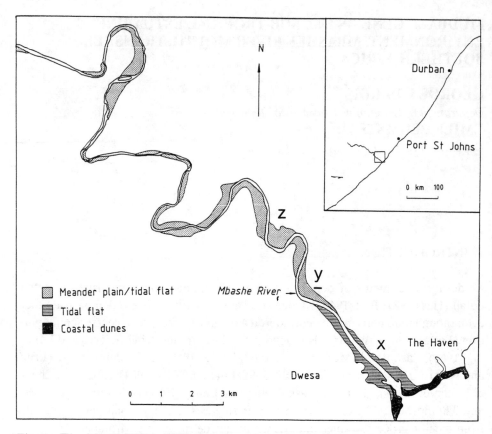

Fig. 1. The Mbashee River Mouth, Transkei, showing the main parts of the river which come under tidal influence.

sected topography is present up to about 800m. Local relief is approximately 200-300m and the general topography is steep and rugged. The river lies within a zone thought to consist of several cyclic erosion surfaces (King, 1972). The study lies within the limits of the supposed Plio-Pleistocene erosion surface, which cut headward from the coast for c. 50km along the main rivers, leading to deep incision. The sites discussed in this paper lie within a 250-300m deep gorge which extends several kilometres upstream from the mouth of the Mbashee River.

The floor of the gorge has a relatively gentle gradient. As a result the river is subject to tidal influences for 11 kilometres upstream from the mouth. Mud ball formation was observed at three main localities in 1985 (Fig. 1), all of which lie with the zone subject to tidal influence. The main site examined in this study occurs on the main tidal flat near the river mouth (Fig. 2). Two additional sites on meander plains/tidal flats approximately 4 km upstream were considered.

The present study is based on observations made between February and April 1985, at the end of the late-summer wet period. This period was characterised by a number of high magnitude floods on the Mbashee River. Mud flats were built up along the banks of the river close to the river mouth. Silt and clay sedimentation

Fig. 2. Location of the main study site (x) at the mouth of the Mbashee River.

occurred on the floodplain in the lower reaches of the river, while high-energy sandy flood deposits formed on high terraces on the upper reaches of the river, as a result of bank over-topping at the height of the flood events. Three 100-year floods were recorded on the Mbashee River during the 1984/1985 wet season, producing unusually high rates of silt and clay sedimentation on its lower reaches.

The mudflats described in the present study are presently covered with 15-25 cms of low-angle planar cross-stratification and horizontally-bedded sands. These were deposited from April to July 1985, during the earlier part of the dry season. They formed principally as a result of aeolian sand deposition from winds blowing on-shore, littoral sand deposition at high tides, and limited high-energy fluvial sedimentation. Mudballs are presently forming 6 km upstream from the river mouth, within the region of the tidal influence.

3. MUD BALL GENESIS

Mud balls at site 1 appear to have developed principally in four main ways;

1. Regular block breakage from mud flats (Fig. 3a). Prolonged dessication during winter dry season has left extensive mud flats exposed at or above the high water

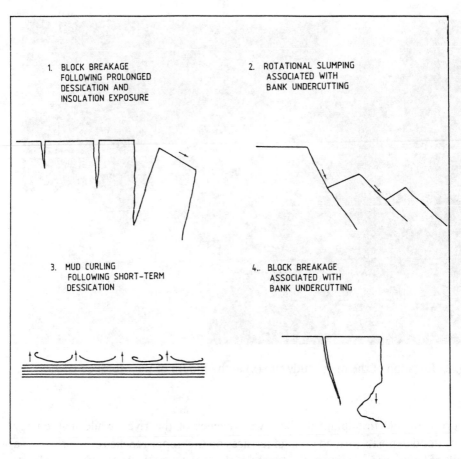

1. BLOCK BREAKAGE
   FOLLOWING PROLONGED
   DESSICATION AND
   INSOLATION EXPOSURE

2. ROTATIONAL SLUMPING
   ASSOCIATED WITH
   BANK UNDERCUTTING

3. MUD CURLING
   FOLLOWING SHORT-TERM
   DESSICATION

4. BLOCK BREAKAGE
   ASSOCIATED WITH
   BANK UNDERCUTTING

Fig. 3. Summary of processes of block breakage at Mbashee River Mouth.

tide mark (Fig. 4). These deposits tend to undergo extensive periodic drying. This results in regular, deep (1 - 15 cm) sun-cracks developing on the mud flat. Where they occur close to the river bank, they produce large cuboid mud blocks (Fig. 5).

2. Irregular block breakage through rotational slumping at the river bank edge (Fig. 3b).

3. Mud curling following short-term dessication (Fig. 3c). Though mud balls do not form by this process, mud pebbles may develop in this manner.

4. Irregular block breakage associated with river bank undercutting (Fig. 3d). This process is relatively common where the tidal flat consists of flaser-type bedding (Reineck and Singh, 1975), characterised by alternating ripple-bedded sand and mud (silt and clay) laminae. Sand laminae are normally laterally discontinuous and may produce block, wedge or irregular shaped mud clasts on breakage.

Clasts examined at site 1 varied in size from 1 cm to 15 cm in diameter. Most of the clasts are unarmoured, with less than 10 % of the clasts showing traces of partial sand grain armouring (Fig. 6). Irregular clasts occur intermixed with shaped

438

Fig. 4. Erosion of mud flats, Mbashee River Mouth. Note the breakage of blocks and progressive rounding of mud blocks away from the bank.

Fig. 5. Influence of surface cracking on mud block shape, Mbashee River Mouth. Regular crackes tend to produce cuboid mud blocks.

clasts at or near the point of origin of clasts. Clasts which have been deposited on the sand bank within a few metres of the point of clast breakage are generally well-rounded with high sphericities. These clasts appear to have undergone rapid abrasion (cf. Smith, 1972), but do not appear to have developed any great degree of armouring (cf. Bell, 1940; Hall and Fritz, 1984; Kugler and Saunders, 1959; Stanley,

Fig. 6. Mud balls showing partial sand armouring (principally in indentations on the surface of the clast), Mbashee River Mouth.

1969). Rounding seems to have occurred largely by wash-over processes (Fig. 6), though slight rounding may result from rolling action. The general absence of armouring is most likely associated with relatively little rolling action.

Mud ball formation commences close (i.e. less than 100 m) to the region of block breakage (Fig. 4), through a combination of rolling, and *in situ* differential attrition. No preferred shape is apparent in site 1. Clasts first form as cuboid forms at the point of breakage (Fig. 5). The majority of clasts in distal localities are spheroidal, elliptical or discoidal in shape, which suggests considerable variation in shaping forces. Hall and Fritz (1984) have shown that marine beach and lake shoreline environments tend to produce a high proportion of triaxial ellipsoids than spherical mud clasts. The mud balls examined in the present study appear to demonstrate both marine and fluvial influences in terms of shaping processes. The geometry of the mud clasts is comparable to descriptions of mud balls from other environments. Hall and Fritz (1984) found that, in intertidal beach environments, sphericities range from 0.4 - 0.9, with 63.3 % having a sphericity between 0.6 - 0.8. Bell (1940) reported greater than 0.95 sphericity in 51 % of mud balls formed in a fluvial environment.

4. DISCUSSION

These observations supplement previous studies of mud ball development and

440

Fig. 7. Differential accumulation of mud balls in depressions in fluviatile sands, Mbashee River Mouth.

allow some general conclusions to be drawn regarding the occurrences of armoured and unarmoured mud balls;

1. Armoured balls appear to form in environments where clasts are subject to a considerable degree of rolling. They are particularly prevalent in wash-backwash zones (i.e. marine or lacustrine beaches or other inter-tidal regions), where clasts are transported to the upper foreshore, where they may be deposited or roll back down the intertidal zone.

2. Unarmoured mud balls appear to form in environments where clasts are subject to minimal rolling action. They are particularly common in fluvial environments.

3. The Mbashee mud balls contain significant proportions of both armoured and unarmoured mudballs, which suggests that both fluviatile processes and wash-backwash processes were active in their development. However, it also seems probable that partial armouring can occur in predominantly fluviatile environments where sand or other materials are washed over the mud balls (cf. Fig. 6). These varieties are relatively easy to identify as they are partially armoured only on the exposed surface on the mud ball. This appears to have occurred in the case of the Mbashee mudballs. These tend to be distributed in inter-sand hollows (Fig. 7).

5. SUMMARY

Processes of mud ball formation and clast evolution are described from a subtropical perennial stream-tidal flat environment at the mouth of the Mbashee River, Transkei, Southern Africa. Dominant processes include wetting-drying cycles, sun-cracking, gravitational clast breakage, clast rolling, clast attrition and resedimentation during periods of flooding. These deposits, though associated with fluviatile and eolian lithofacies, only occur within the zone of tidal influence. This suggests that wetting and drying cycles play a predominant role in mud ball genesis.

6. ACKNOWLEDGEMENTS

The research was funded by the Research Committee of the University of Transkei, which is gratefully acknowledged.

7. REFERENCES

Bell, H.S. 1940. Armoured mud balls: their origin, properties, and role in sedimenta-tion. Journal of Geology, 48, 1 - 31.

Dickas, A.B. and Lunking, W. 1968. The origin and destruction of armoured mud balls in a freshwater lacustrine environment, Lake Superior, Journal of Sedimentary Petrology, 38, 1366-1370.

Eksteen, Van Der Valt, and Missen 1969. Report on the   General Hydrological Properties of the Transkei, Department of Agriculture and Forestry, Transkei.

Fritz, W.J. and Harrison, S. 1983. Giant armoured mud boulder from the 1982 Mount St. Helens mudflows. Journal of Sedimentary Petrology, 53, 131-133.

Hass, W.H. 1927. Formation of clay balls. Journal of Geology, 35, 150-157.

Hall, A.N. and Fritz, W.J. 1984. Armoured mud balls from   Cabretta and Sapelo Bar-rier Islands, Georgia. Journal of Sedimentary Petrology, 54, 8310-8350.

Karcz, I. 1969. Mud pebbles formed in a flash flood   environment. Journal of Sedimentary Petrology, 39, 333-337.

Karcz, I. 1972. Sedimentary structures formed by flash floods in southern Israel. Sedimentary Geology, 7, 161-182.

King, L.C. 1972. The Natal Monocline:explaining the origin and scenery of Natal, South Africa, University of Natal Press, Pietermaritzburg, 134pp.

Kugler, H.G. and Saunders, J.B. 1959. Occurrence of armoured mud balls in Trinidad, West Indies. Journal of Geology, 67, 563-565.

Leney, G.W. and Leney, A. 1957. Armoured till balls in   Pleistocene outwash of southeastern Michigan. Journal of   Geology, 65, 105-107.

Little, R.D. 1982. Lithified armoured mud balls of the   lower Jurassic Turner Falls sandstone, north central   Massachussetts. Journal of Geology, 90, 203-207.

McKee, E.D. Crosby, E.J. and Berryhill, H.L. 1967. Flood   deposits, Bijou Creek, California, June 1965. Journal of Sedimentary Petrology, 37, 829-851.

Ojakangas, R.W. and Thompson, J.A. 1977. Modern armoured and mud balls in an urban environment. Journal of Sedimentary Petrology, 47, 1630-1633.

Richter, R. 1926. Die Entstehung von Tongerollen und Tongallen under Wasser. Senckenbergiana, 8, 305-315.

Smith, N.D. 1972. Experiments on the durability of mud clasts. Journal of Sedimentary Petrology, 42, 378-383.

Stanley, D.J. 1964. Large mudstone nucleus spheroids in submarine channel deposits. Journal of Sedimentary Petrology, 34, 672-676.

Stanley, D.J. 1969. Armoured mud balls in an intertidal environment, Minas Basin, southeast Canada. Journal of Geology, 77, 683-693.

Zingg, Th. 1935. Beitrage zur schotteranalyse. Schweiz. Mineralog. Petrog. Mitt., 15, 39-140.

SUBSURFACE STORMFLOW RESPONSE TO RAINFALL ON A HILLSLOPE PLOT WITHIN THE ZULULAND COASTAL ZONE

G.J.MULDER
Department of Geography, University of Zululand

1. INTRODUCTION

Runoff contributions to stormflow production vary from one catchment to another of which the mechanisms involved are as yet not fully understood. Horton (1933) implied that most rainfall events exceed infiltration and stormflow being the result of surface runoff. Betson (1964), Dunne and Black (1970), Harr (1977) and others have found however that rainfall capacities seldom exceeded infiltration capacities.

Ring infiltrometer studies on soils under natural grassland in the Zululand coastal zone also indicated this (Mulder and Harmse, 1987). Minimum infiltration values varied from 7 to 760 mm/hr with a positively skewed distribution and a geometric mean of 44.7 mm/hr.

The maximum daily 30 mm rainfall intensities in Fig. 1 indicate that very few events exceed 40 mm/hr (2.1 %) for a 30 minute duration. Infiltration capacities also recover rapidly from the mimimum rate to a higher value during dry intervals and periods of lower rainfall intensity. The infiltration values could however have been over or under estimated due to the limitations of the ring infiltrometer. It seems however that there is very little chance of surface flow occurring except in saturated or highly compacted areas.

The variable source area concept (Hewlett and Hibbert, 1963) gives the most likely explanation for stormflow production in this area. Variable source areas are not stationary but expand and contract and are saturated from below due to subsurface or groundwater flow. Subsurface flow is considered too slow to be able to contribute to stormflow. This was also confirmed by Dunne and Black (1970) and Bevon and Kirkby (1979) but Harr (1977) found however that subsurface flow and channel interception averaged 97 % and 3 % respectively of the stormflow without any overland flow. Previously stored water could however move very quickly downwards through the soil profile towards the expanded stream area through the process of displacement or translatory flow (Hewlett and Hibbert, 1963) and thus contribute to the stormflow hydrograph.

Geomorphological Studies in Southern Africa, G.F.Dardis & B.P.Moon (eds)
© 1988 Balkema, Rotterdam. ISBN 90 6191 831 6

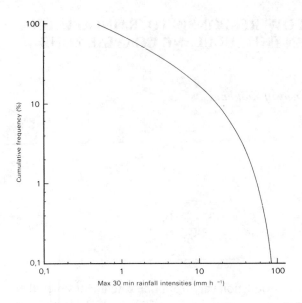

Fig. 1. Cumulative relative frequency of daily maximum 30 mm rainfall intensities (University of Zululand weather data).

2. STUDY AREA

The study area (Fig. 2) is situated close to Empangeni, 160 km north of Durban, and is situated in a physiographically homogenous region to the north-eastern side of the Ongoye hills. These hills have developed from a horst, bound north and south by post Karoo faults (Beater and Maud, 1960). The horst forms part of the Tugela complex (Geological series, 1975), consisting mainly of biotite granite gneiss and less resistent biotite quartz-feldspathic schists.

The soils in this area have been mapped (Hope and Mulder, 1979) according to the Binomial system (MacVicar et al., 1977), with the Glenrosa form being dominant on the interfluves and Katspruit and Fernwood dominating the bottom-land areas.

The climate is subtropical, with the maximum rainfall occurring during the summer. The average rainfall is 1344 mm/yr. which is very high compared to the rest of the country.

3. THE RUNOFF PLOT

The trench is situated above the sedge area and the stormflow produced from the

446

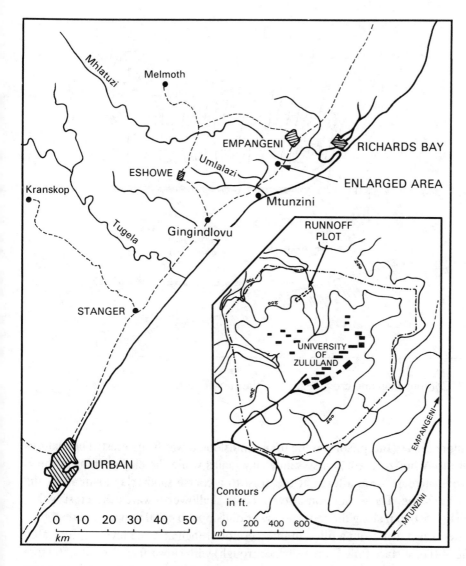

Fig. 2. Position of the runoff plot in relation to the Zululand coastal zone.

saturated bottomland area has therefore been excluded from the study.

The drainage trench is sealed off at the bottom on a solid granite-gneiss rock surface. The underground rock surface along the hillslope is slightly concave as the depth of weathering along the midslope is deeper than at the top or bottom of the slope. This became evident after a number of holes were augered to the rock surface for determination of groundwater levels.

The soils have been classified according to the Binomial system (MacVicar et al., 1977) as follows (USDA soil order equivalent in brackets). The Swartland series (Alfisol) on the foot slope changes gradually to the Sibasa series (Alfisol) on the lower midslope with another gradual transition from a deep to shallow face of the

447

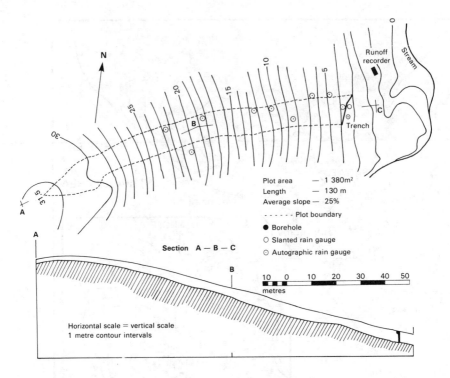

Fig. 3. Contour map and cross-section of the runoff plot.

Robmore series (Inceptisol) on the upper midslope towards the crest. The textures of all the A horizons are coarse sandy clay loams while the clay content in the B horizons increase from a loam at the crest to a coarse sandy clay loam along the lower midslope. The A horizon also becomes shallower towards the crest with a depth of c.500 mm along the foot and lower midslope to c.200 mm at the crest.

Water was drawn from a filled in interception trench at three levels as indicated in Fig. 4 (i.e. surface flow from a surface trough) with subsurface flow at c.500 mm on the B horizon (having a substantially higher clay content) and groundwater from the rock surface which is c.2 m deep. The surface trough was covered with a galvanized roof to eliminate channel interception.

Three tipping bucket flow recorders (built according to specifications by Chow, 1977) measuring 1.5 litres per tip were installed in a recorder hut. This type of recorder is ideal when recording low flows and has an accuracy tolerance of 5 % (Fig. 5). The volume per count increases steadily with an increase in discharge and exceeds the 5 % tolerance limit when the rate becomes less than two minutes per tip and created some level of under estimation when recording surface runoff.

Daily records over a year and a half showed virtually no subsurface flow along the A horizon while the surface flow was only recorded during very extreme events. Magnetic tape counters that record on a continuous basis were then subsequently installed during 1982 to record only the groundwater and surface flow.

448

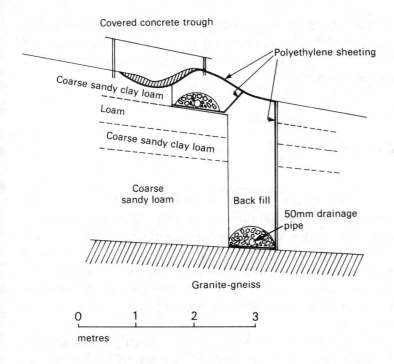

Fig. 4. Cross-section of the interception trench.

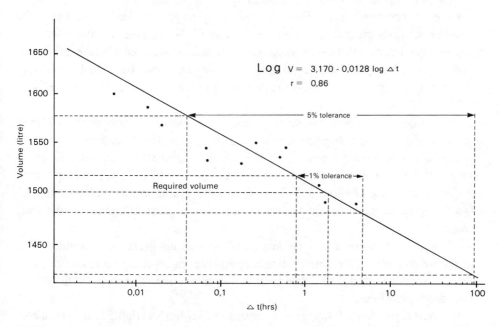

Fig. 5. Calibration curve for the 1.5 litre tipping bucket recorder.

An autographic raingauge, check gauge and slanted gauge were installed next to the trench. The aperture of the slanted gauge was set to the same slope angle and aspect as that of the hillslope plot. This was done in order to monitor the influence of slope angle and aspect to rainfall.

4. RESULTS AND DISCUSSION

The experiment started during an abnormally dry period of 1980. Table 1, which gives a summary of the daily record of flow for the 1980-1981 hydrological year, clearly indicates groundwater flow as being the main contributor to hillside runoff. The subsurface flow is not indicated as no flow at that level was recorded during this period. A small amount of flow was, however, recorded the previous year during a period of extremely high rainfall intensity. The groundwater table extended on average about 25 m upslope from the interceptor trench during average summer conditions.

The total response of 8 % is low compared to the average yearly streamflow response of 30 % in this region as determined by Whitmore (1970). The low antecedent rainfall and subsequent soil moisture conditions during mid 1980 are the reason for the low response from October to December 1980. Conditions did improve, however, with the maximum being recorded from May to June 1981. The high rainfall figure during May was recorded at the end of the month causing the increase in flow during June, thus explaining the abnormally high response of 63 %.

An event recorded during February 1982 (Fig. 6) gives a clear indication of the rapid reaction of groundwater seepage to rainfall. No surface or subsurface flow was recorded during this event. A standard procedure was used whereby the quick-flow is separated from the delayed flow with a line projected from the beginning of the storm hydrograph increasing the delayed flow by 1.13 mm/day/yr.(Hewlett and Hibbert, 1965).

The influence of aspect and slope on the actual amount of precipitation is indicated in Fig. 7. The autographic raingauge therefore slightly underestimates the precipitation on the hillside plot (e.g. 38.7 mm measured with an autographic gauge would give a true average value of 40 mm by means of the regression equation), giving a difference of about 3 %. The slight underestimation of precipitation with the autographic gauge could therefore not have a great influence on the final results.

The low response could be attributed to certain practical limitations like problems concerning the determination of actual area contributing to groundwater flow, as seepage through and diversion of flow along quartz veins could distort the downslope flow lines.

The best explanation however is by means of Hewlett's variable source area concept. He ascribes the main source of stormflow to the production of surface flow

Table 1. Monthly rainfall-runoff plot recorded during the 1980-1981 hydrological year.

MONTH	RAIN-FALL (mm)	RUNOFF (in mm)		Groundwater Contribution (%)	RUNOFF RESPONSE TO RAINFALL (%)
		Surface	Groundwater		
October	38.1	0	0.678	100	1.78
November	69.3	0	0.329	100	0.47
December	47.7	0	0.038	100	0.08
January	205.0	0.209	4.359	95	2.23
February	179.7	0.190	7.564	97.5	4.31
March	68.3	0	3.793	100	5.55
April	95.6	0.003	0.467	99.3	0.49
May	292.4	0.114	28.331	99.6	9.73
June	29.0	0.058	18.083	98.7	62.56
July	60.6	0.069	1.930	96.5	3.30
August	102.1	0.387	13.491	97.2	13.59
Sept	135.3	0.325	25.262	98.7	18.88
TOTAL	1322.6	1.355	104.325	98.5	7.99

due to low infiltration capacities on saturated bottomland areas. Weyman (1970) describes the upslope groundwater stormflow response to rainfall as the result of rapid increases in throughflow output as saturated areas extend upslope. The quick response is explained by Hewlett through the process of translatory flow. There is still some uncertainty about the extent of saturation above the drainage trough, but it does not look as if the soil ever gets saturated right up to the surface as very little flow has been generated at B-horizon level. The push through mechanism of unsaturated flow due to the process of displacement of water from previous rain events seems to be much faster than was supposed by Bevon and Kirkby (1970) and Dunne and Black (1970).

Daily surface flow values were compared with certain precipitation values to get an indication of the influence of antecedent soil moisture conditions on hillside stormflow runoff volumes. The best correlations were between daily runoff values

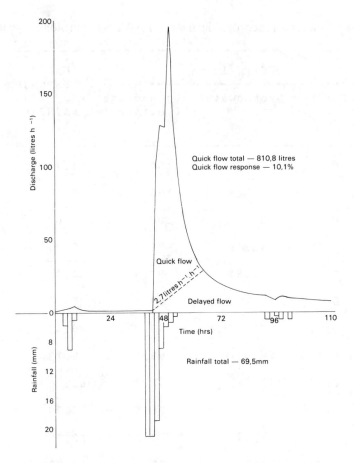

Fig. 6. A hydrograph indicating groundwater seepage in relation to hourly rainfall intensities of an event during February, 1986.

and daily precipitation before the maximum 30 minute intensity is reached ($r = 0.87$) and daily precipitation values ($r = 0.72$).

5. CONCLUSIONS

There is a large contribution of groundwater flow towards the stormflow hydrograph. The surface flow contribution however, is very low due to the fact that the infiltration capacities usually exceed rainfall intensities. The low hillside runoff response compared to the average response of the catchment basin is probably due to the fact that most of the stormflow is generated through surface runoff from the bottomland variable source areas.

452

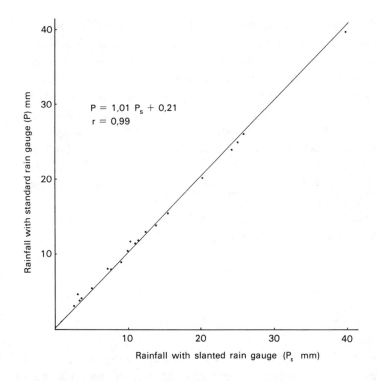

$$P = 1{,}01\, P_s + 0{,}21$$
$$r = 0{,}99$$

Fig. 7. The difference between rainfall data collected with a standard gauge and corrected data from a guage set at an angle.

Significant correlation between runoff and antecedent rainfall data again indicates the importance of antecedent soil moisture conditions in the prediction of hillslope runoff. The rapid groundwater stormflow response to rainfall should be investigated further as it is not known whether the stormflow contribution is only from the footslope or through translatory flow from the entire slope.

6. SUMMARY

Runoff contributions to stormflow production vary from one catchment to another of which the mechanisms involved are yet not fully understood. The aims of the study was to determine to what extent surface, subsurface and groundwater flow along the hillslope, in the Zululand coastal zone, contribute to the stormflow hydrograph. A runoff plot measuring the three flow components was chosen to represent average hillslope conditions prevailing in a physiographically homogenous region north-east of the Ngoye hills. This area is close to Empangeni, in the Zululand coastal zone. The climate is subtropical with an annual rainfall average of

1344 mm which mainly occurs during the summer. The slope on which the runoff plot is situated is slightly convex with a short footslope adjoining a flat bottomland sedge area. This type of bottomland area is found along most first and second order perennial streams in this area and serves as a permanent partial or variable source area for producing stormflow. Hillslope runoff data showed that over a period of two years groundwater flow contributed most to the stormflow hydrograph with hardly any surface flow. Significant correlations between runoff and antecedent rainfall data before maximum rainfall intensity is reached again indicates the importance of antecedent soil moisture conditions in the prediction of hillslope runoff. A low response (8 % of the total rainfall), compared to an average response of 30 % for catchments in this region suggests that a large percentage of the stormflow from first and second order catchments in this region is generated mainly through surface runoff from the bottomland sedge areas.

7. REFERENCES

Beater, B.E. and Maud, R.R. 1960. The occurrence of an extensive fault system in S.E. Zululand and its possible relationship to evolution of part of the coastline of Southern Africa. Transactions of the Geological Society of South Africa, 63, 51-54.

Betson, R.P. 1964. What is watershed runoff? Journal of Geophysical Research, 69, 1542-1552.

Bevon, K.J. and Kirkby, M.J. 1979. A physically based, variable contributing area model of basin hydrology. Hydrological Sciences Bulletin, Wallingford, 24, 43-69.

Chow, T.L. 1977. A low cost tipping bucket flow meter for overland and subsurface stormflow studies. Canadian Journal of Soil Scince, 56, 197-202.

Dunne, T. and Black, R.D. 1970. An experimental investigation of runoff production in permeable soils. Water Resources Research, 6, 478-490.

Geological series, 1975. 2831 DD Felixton. Government Printer, Pretoria.

Harr, R.D. 1977. Water flux in soil and subsoil on a steep forested slope. Journal of Hydrology, 33, 37-58.

Hewlett, J.D. and Hibbert, A.R. 1963. Moisture and energy conditions within a sloping mass during drainage. Journal of Geophysical Research, 68, 1081-1087.

Hewlett, J.D. and Hibbert, A.R. 1967. Factors affecting the response of small watersheds to precipitation in humid areas. In: Sopper, W.E. and Lull, H.W. (eds.), Forest Hydrology, Pergamon, Oxford, 275-290.

Hewlett, J.D. and Troendle, C.A. 1975. Non-point and diffused water sources: a variable source area problem. Symposium Utah State University, Logan, Utah, 11-13 Aug. 1975, American Society of Civil Engineers, 21-45.

Hope, A.S. and Mulder, G.J. 1979. Hydrological investigations of small catchments in the Natal Coastal Belt and the role of physiography and landuse in the raifall-runoff process. Report No. 1, series B, University of Zululand.

Horton, R.E. 1933. The role of infiltration in the hydrologic cycle. Transactions of the American Geophysical Union, 14, 446-460.

MacVicar, C.N., De Villiers, J.M., Loxton, R.F., Vester, E. Lambrechts, J.J.N., Merryweather, F.R., Le Roux, J., Van Rooyen, T.H. and Harmse, H.J. von M. 1977. Soil classification. A binomial system for South Africa. Department of Agriculture Technical Services, Pretoria.

Mulder, G.J. and Harmse, H.J. von M. 1987. The influence of selected physical variables of soils in the Ntuze catchment on the infiltration capacity (Zululand coastal zone). Water SA, 13, 43-48.

Weyman, J.S. 1970. Throughflow on hillslope and its relation to the stream hydrograph. Bulletin of the International Association of Scientific Hydrologists, 15, 25-32.

Whitmore, J.S. 1970. The Hydrology of Natal. Symposium on Water Natal, Paper 1, Durban.

GEOMORPHOLOGY OF THE ESIKHALENI MASS MOVEMENT COMPLEX, TRANSKEI, SOUTHERN AFRICA: PRELIMINARY OBSERVATIONS

HEINRICH R. BECKEDAHL, PATRICIA M. HANVEY, GEORGE F. DARDIS
Department of Geography, University of Transkei.

1. INTRODUCTION

Landslides and related mass movement phenomena are relatively common features in tropical and sub-tropical areas of high relief, and are often associated with areas of tectonic uplift (Crozier, 1986). The study of landslides in southern Africa has been largely neglected (cf. Brink, 1982), despite clear evidence, for example, during the high intensity rainfall events of January 1984 and November 1987 in Natal, of climatic and/or seismic triggering of landslides, in many instances causing considerable damage in residential areas. Relatively little is known regarding the nature or distribution of landslides in southern Africa, and few studies have attempted to map landslide phemomena in detail, monitor their rate(s) of development, or establish (if possible) the main causes of instability. A number of small-scale engineering studies of slope instability have been undertaken in South Africa (Knight et al., 1977; Sudgen et al., 1977; Smedley and Nowlan, 1978; Garland, 1978; Webb, 1983; Venter, 1983; Brink, 1981,1983), which demonstrate that instability is common, particularly in coastal regions underlain by Karoo mudrocks, reflecting combinations of weathering, climate and geotechnical properties of the bedrock. Venter (1983) has identified three main types of mass movement on Karoo mudrocks in the Durban area; rotational slides, translational slides and rockfalls. Some caution must, however, be exercised in using particular genetic terms such as these to describe mass movement phenomena.

The problem of terminology/nomenclature has frequently been addressed in the literature (cf. Sharpe, 1938; Hutchinson, 1968; Coates, 1977; Varnes, 1978; Hansen, 1984). The term 'landslide' has generally been used to denote the rapid downward movement of a mass of rock and soil on a slope under the influence of gravity. The term is, however, viewed as a misnomer, in that the most commonly occurring types of movement are related (at least in part) to flowage rather than to actual sliding (Skempton and Hutchison, 1969; Chorley et al.,1984). This has resulted in various alternative names being proposed, viz. slope failure (Terzaghi, 1950), debris slides,

Geomorphological Studies in Southern Africa, G.F.Dardis & B.P.Moon (eds)
© 1988 Balkema, Rotterdam. ISBN 90 6191 831 6

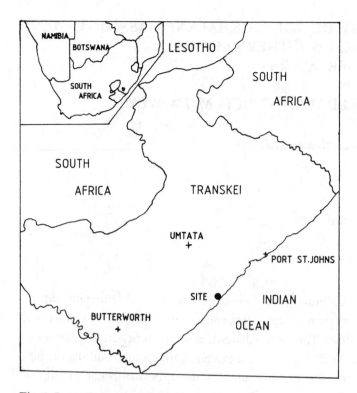

Fig. 1. Location of the Esikhaleni site in Transkei.

debris avalanches and debris flows (Varnes , 1958; Blong, 1973b), mass movement (Hutchison, 1968), and slope movement (Varnes, 1978). The wide diversity of classification systems is due both to the broad spectrum of interest in problems of slope instability, and difficulties in categorizing transitional forms of mass movement (Hutchinson, 1968; Varnes, 1978; Hansen, 1984). The criteria upon which classifications have been based include the type of movement, the nature and quality of the material, morphometric characteristics, the degree of displacement, degree of disruption or disturbance of the material, the rate of movement, water content, cause and degree of potential hazard (Sharpe, 1938; Varnes, 1958, 1978; Blong, 1973; Scheidegger, 1984). Of these, the most common criteria have been the type and rate of movement, and the moisture content of the material prior to failure (Carson and Kirkby, 1972; Nemcok et al., 1972; Garland, 1978; Chorley et al., 1984; Crozier, 1986).

In this paper, preliminary observations of a mass movement complex, identified at Esikhaleni (lat. 29°06'S;long. 32°02'E) on the Transkeian coast (Fig. 1), are outlined. Slope failure has occurred on hillslopes on the southern flank of the Mpako River, in an area underlain by dolerite-capped, fragmented shales and mudstones of the Ecca Group (Tankard et al., 1982). It consists of a series of landslides, minor rock falls and soil creep, extending over an area of approximately 1.5 sq. km. The volume of material

458

involved in the main instability is estimated at 2.5 x 10^6 cubic metres.

2. METHODS

The Esikhaleni region (Fig. 2) was mapped using 1:10000 aerial photographs, taken in 1982, and high-oblique aerial photographs of varying scales (1:1000-1:5000) taken in November 1987. The present paper focuses primarily on initial investigations undertaken in the main area of instability at Esikhaleni (Fig. 2). More extensive instability has been uncovered in the general area which is not discussed here. The area was mapped on the ground in September and November 1987. A transect (Fig. 3) was surveyed through one of the major lobes using a dumpy level and staff. Major lobes, interlobate boundaries, slip scars and areas of undifferentiated mass movement were recorded (Fig. 2). This allowed simple morphometric parameters (cf. Crozier, 1973; 1986) to be calculated (Table 1), to facilitate categorisation of the type (or types) of mass movement at Esikhaleni.

The morphometric parameters used in this study are based on indices defined by a number of workers (Brunsden, 1973; Crozier, 1973; Cooke and Doornkamp, 1974; Hansen, 1984; Crozier, 1986). As the actual event of slippage is seldom witnessed, or formative processes either observed or measured, at the time of failure, a reconstruction of the processes is often only possible through detailed morphometric analysis of the resultant forms. Although Crozier's (1973) classification provides a useful starting point, a certain degree of uncertainty remains, due to the overlap in the standard deviations of the different types of instability (Cooke and Doornkamp, 1974; Hansen, 1984). The difficulties of adequate classification of mass movements is demonstrated by Varnes' (1958) study of 92 landslides, of which 78 were classified as 'complex slope failures', involving two or three different types of slope failure (Blong, 1973b).

The morphometric indices established from the transect (Fig. 3) (in particular the depth [D/L x 100% = 8.78] and the viscous flow [Lt/D = 1.058] indices) suggests that the mass movement complex is predominantly of earth flow type (by viscous flow processes) (Sharpe, 1938; Crozier, 1973). These must, however, be regarded as tentative until more detailed analysis of individual lobes has been undertaken.

3. GEOMORPHOLOGY

Eight major lobes and three major tensional zone slip scars (1-3, Figs. 2, 4) have been identified in the mass movement complex. The main scar (1, Figs 2, 4) extends to the top of the ridge and is c. 1070 m in length. It varies from 1-15 m in height. The second major slip scar (2, Figs. 2, 4) is c.550 m in length and varies from 0.5-7.0 m in height.

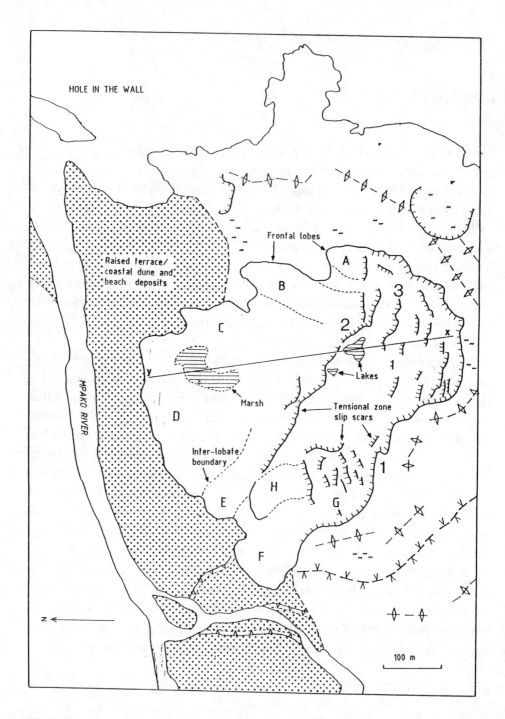

Fig. 2. Geomorphology of the Esikhaleni mass movement complex, showing a series of lobes (A-H), associated with a number of tensional zone slip scars. Lobes C-F have over-flowed onto raised terrace and coastal dunes sediments. (   ) denotes areas of undifferentiated instability. X-Y denotes the line of transect shown in Fig. 3.

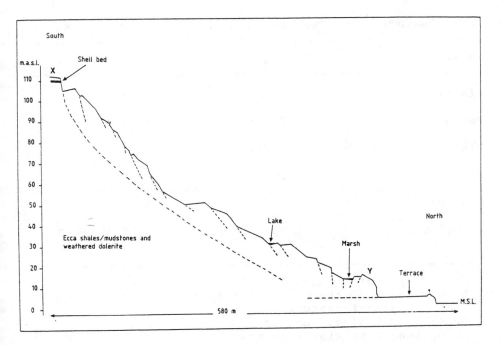

Fig. 3. Transect through lobe D (see Fig. 2).

The third major slip scar (3, Figs. 2, 4), which lies between slip scars 1 and 2, is c.150 m in length and varies from 4-25 m in height (Figs, 5, 6). These are crude estimates as, in reality, the tensional zone slip scars tend to have a relatively complex morphology, with each "scar" characterised by a series of discontinuous, cresent-shaped scars of varying width and depth (Fig. 6, 7). Second and third degree slip scars may also form within individual lobes (zones w, Fig. 6).

While no measurements have yet been made, it seems likely that the length of individual lobes is related primarily to the water content of the lobes and slope angle. The longer, more tenuous lobes (B, C, Fig. 2) commonly contain ponded water bodies (Figs. 2, 6), and marshes have developed along the margins of these lobes resulting from groundwater seepage from the lobes. These are also commonly found at inter-lobate boundaries.

Whereas lateral boundaries of individual lobes are often difficult to distinguish (for example in lobe A, Fig. 6), the frontal zones of the lobes are quite distinct. They are characterised by a head, varying from 5-20 m in height (figs. 8, 9) and often show evidence of secondary slippage and terracette formation on their flanks (Fig. 8). The frontal lobe of lobe C is characterised by marshy areas developed both in front and behind the lobe, and forms a major elongate ridge, c. 200 m long, 5-20 m in height, and 10-50 m wide (Fig. 8).

Analysis of field data suggests that failure appears to have occurred retrogressively away from the Mpako River. Although the river's course has been marginally affected

461

Table 1. Morphometric indices for the Esikhaleni mass movement complex.

INDEX	PARAMETER	VALUE
Classification	D/L *	0.0878
Depth	D/L x 100%	8.78
Displacement	Lr/Lc	0.83
Tenuity	Lm/Lc	0.955
Viscous flow	Lf/D	1.058
Dilation	Wx/Wc	0.76 (Lobe B) 0.91 (Lobe H)

```
D  = Depth of rupture
L  = Length of the mass movement from crown to toe
Lr = Length of rupture
Lc = Length of concave scar
Lm = Length of displaced mass
Lf = Length of the head of the landslip (that part
     of the displaced mass occurring below the original
     surface)
Wx = Width of displaced mass
Wc = Width of surface of rupture
```

by lobe F (Fig. 3), the abandoned channel and ox-bow lakes occurring on the raised terrace indicate that the initial slope instability may have resulted from fluvial under-cutting of the slope. Wave-induced vibration has been sufficient to realise latent slope instability elsewhere (Varnes, 1958), and it is likely that slope failure at the Esikhaleni site may have been triggered by the same processes once the threshold for failure had been reached.

Some evidence exists to support Hansen's (1984) observation that creep processees preceed other, larger scale failure phenomena. This is illustrated clearly by the eastern sector of
the main complex, where a comparison of the present with 1982 aerial survey shows soil creep upslope of a small scale viscous flow.

The morphology of the section (Fig. 3) also suggests retrogressive failure (cf. Brunsden, 1987; Van Asch et al., 1984). Evidence in support of retrogressive failure may be seen from the nature and composition of the debris lobes themselves. The slope consists of contact metamorphosed, fractured shales and mudstones that have been intruded by doleritic material which also constitutes a sill at the crest of the slope. The basal debris lobes (C,D,F, Fig. 4) consist predominantly of weathered shale and disturbed regolith, whereas the upper lobes, which in several instances partially over-run the basal lobes, consist of a high percentage of weathered doleritic material to the

Fig. 4. High-oblique aerial photograph of the eastern side of the Esikhaleni mass movement complex (taken in November 1987), showing major slip scars (1-3) and lobe B. Note the interlobate marsh (x) in lobe C.

Fig. 5. Surface view of lobe B, taken on the raised terrace, and showing the main slip scars (1-3) in the complex. (Photo taken in September 1987).

463

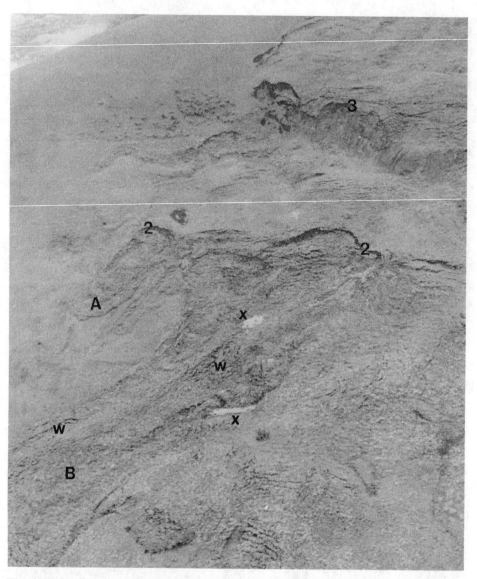

Fig. 6. High-oblique aerial photograph of lobes A and B, which are associated with slip scar 2. Slip scar 3 lies directly behind these lobes. Note the ponded water bodies (x) and tensional slip scars (w) occurring within lobe B (Photograph taken November 1987).

virtual exclusion of the shales. It would thus appear that slippage and subsequent flow first occurred in the shales to a depth determined by the weathering front. Subsequent failure of the headscar resulted in flowage of material, partly burying the previous lobes. Examination of the head scar suggests that failure within the dolerites was related primarily to the prevailing weathering front.

The gradient of the failed slope approximates 15°, and field observations suggest

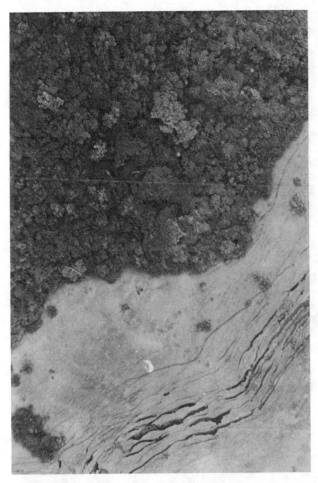

Fig. 7. Discontinuous tensional zone slip scars, formed at the top of the ridge on which the Esikhaleni mass movement complex is situated. (Photograph taken in November 1987).

that the original slope did not exceed 20°. These values are significant in that they accord with observations elsewhere, that, where slopes are relatively gentle, slipped blocks will show back-tilting (Skempton and Hutchinson, 1969). The values also accord with the findings of Campbell (1975) that earth flows on slopes of less than 27° will not attain avalanche speeds. That this has not occurred may be seen from the lack of spreading of the toe section onto the flood plain adjacent to the Mpako river.

4. DISCUSSION

A number of points remain unanswered; (1) what events or factors triggered initial movement in the complex; (2) what factors lead to recurrence of flow; and (3) when will the complex reach quasi-equilibrium?

465

Fig. 8. High-oblique aerial photograph of the frontal region of lobe C, showing ter-racettes and slip scars developed in the frontal zone. Note the marshy region behind the frontal zone (top left of photo) and in front (left) of the lobe. (Photograph taken in November 1987).

Fig. 9. Surface view of the frontal terminus region of lobe C, showing secondary mass movements resulting from oversteepening during forward movement of the lobe.(Photograph taken September 1987).

4.1. Triggering mechanisms

Dating of the shell layer at the top of the complex (Fig. 3) is underway, to place a mimimum age on the mass-movement complex. Though detailed analysis has not been undertaken, it seems clear that the lobes in the complex are of several ages, and there is clear evidence of superimposition of flows. The mass movement complex may be quite old (i.e. Mid- or Late-Holocene in age), but does not appear to be triggered by large-scale climatic changes, which suggests that initial movement in the complex may have resulted from other processes.

It seems likely that the initial instability is associated with changes in the course of the Mpako River, leading to under-cutting of the bedrock slope. Flow lobes F and H (Fig. 2), for example, which are relatively recent superimposed flows, appear to have formed as a result of cutting of a lower terrace on the south-western flank of the Mpako River (Fig. 2). Changes in the course of the river were most likely induced by Holocene changes in sea level (Dardis and Grindley, 1988; Beckedahl, 1988), but may also arise as a result of high stage flood events. There is, however, no recent evidence of dissection of the mass movement complex as a result of flooding. The principal scars lie approximately parallel to the direction of major regional slope in the area. This supports Ai and Scheidegger's (1984) findings that there is a direct correlation between the incidence of slope instability, slope orientation and the direction of principal tectonic stress. While this will determine the configuration of the flow it has little effect on triggering mechanisms (other than setting a limiting threshold for movement).

4.2. Flow recurrence

Field evidence indicates that the head scars may fail further, and also that the debris lobes are prone to movement by creep and slump processes. Potential instability is also suggested by the displacement index value of 0.82 for the section (Table 1). Where the slope is inherently unstable, flow recurrence may result from a number of processes. An apparent correlation has been recorded between flows of the soil-slip origin and high intensity precipitation events (So, 1974; Campbell, 1975; Caine, 1980). When infiltration through the regolith exceeds the transmissive capacity of the underlying strata, an accumulation of water results in saturation of the regolith, with concomitant reduction in shearing resistance (Varnes, 1978; Chorley et al., 1984). The moisture content of the regolith is a primary factor governing stability, both for failure (cf. Shreve, 1968; Campbell, 1975; de Ploey and Cruz, 1979; Quigley et al., 1980; Bryan and Rice, 1980), and subsequent remobilisation of the debris (Leighton, 1974). In this, the role of high magnitude rainfall events has been indicated by several researchers (cf. Caine, 1980; Hsu, 1987; Grainger and Kalaugher, 1987). Small-scale flow recurrences of 0.5-1.5 m displacement have been observed in the Esikhaleni mass movement complex between September and November 1987, during a period in which a number of high-intensity rainfall events occurred.

Movement is initiated when the shearing resistance is less than the shear stress. The actual trigger setting off instability is frequently a vibration, either resulting from an earthquake, or the pounding from waves or from cattle moving over the slope (Sharpe, 1938; Varnes, 1978). Minor secondary failures were triggered at Esikhaleni by an earth tremor which took place in October, 1986. However it is not clear to what extent these are responsible for large-scale movements in the complex.

4.3. Flow stabilisation

The displacement index value (Lr/Lc = 0.83) suggests that the potential for further reactivation of the complex exists. Evidence is the secondary slumping and creep suggests that only minor movements are presently underway. However, the possibility of largescale catastrophic failure cannot be ruled out.

5. FUTURE WORK

Further work is planned at this site, to; (1) establish the age of the complex, (2) to monitor present rates and types of mass movement, (3) to map individual lobes in detail, to establish age relationships between individual lobes and provide more detailed morphometric data on the complex, (4) to examine the internal properties of the flow, to identify zones of potential instability and planes of weakness, and (5) to establish likely formative processes.

6. ACKNOWLEDGEMENTS

We wish to thank Peter Storey, Tescor, for invaluable helicopter support during the course of fieldwork.

468

7. REFERENCES

Ai, N.S. and Scheidegger, A.E. 1984. On the connection between the neotectonic stress field and catastrophic landslides, Proceedings of the 24th International Geological Congress, Moscow, 180-9.

Blong, R.J. 1973a. Relationships between morphometric attributes of landslides, Zeitschrift für Geomorphologie, Supplement Band 18, 66-77.

Blong, R.J. 1973b. A numerical classification of selected landslides of the debris slide-avalanche-flow type, Engineering Geology, 7, 99-114.

Brink, A.B.A. 1981. Engineering Geology of Southern Africa, Vol. 2. Building Publications, Silverton, 255pp.

Brink, A.B.A. 1983. Engineering Geology of Southern Africa, Vol. 3. Building Publications, Silverton, 255pp.

Brunsden, D. 1973. The application of systems theory to the study of mass movewment, Geologica Applicata e Idrolgeologia, 8, 1, 185-207.

Brunsden, D. 1979. Mass movements. In: Embleton, C. and Thornes, J.B., (eds.), Processes in Geomorphology, Edward Arnold, London, 130-86.

Brunsden, D. and Jones, D.K.C. 1980. Relative time scales and formative events in coastal landslide systems. Zeitschrift für Geomorphologie, Supplement Band 34, 1-19.

Bryan, R.B. and Price, A.G. 1980. Recession of the Scarborough Bluffs, Ontario, Canada. Zeitschrift für Geomorphologie, Supplement Band 34, 48-62.

Caine, N. 1980. The rainfall intensity - duration control of shallow landslides and debris flows. Geografiska Annaler, 62A, 23-27.

Campbell, D.A. 1951. Types of erosion prevalent in New Zealand; Association Internationale d'Hydrologie Scientifique, Assemblee General de Bruxells, Tome II, 82-95.

Campbell, R.H. 1975. Soil slips, debris flows and rainstorms in the Santa Monica Mountains and vicinity, southern California, US Geological Survey, Professional Paper 851.

Carson, M.A. and Kirkby, M.J., 1972: Hillslope form and process, Cambridge University Press, London.

Carson, M.A. 1976. Mass-Wasting, slope development and climate, In: Derbyshire, E.,(ed.), Geomorphology and Climate, Wiley and Son, Chichester, 101-36.

Chorley, R.J., Schumm, S.A. and Sugden, D.E. 1984. Geomorphology, Methuen and co. London.

Coates, D.R. 1977. Landslide perspectives, in: Coates, D.R. (ed.), Landslides, Geological Society of America, 3-28.

Cooke, R.U. and Doornkamp, J.C. 1974. Geomorphology in Environmental Management: An Introduction, Oxford University Press, Oxford.

Crozier, M.J. 1973. Techniques for the morphometric analysis of landslips, Zeitschrift für Geomorphologie, 17, 78-101.

Crozier, M.J. 1986. Landslides: Causes, Consequences and Environment, Croom Helm,

Sydney.

Dardis, G.F. and Grindley, J.R. 1988. Coastal geomorphology. In: Moon, B.P. and Dardis, G.F. (eds.), The Geomorphology of Southern Africa, Southern Book Co., Johannesburg (in press).

De Ploey, J. and Cruz, Z. 1979. Landslides in the Serra do Mar, Brazil, Catena, 6, 111-122.

Garland, G.G. 1978. An embankment landslide in Berea Red Sands, Durban, South African Geographical Journal, 60, 63-70.

Grainger, P. and Kalaugher, P.G. 1987. Intermittent surging movements of a coastal landslide. Earth Surface Processes and Landforms, 12, 597-603.

Hansen, M.J. 1984. Strategies for classification of landslides. In: Brunsden, D. and Prior, D.B. (ed.), Slope Instability, John Wiley and Sons, New York, 1-25.

Hutchison, J.N. 1973. The response of London Clay cliffs to differing rates of toe erosion, Estratto da Geologica Applicata e Idrogeologia, VII pt 1, 221-39.

Hutchinson, J.N. 1968. Mass movement. In: Fairbridge, R.W., (ed.), Encyclopaedia of Earth Sciences, Reinhold, New York, 688-695.

Innes, J.L. 1984. Debris Flows, Progress in Physical Geography, 7, 469-501.

Knight, K., Everitt, P.R. and Sudgen, M.B. 1977. Stability of shale slopes on the Natal coastal belt. Proceedings of the Fifth South-East Asian Conference on Soil Engineering, Bangkok, 201-212.

Leighton, F.B. 1974. Landslides and hillslope development. In: Coates, D.R. (eds.), Environmental Geomorphology and Landscape Management, Volume 2, Benchmark Books,Dowden,Hutchinson and Ross Incorporated,Stroudsburg, Pennsylvania.

Nemcok, A.; Pasek, J. and Rybar, J. 1972. Classification of landslides and other mass movements, Rock Mechanics, 4, 71-8.

Prior, D.B., Rouge, B. and Renwick, W.H. 1980. Landslide morphology and processes on some coadstal slopes in Denmark and France. Zeitschrift für Geomorphologie, Supplement Band 34, 63-86.

Quigley, R.M. and Di Nardo, L.R. 1980. Cyclic instabilitry modes of eroding clay bluffs, Lake Erie Northshore Bluffs at Port Bruce, Ontario, Canada, Zeitschrift für Geomorphologie, Supplement Band 34, 39-47.

Selby, M.J. 1982. Hillslope Materials and Processes, Oxford University Press, Oxford.

Scheidegger, A.E. 1984. A Review of recent work on mass movements on slopes and on rock falls. Earth Science Reviews, 21, 225-49.

Sharpe, C.F.S. 1938. Landslides and Related Phenomena, Columbia Univ. Press, New York.

Skempton, A.W. and Hutchison, N.J. 1969. Stability of natural slopes and embankment foundations. 7th International Conference on Soil Mechanics and Foundation Engineering, Mexico, 291-340.

Skempton, A.W. and Hutchison, J.N. (eds.) 1976. A discussion on valley slopes and cliffs in southern England: Morphology, mechanics and Quaternary history. Philosophical Transactions of the Royal Society of London, Series A, 283, 421-631.

Smedley, M.I. and Nowlan, P.H. 1978. A geotechnical investigation of a landslide on the Broodsnyersplaas to Richards Bay coal line. The Civil Engineer in South Africa, 20(7), 169-173.

So, C.L. 1974. Mass movements associated with the rainstorm of June 1966 in Hong Kong. In: Coates, D.R. (ed.), Environmental Geomorphology and Landscape Management, Volume 2, Benchmark Books,Dowden,Hutchinson and Ross Incorporated,Stroudsburg, Pennsylvania, 244-254.

Sudgen, M.B., Van Wieringen, M. and Knight, K. 1977. Slip failure in bedded sediments. Proceedings of the Ninth International Conference on Soil Mechanics and Foundation Engineering, Tokyo, vol. 2, 155-160.

Tankard, A.J.; Jackson, M.P.A.; Eriksson, K.A.; Hobday, D.K.; Hunter, D.R. and Minter W.E.L. 1982. Crustal Evolution of Southern Africa: 3.8 Billion Years of Earth History, Springer Verlag, New York.

Terzaghi, K. 1950. Mechanism of landslides. Geological Society of America, Berkey Volume, 83-123.

Van Asch, W.J., Brinkhorst, W.H., Buist, H.J. and Vessem, P.V. 1984. The development of landslides by retrogressive failure in varved clays. Zeitschrift für Geomorphologie, Supplement Band 40, 165-181.

Varnes, D.J. 1958. Landslide types and processes. In: E.B. Eckel (Ed.): Landslides and Engineering Practice, (Highway Research Board, Washington), Special Publication 29, NAS-NRC Publication 544.

Varnes, D.J. 1978. Slope movements and types of processes. In: Landslides: Analysis and Control. Transportation Research Board National Academy of Science, Washington Special Report, 176, 11-33.

Venter, J.P. 1983. Karoo mudrock. In: Brink, A.B.A. (ed.), Engineering Geology of Southern Africa, Vol. 3, Building Publications, Silverton, 73-106.

Webb, D.L. 1983. Slope failure on shale of the Pietermaritzburg Formation. In: Brink, A.B.A. (ed.), Engineering Geology of Southern Africa, Building Publications, Silverton, 107-111.

FLOOD GEOMORPHOLOGY AND PALAEOHYDROLOGY OF BEDROCK RIVERS

VICTOR R. BAKER
Department of Geosciences, University of Arizona

1. INTRODUCTION

In studying the response of river channels to flood flows an important distinction must be made with regard to the resistance of the channel boundaries to imposed stresses. Alluvial rivers have highly deformable boundaries, while non-alluvial rivers do not.

Until recently, modern quantitative fluvial geomorphological research has focused on studies of alluvial rivers. The channel characteristics of such rivers are established by variations in the discharge of water and sediment in flows of intermediate magnitude and frequency (Leopold and Maddock, 1953; Wolman and Miller, 1960). Alluvial river channels are self-formed by the independent adjustment of morphological variables to the water and sediment discharges. The adjustment is characterized by a tendency toward equilibrium or regime. Moreover, as shown in the writings of Luna B. Leopold (e.g. Leopold et al., 1964), measurements of both the process variables and the response variables can be organized into empirical and theoretical explanations of fluvial phenomena. By systematizing a great body of quantitative measurement Leopold and his colleagues discovered fundamental properties of alluvial rivers from which to predict their behavior. Thus, geomorphology has been especially successful in advancing the understanding of those rivers that Leopold and Langbein (1962, p. A11) termed, "...authors of their own geometries".

As in all good science, the rise of a comprehensive theory also engenders the appearance of anomalies, that is, phenomena which are inconsistent with the theory. Where bed and bank resistance are extreme, as in the case of many bedrock rivers, deformation does not occur for even very large and rare floods. Moreover, when geomorphic and climatic factors allow the occurrence of very large floods, phenomenal sediment transport and erosion occur as a result of exceeding high thresholds of resistance (Baker, 1977). Bedrock rivers subject to such floods comprise a class of fluvial systems that have only recently been analysed in quantitative terms. Their study has been facilitated by the fact that under certain conditions,

Geomorphological Studies in Southern Africa, G.F.Dardis & B.P.Moon (eds)
© *1988 Balkema, Rotterdam. ISBN 90 6191 831 6*

Fig. 1. Oblique aerial view of the Finke River gorge in central Australia. The river is entrenched into resistant sandstone and transports a heavy sand load derived from desert basins (Baker et al., 1983a). Palaeoflood sediments are preserved along channel margins allowing reconstruction of floods over the last few millennia (Baker et al., 1987a).

flood-dominated bedrock rivers may preserve a detailed and accurate record of their history of major flow events. As stated by Baker and Pickup (1987, p. 645) this class of rivers can be considered "...the chroniclers of their own cataclysms". In contrast to the self-forming or self-regulating tendencies of alluvial rivers, this special class of bedrock rivers has the scientifically intriguing state of being self-gauging.

Methodologies for the recognition and analysis of self-gauging bedrock rivers originated in studies in the southwestern United States (Baker, 1975; Baker et al., 1979). In subsequent work, to be reviewed in this paper, the following regional factors were found to produce ideal conditions for the occurrence of self-gauging rivers; (1) confined canyons or gorges developed in resistant geological materials, (2) adequate concentrations of sand, silt, and coarser materials in transport, and (3) resistant channel beds not subject to aggradation. These criteria are met in many of the ancient shield landscapes and adjacent highland terrains of the southern hemisphere. Work in central and northern Australia (Fig. 1) revealed self-gauging rivers in tropical, savanna, and desert settings for rivers flowing through fold belts and plateau uplands (Baker et al., 1983b; Baker and Pickup, 1987). Numerous analogous settings in southern Africa afford excellent potential study sites.

Fig. 2. Flood-transported boulders comprising a 100 m by 300 m boulder bar in the Katherine Gorge, northern Australia (Baker and Pickup, 1987). The largest transported boulder measured 4m by 3m by 1.5 m.

2. EFFECTS OF RISING FLOOD STAGE

In an alluvial river the rising stages of flood flow produce a series of equilibrium or regime adjustments among sediment capacity, water discharge, and morphological variables. In the channel, velocity and sediment transport capacity rise until overbank conditions occur. If the stream has excess sediment transport capacity, it adjusts by eroding its bed or banks. If the sediment concentration rises too fast, the stream deposits this sediment as natural levees. Each response provides a means of restoring equilibrium (Maddock, 1976).

In resistant-boundary bedrock streams, these adjustments cannot occur. The systems are sediment-limited, and equilibrium or regime responses are not possible. In the most resistant rock types, narrow-deep cross sections are common (Baker, 1984). As flood flows rise, most of the increased discharge is accommodated by stage increases.

Narrow-deep cross sections in bedrock channels profoundly influence the velocities, bed shear stresses, and stream powers per unit boundary areas generated by great floods (Baker and Costa, 1987). The combination of steep water-surface gradients and inordinately great flow depths insures that each incremental increase in discharge produces immense increases in the variables responsible for erosion and sediment transport. However, because resistant-boundary channels are sediment-limited, the excess energy must be dissipated as remarkable intense turbulent phenomena (Baker, 1978). Sand and even gravel, may be transported in suspension (Baker and Costa, 1987). Boulders are easily moved (Fig. 2), and immense erosion can be achieved (Baker, 1988; Patton and Baker, 1977; Nanson, 1986).

Ironically, the intense flood flow phenomena of narrow-deep bedrock channels also provide a phenomenal opportunity to reconstruct past flood chronologies. The cataclysmic flood phenomena that defy direct study are responsible for the emplacement of a long-term record that can be studied indirectly. The remainder of this paper will illustrate the analysis of self-gauging rivers.

3. PALAEOFLOOD EVIDENCE

Palaeohydrological studies of bedrock streams follow in a long tradition of study (Costa, 1986; Patton, 1987). However, until recently much of that tradition applied to the studies of alluvial rivers. Various general techniques have been advocated for fluvial palaeohydrological studies, or palaeofluminology (Baker, 1983a), as reviewed by Foley (1984), Maizels (1983, 1987), Starkel and Thornes, (1981), and Williams (1984). Among the categories of analysis are regime-based palaeoflow estimates (RBPE), methods that relate palaeohydraulic factors (usually shear stress, velocity, or stream power) to the maximum particle sizes transported by a river, and methods that determine palaeostages from various indicators.

RBPE studies involve empirically derived relationships that relate relatively high-probability flow events, such as the mean annual flood or bankfull discharge, to palaeochannel dimensions, sediment types, palaeochannel gradients and other field evidence. RBPE studies are described in a large scientific literature, summarized by Dury (1976), Ethridge and Schumm (1978), and Williams (1984).

RBPE studies apply to alluvial channels, which are self-formed by the interaction of water and sediment. During floods such channels adjust their width, depth, and slope to the prevailing discharge. Relatively large flows, exceeding bankfull, spread over floodplains. Because of floodplain storage, large increases in flood discharge produce only small increases in stage for such overbank flows. Because nearly all RBPE relationships apply to relatively frequent floods, they are of minimal use in the evaluation of rare, extreme floods.

Maximum particle size studies involve regression expressions and theoretical considerations that determine shear stress, velocity or stream power (SS-V-SP). The palaeohydrologic applications assume that the SS-V-SP estimates apply to the maximum foods that transported the particles. Whether based on theoretical (Baker, 1974; Komar, 1988) or empirical procedures (Costa, 1983; Williams, 1983) key data for SS-V-SP studies are palaeochannel dimensions, slopes, and maximum particle sizes. SS-V-SP studies generally produce only a single estimate of the largest flood experienced in a give time period. Particle sizes sufficiently large to have been transported near the competence limit of that flood need to be present

476

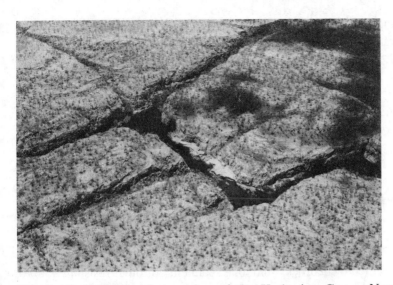

Fig. 3. Aerial view of a portion of the Katherine Gorge, Northern Territory, Australia. Flood flows proceed from left (east) to right (west) in this scene, which was photographed during dry-season low flows in October, 1979. The prominent rectangular pattern of slot-like canyons results from vertical joints in the resistant sandstone bedrock. The tributary at the top center (Butterfly Gorge) contains slackwater deposits described by Baker et al. (1985) and by Baker and Pickup (1987).

in the study reach. Accuracy of the method is low, as discussed by Church (1978).

The most accurate and detailed reconstructions of large palaeofloods are achieved by studies of stable-boundary fluvial reaches characterized by slackwater deposits and palaeostage indicators (SWD-PSI) (Baker et al., 1979, 1983, Kochel and Baker et al., 1988; Baker and Kochel, 1988). Slackwater deposits consist of sand and silt (occasional gravel) that accumulate relatively rapidly from suspension during major floods, particularly where flow boundaries result in markedly reduced local flow velocities. For palaeoflood studies, a slackwater sedimentation site should be optimum for both the accumulation and preservation of the relatively fine-grained flood sediments carried high in flood flows at maximum stage. Assuming available bed material load for transport, local sites of slackwater deposition may be of several types, including mouths of tributaries, upstream of abrupt channel constructions, flow-separation zones in abrupt channel expansions, caves and rock shelters on canyon walls, downstream of large boulders and other flow obstructions, and eddy zones or ineffective flow areas associated with channel bands or valley-side alcoves. The dynamics of flood slackwater sedimentation at such sites has been described by Baker et al. (1983a), Baker and Kochel (1988), and Kochel and Ritter (1978).

Tributary mouth sites are very common (Fig. 3). They occur because relatively small tributaries debouch their peak flows before mainstem flood peaks. The

477

mainstem flooding may then backflood the tributary up to a level nearly equivalent to the mainstem flood stage. Bedrock caves are less common, but they are especially valuable for the long-term preservation of flood slackwater sediments. Caves or rockshelters with large openings readily accumulate slackwater sediments if located at optimum levels in relation to flood stages. Moreover, the cave environment prevents subsequent rainwater and tributary flow erosive effects. Reduced biological activity preserves deposit stratigraphy.

4. SWD - PSI PALAEOFLOOD HYDROLOGY

The best SWD-PSI palaeoflood sites are located in channel reaches with flow boundaries constrained by bedrock, immobile sediment , or other resistant boundary materials. Such channels produce relatively large stage changes for changes in flood discharge (Baker, 1977, 1984). Moreover, they do not change their cross sections appreciably during major floods, as commonly occurs in alluvial channels (Leopold and Maddock, 1953).

Field surveys are required to identify and describe various palaeostage indicators along a SWD-PSI study reach, date the flood deposits, and survey the hydraulic geometry. Hydraulic geometry surveys generally follow established engineering principles, but the dating of flood deposits requires stratigraphic analysis. Individual flood sedimentation units are recognized (Fig. 4), discriminated, and correlated between multiple sites along a palaeoflood study reach (Kochel et al., 1982). Most important is the use of geochronologic procedures to provide absolute dates of the individual flood events.

Radiocarbon dating is the most common geochronologic tool used in palaeoflood studies. Floods in the period 1950 to present can be dated essentially to the calendar year (Baker et al., 1985). Floods in the period 1650 to 1950 require supplemental dating by historical documentation, archaeology, dendrochronology, or other means. Floods in the periods 10000 to 350 years B.P. (before present) can generally be dated to high accuracy by conventional radiocarbon procedures. The new procedure of tandem-accelerator mass spectrometry (Taylor et al., 1984) makes possible the dating of even 1 to 2 mg of carbon.

Other geochronology tools may be used in some applications. Wohl et al.(in press) found that thermoluminescence dating was useful in a study of palaeoflood slackwater deposits in northern Australia. Tree ring studies have also proven especially useful for the dating of palaeofloods in the southwestern United States (Clark, 1986; Smith and McCord, 1986).

After palaeoflood stages have been established by analysis of various slackwater deposits and palaeostage indicators, the investigator must transform those data into palaeodischarge estimates. These can be accomplished by several hydraulic procedures, including the slope-area method and step-backwater methods. The former

Fig. 4. Sandy slackwater deposits and intercalated tributary alluvium (coarse gravel) at the mouth of Cave Creek, a tributary to Aravaipa Canyon in southeastern Arizona. Note the prominent organic layer being pointed to by the student.

was utilized in the first SWD-PSI studies (Baker et al., 1979, 1983; Kochel and Baker, 1982; Kochel et al., 1982), but more recent work has utilized step-backwater analysis (Ely and Baker, 1985; O'Connor et al., 1986; Partridge and Baker, 1987; O'Connor and Webb, 1988; Webb et al., 1988). A significant advantage of step-backwater analysis is that the hydraulic calculations can be performed independently of the high-water indicator survey. Thus, problems in the geomorphology of palaeoflood indicators can be separated from problems in specifying the hydraulics of the study reach. Recent advances in computer flow modelling (Feldman, 1981) have greatly facilitated the latter. Of prime importance to accurate flow modelling is a precise characterisation of channel geometry. In stream channels with non-deformable boundaries (the preferred application) the survey cross sections should be chosen to account for the major energy losses that occur at channel expansions and construction.

Water-surface profiles produced by modern step-backwater flow-modelling procedures require input parameters of energy-loss coefficients, stage, and discharge. The initiating stage choice is of little consequence, since profiles generated at equal discharge will converge after a few computed steps. Moreover, several studies have demonstrated that step-backwater flow modelling of relatively deep, rare floods is remarkably insensitive to reasonable errors in estimating either Manning's coefficient or the expansion/contraction coefficients (Dawdy and Montayed, 1979; O'Connor et al., 1986; Sauer et al., 1985). These observations apply to relatively narrow, deep flood channels with resistant boundaries. Alluvial reaches, subject to appreciable scour and filling, are generally not appropriate for palaeoflood hydrologic studies.

Discharge estimation error in palaeoflood hydrologic analysis can be reduced in several ways. The categories of error reduction include; (a) palaeoflood cross-sectional stability, (b) palaeoflood water-surface estimation, and (c) palaeoflood flow coefficients. An important strategy is to separate the problems of categories (a) and (b), which depend on field conditions, from catergory (c), which depends on calculation procedure. This separation is not possible in the slope-area calculation procedure, since it requires that cross sections used in the analysis be located at sites where palaeoflood highwater indicators are present. Discharge estimation error in palaeoflood studies is somewhat analogous to estimation error in historic flood studies. However, several aspects of palaeoflood studies allow more precise controls on error than possible in historic flood studies or even some systematic flood studies. The main reason for this is the preservation of the best palaeoflood information at exceptionally stable geological sites. Methods of reducing palaeodischarge estimation error in the various categories are discussed by Kochel et al. (1982) and Baker (1987).

5. MAGNITUDE AND FREQUENCY OF PALAEOFLOODS

SWD-PSI studies can yield varying amounts of information on the palaeostages and ages associated with ancient flow events. The "worst case" example is a single flood stage indicated by a high-water indicator and a single date on the event. A variant of this "worst case" is a single vertically-stacked sequence of slackwater deposits (Fig. 4). Since a new flood deposit cannot be emplaced unless the responsible flood state exceeds the elevation of a previous deposit, the record flood series is censored. The censoring level (CL) is the elevation of the top of the deposit by the succeeding large flood. Note that CL increases when exceedences occur. However, the exact magnitude of exceedence is unknown, since various depths of flood water above the CL are capable of emplacing a deposit. Note also that floods smaller than this upward moving CL are not recorded in this example.

In practice, it is rare that a SWD-PSI study will only yield a single indicator. More commonly, the problems of the single-site, vertically-stacked sequence are minimized by correlation and tracing through stratigraphic analysis (Kochel et al., 1982). Deposits of a given palaeoflood are traced laterally to their highest levels as they thin upstream along a tributary. Additional checks on the maximum flood level are provided by scour lines that can be traced to dated deposits. Moreover, inset deposits generally accumulate below vertical slackwater depositional stacks. Inset relationships allow the identification of smaller floods, because a new, lower threshold is established by the river itself.

In the general case, characterizing most SWD-PSI studies, numerous vertical stacks, which began accumulating at different CL positions, are correlated along a study reach. Scour lines or other direct high-water indicators are incorporated into the survey. Such studies rely on the highest deposit or mark of a given flood to define the magnitude of that flood. Thus, greater accuracy is achieved when the SWD-PSI study attempts to document as many sites as possible in an appropriate study reach. Floods that fail to leave deposits at one site, because they fail to exceed the local CL, will be preserved at other, lower sites. Sites posing a very high CL for a given flood will preserve the very highest deposits emplaced by that flood, thereby establishing its stage. For engineering studies the value of the increased accuracy achieved by such studies must be balanced by the increased expense of prolonging the fieldwork.

One way to insure an appropriate palaeoflood record for statistical analysis is to assert the goal of reconstructing a complete catalogue of palaeodischarges exceeding or nonexceeding various censoring levels over specified time periods. This results in a type of systematic data set that can be subjected to flood-frequency analysis (Stedinger and Cohn, 1986; Stedinger and Baker, 1987). Baker and Pickup (1987) illustrate this analytical procedure for the Katherine Gorge, Northern Territory, Australia.

6. APPLIED PALAEOFLOOD HYDROLOGY

In many countries, including those of southern Africa, there are numerous local opportunities to generate excellent SWD-PSI palaeoflood records. The expense of such studies is minor in relation to planning costs for major high-risk projects such as large dams. At present these opportunities are largely being ignored. Valuable palaeoflood records are actually being destroyed by reservoir construction. There is a critical need to generate palaeoflood data to optimum SWD-PSI study sites.

A major reason for lack of use of SWD-PSI methodology is its multidisciplinary complexity. SWD-PSI studies require expertise in geology and geomorphology as well as in hydrology and hydraulics. Lack of familiarity with concepts and terminology have hindered the adoption by engineers of palaeoflood methodologies, but the potential contribution of palaeoflood data should outweigh this artificial barrier. It is imperative that the design engineering community be made aware of modern advances in palaeoflood hydrology.

At a minimum, the physical evidence of large palaeofloods should be considered to provide objective evidence of the likelihood and frequency of larger floods than can be documented in systematic or historic flood records. Palaeoflood data should be used to set the credibility of calculated probable maximum flood values used in the design of high-hazard dams. For critical projects the palaeoflood data should at least be collected, appropriately weighted, and considered in the overall decision process leading to design.

The most extensive application of SWD-PSI palaeoflood hydrology to a practical problem has been the dam-risk evaluation of Salt River System in Arizona (Baker et al., 1987a). Work by Ely and Baker (1985) and Partridge and Baker (1987) showed that the largest floods to occur at critical dam sites in the past 1000 to 2000 years were only 33 - 25 % of the probable maximum flood values applied theoretically to those sites. Moreover, a statistical analysis of palaeoflood, historical and systematic data for the sites was performed using maximum likelihood methods (Stedinger et al., 1988). It was found that occurrences of floods substantially larger than those in the palaeoflood record are very unlikely.

7. SUMMARY

Narrow-deep bedrock streams display an enhanced response to floods. Where bed and bank resistance are extreme, and where climatic and geomorphic factors allow, fluvial systems may be dominated by rare, great floods that produce extraordinary values of boundary shear stress and stream power per unit area of stream bed. Until recently fluvial geomorphologists have emphasized studies of alluvial rivers, which adjust their form to relatively frequent and measurable flow parameters. Such rivers, characterised by Luna Leopold and Walter Langbein as the "...authors of their own geometries...", proved most amenable to quantitative analysis. Because of their rarity and difficulty of measurement, cataclysmic floods remained under appreciated as agents of landscape change.

The emerging science of palaeoflood hydrology allows the accurate, quantitative reconstruction of flood histories in appropriate settings. Narrow deep bedrock canyons with stable channel geometries and appropriate sediment characteristics may preserve remarkably detailed sequences of ancient flood slackwater deposits and palaeostage indicators (SWD-PSI). Recent advances in geochronology, flow modelling, and statistical analysis of palaeoflood data have greatly increased the ability to extract useful hydrologic information from SWD-PSI studies. SWD-PSI investigations can provide reconstructions of discharges and magnitudes for multiple palaeofloods with remarkably high accuracy over time scales of centuries and millennia. In essence, rivers displaying well-preserved SWD-PSI are the chroniclers of their own cataclysms. The ancient shield landscapes of the southern hemisphere offer numerous opportunities for the application of SWD PSI palaeoflood hydrology. Work in the Northern Territory, Australia, demonstrates the technique in tropical, savanna, and desert settings for rivers flowing through fold belts and plateau uplands. Numerous analogous settings in southern Africa afford excellent potential study sites.

8. REFERENCES

Baker, V.R. 1974. Paleohydraulic interpretation of Quaternary alluvium near Golden, Colorado. Quaternary Research, 4, 95-112.

Baker, V.R. 1975. Flood hazards along the Balcones Escarpment in central Texas: Alternative approaches to their recognition, mapping and management. University Texas, Bureau of Economic Geology Circular 75-5, 22p.

Baker, V.R. 1977. Stream-channel response to floods with examples from central Texas. Geological Society of America Bulletin, 88, 1057-1071.

Baker, V.R. 1978. Paleohydraulics and hydrodynamics of scabland floods. In: Baker, V.R. and Nemmendal, D. (eds.), The Channeled Scabland, National Aeronautics and Space Administration, Washington, D.C., 59-79.

Baker, V.R. 1983a. Large-scale fluvial palaeohydrology. In: Gregory, K.J. (ed.), Background to Palaeohydrology: A Perspective, John Wiley and Sons, Chichester, 453-478.

Baker, V.R. 1983b. Paleoflood hydrologic techniques for the extension of stream flow records. Transportation Research Record, 922, 18-23. (National Research Council, Washington, D.C.).

Baker, V.R. 1984. Flood sedimentation in bedrock fluvial systems. In: Koster, E.H. and Steel, R.J. (eds.), Sedimentology of Gravels and Conglomerates, Canadian Society of Petroleum Geologists Memoir, 10, 87-98.

Baker, V.R. 1987. Palaeoflood hydrology and extraordinary flood events. Journal of Hydrology (in press).

Baker, V.R. 1988. Flood erosion. In: Baker, V.R., Kochel, R.C. and Patton, P.C. (eds.), Flood Geomorphology, John Wiley and Sons, New York (in press).

Baker, V.R. and Costa, J.E. 1987. Flood power. In: Mayer, L. and Nash, D.B. (eds.), Catastrophic Flooding, George Allen and Unwin, London, 1-24.

Baker, V.R., Ely, L.L., O'Connor, J.E. and Partridge, J.B. 1987a. Palaeoflood hydrology and design applications. In: Singh, V.P. (ed.), Flood Frequency and Risk Analysis, D. Reidel Publ. Co., Dordrecht, The Netherlands, (in press).

Baker, V.R. and Kochel, R.C. 1988. Flood sedimentation: Bedrock fluvial Systems. In: Baker, V.R., Kochel, R.C. and Patton, P.C. (eds.), Flood Geomorphology, John Wiley and Sons, New York (in press).

Baker, V.R., Kochel, R.C. and Patton, P.C. 1979. Long-term flood frequency analysis using geological data. International Association of Hydrological Sciences Publication, 128, 3-9.

Baker, V.R. , Kochel, R.C., Patton, P.C. and Pickup, G. 1983a. Palaeohydrologic analysis of Holocene flood slack-water sediments. In: Collinson, J. and Lewin, J. (eds.), Modern and Ancient Fluvial Systems, International Association of Sedimentologists Special Publication No. 6, 229-239.

Baker, V.R. and Pickup, G. 1987. Flood geomorphology of the Katherine Gorge, Northern Territory, Australia. Geological Society of America Bulletin, 98, 635-646.

Baker, V.R., Pickup, G. and Polach, H.A. 1983b. Desert palaeofloods in central

Australia. Nature, 301, 502-504.

Baker, V.R., Pickup, G. and Polach, H.A. 1985. Radiocarbon dating of flood events, Katherine Gorge, Northern Territory, Australia. Geology, 13, 344-347.

Baker, V.R., Pickup, G. and Webb, R.H. 1987b. Palaeoflood hydrologic analysis at ungaged sites, central and northern Australia. In: Singh, V.P. (ed.), Flood Frequency and Risk Analysis, D. Reidel Publ. Co., Dordrecht, The Netherlands (in press).

Church, M. 1978. Palaeohydrological reconstructions from a Holocene valley fill. In: Miall, A.D. (ed.), Fluvial Sedimentology, Canadian Society of Petroleum Geologists Memoir, 5, Calgary, Alberta, 743-772.

Clark, S. 1986. Potential for cottonwoods as indicators of past floods. In: Proceedings, International Symposium on Ecological Aspects of Tree Ring Analysis, Columbia University, N.Y., 17-21.

Costa, J.E. 1983. Paleohydraulic reconstruction of flash-flood peaks from boulder deposits in the Colorado Front Range. Geological Society of America Bulletin, 94, 986-1004.

Costa, J.E. 1986. A history of paleoflood hydrology in the United States. EOS, 67, 425-430.

Dawdy, Dr. R. and Motayed, A.K. 1979. Uncertainties in determination of flood profiles. In: Inputs for Risk Analysis in Water Resources, Water Resources Publications, Fort Collins, Colorado, 193-208.

Dury, G.H. 1976. Discharge prediction, present and former, from channel dimensions. Journal of Hydrology, 30, 219-245.

Ely, L.L. and Baker, V.R. 1985. Reconstructing paleoflood hydrology with slackwater deposits: Verde River, Arizona. Physical Geography, 6, 103-126.

Ethridge, F.G. and Schumm, S.A. 1978. Reconstructing paleochannel morphologic and flow characteristics: methodology, limitations and assessment. In: Miall, A.D. (ed.), Fluvial Sedimentology, Canadian Society of Petroleum Geologists Memoir, 5, Calgary, Alberta, 703-721.

Feldman, A.D. 1981. HEC models for water resources system simulation, theory and experience. In: Chow, V.T. (ed.), Advances in Hydroscience, Academic Press, New York, 12, 297-423.

Foley, M.G., Doesburg, J.M. and Zimmerman, D.A. 1984. Paleohydraulogic techniques with environmental applications for siting hazardous waste facilities. In: Koster, E.H. and Steel, R.J. (eds.), Sedimentology of Gravels and Conglomerates, Canadian Society of Petroleum Geologists Memoir, 10, Calgary, Alberta, 99-108.

Kochel, R.C. 1988. Extending stream records with slackwater paleoflood hydrology: examples from west Texas. In: Baker, V.R., Kochel, R.C. and Patton, P.C. (eds.), Flood Geomorphology, John Wiley and Sons, New York. (in press).

Kochel, R.C. and Baker, V.R. 1988. Paleoflood analysis using slackwater deposits. In: Baker, V.R., Kochel, R.C. and Patton, P.C. (eds.), Flood Geomorphology, John Wiley and Sons, New York (in press).

Kochel, R.C., Baker, V.R. and Patton, P.C. 1982. Paleohydrology of southwestern Texas. Water Resources Research, 18, 1165-1183.

Kochel, R.C. and Ritter, D.F. 1987. Implications of flume experiments on the interpretation of slackwater paleofloods sediments. In: Singh, V.P. (ed.), Flood Frequency and Risk Analysis, D. Reidel, Publ. Co., Dordrecht, The Netherlands (in press).

Komar, P.D. 1988. Sediment transport by floods. In: Baker, V.R., Kochel, R.C. and Patton, P.C. (eds.), Flood Geomorphology, John Wiley and Sons, New York (in press).

Leopold, L.B. and Langbein, W.B. 1962. The concept of entropy in landscape evolution. U.S. Geological Survey Professional Paper 500-A, A1-A20.

Leopold, L.B. and Maddock, T. Jr. 1953. The hydraulic geometry of stream channels and some physiographic implications. U.S. Geological Survey Professional Paper 252, 1-57.

Leopold, L.B., Wolman, M.G. and Miller, J.P. 1964. Fluvial Processes in Geomorphology. Freeman, San Francisco, 522p.

Maddock, T. Jr. 1976. A primer on floodplain dynamics. Journal of Soil and Water Conservation, 31, 44-47.

Maizels, J.K. 1983. Palaeovelocity and palaeodischarge determination for coarse gravel deposits. In: Gregory, K.J. (ed.), Background to Palaeohydrology: A Perspective, John Wiley and Sons, New York, 101-139.

Maizels, J.K. 1987. Large-scale flood deposits associated with the formation of coarse-grained, braided terrace sequences. In: Ethridge, F.G., Flores, R.M. and Harvey, M.D. (eds.), Recent Developments in Fluvial Sedimentology, Society of Economic Paleontologists and Mineralogists, Special Publication 39, 135-148.

Nanson, G.C. 1986. Episodes of vertical accretion and catastrophic stripping: a model of disequilibrium flood-plain development. Geological Society of America Bulletin, 97, 1467-1475.

O'Connor, J.E. and Webb, R.H. 1988. Hydraulic modeling for paleoflood analysis. In: Baker, V.R., Kochel, R.C. and Patton, P.C. (eds.), Flood Geomorphology, John Wiley and Sons, New York (in press).

O'Connor, J.E., Webb, R.H. and Baker, V.R. 1986. Paleohydrology of pool and riffle pattern development, Boulder Creek, Utah. Geological Society of America Bulletin, 97, 410-420.

Partridge, J.B. and Baker, V.R. 1987. Paleoflood hydrology of the Salt River, Arizona. Earth Surface Processes and Landforms, 12, 109-125.

Patton, P.C. 1987. Measuring the rivers of the past: A history of fluvial palaeohydrology. In: Landa, E.P. and Ince, S. (eds.), The History of Hydrology, American Geophysical Union History of Geophysics Series Vol. 3. (in press).

Patton, P.C. and Baker, V.R. 1977. Geomorphic response of central Texas stream channels to catastrophic rainfall and runoff. In: Doehring, D. (ed.), Geomorphology of Arid and Semi-arid Regions, Publications in Geomorphology, Binghamton, N.Y., 189-217.

Patton, P.C., Baker, V.R. and Kochel, R.C. 1979. Slackwater deposits: a geomorphic technique for the interpretation of fluvial paleohydrology. In: Rhodes, D.D. and Williams, G.P. (eds.), Adjustments of the Fluvial System, Kendall/Hunt Publishing Company, Dubuque, Iowa, 225-253.

Sauer, V.B., Curtis, R.E., Santiago-Rivera, L. and Gonzalez, R. 1985. Quantifying flood discharges in mountainous tropical streams. In: Quinones, F. and Sanchez, A.V. (eds.), International Symposium on Tropical Hydrology and 2nd Caribbean Islands Water Resources Congress, American Water Resources Association, Bathesda, Maryland, 104-108.

Smith, S.S. and McCord, V.A.S. 1986. Tree-ring reconstruction of flood events along Kanab Creek, south-central Utah. Geological Society of America Abstracts with Programs, 18, 755.

Starkel, L. and Thornes, J.B. 1981. Palaeohydrology of River Basins. British Geomorphological Research Group Technical Bulletin 28, Geo Books, Norwich, 107pp.

Stedinger, J.R. and Baker, V.R. 1987. Surface water hydrology: historical and paleoflood information. Reviews of Geophysics, 25, 119-124.

Stedinger, J.R. and Cohn, T.A. 1986. The value of historical and paleoflood information in flood frequency analysis. Water Resources Research, 22.

Stedinger, J.R., Therivel R. and Baker, V.R. 1988. Flood frequency analysis with historical and paleoflood information. In U.S. Committee on large Dams Eighth Annual Lecture Notes, Salt River Project, Phoenix, Arizona. (in press).

Taylor, R.E., Donahue, D.J., Zabel, T.H., Damon, P.E. and Jull, A.J.T. 1984. Radiocarbon dating by particle accelerators: An archaeological perspective. In: Archaeological Chemistry 111, Lambert, J.B. (ed.), American Chemical Society Advances in Chemistry Series, 205, 333-356.

Webb, R.H., O'Connor, J.H. and Baker, V.R. 1988. Paleohydrologic reconstruction of flood frequency on the Escalante River, south-central Utah. In: Baker, V.R., Kochel, R.C. and Patton P.C. (eds.), Flood Geomorphology, John Wiley and Sons, New York (in press).

Williams, G.P. 1983. Paleohydrological methods and some examples from Swedish fluvial environments. 1 - cobble and boulder deposits. Geografiska Annaler, 65A, 227-243.

Williams, G.P. 1984. Paleohydrologic equations for rivers. In: Costa, J.E. and Fleischer, P.J. (eds.), Developments and Applications of Geomorphology, Springer-Verlag, Berlin, 343-367.

Wohl, E.E., Murray, A.S. and East, T.J. 1988. Thermoluminescence dating of fluvial sands, East Alligator River, Australia. Quaternary Research. (in press).

Wolman, M.G. and Miller, J.P. 1960. Magnitude and frequency of forces in geomorphic processes. Journal of Geology, 68, 54-74.

MORPHOLOGY OF THE MOOI RIVER DRAINAGE BASIN

A.B. DE VILLIERS

Department of Geography, Potchefstroom University

1. INTRODUCTION

The aim of this study was to determine the existence of significantly different morphological regions on the basis of a quantitative drainage basin analysis. The Mooi
River drainage basin, situated in the southern Transvaal between 25°54'S and 26°55'S,
and 26°55'E and 27°49'E, was selected for the study. This basin is situated in the area
called "Pre-Karoo Highveld" (Wellington, 1955) or the "South-western Transvaal
Highplain with homoclinal ridges" (Van Zyl, 1985).

A morphological map was compiled by means of aerial photographs and field work.
The quantitative basin analysis, of a sample of drainage basins, confirms the existence
of significantly different morphological regions within the area. These differences can
be explained mainly in terms of the underlying geology.

2. GEOLOGY

As the present study is primarily concerned with the influence of rock type and structure on the morphology of the basin, a map was compiled showing the dominant rock
types present in the area. With the exception of Permo-Carboniferous Ecca
sandstones and shales and Cainozoic alluvium, the majority of rocks are Pre-Cambrian
in age. Lithologies of the different geological formations (e.g. for example, quartzite)
were combined as a single lithological unit on the map. The percentage area covered
by each of the different lithological units are given in Table 1. These units were derived
from vaious maps published by the Geological Survey. The generalized lithology of
the Mooi River drainage basin is shown in Fig. 1.

Four major structural elements are present in the area. These elements are related
to two major geological events, namely the formation of the Vredefort Dome and the
later emplacement of the Bushveld Igneous Complex (Fig.2). The first major structural element is found in the overturned collar rocks, lining the Archean granite core

Geomorphological Studies in Southern Africa, G.F. Dardis & B.P. Moon (eds)
© 1988 Balkema, Rotterdam. ISBN 90 6191 831 6

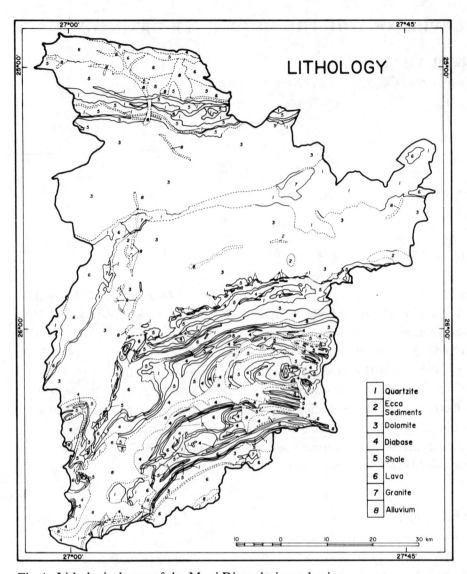

Fig. 1. Lithological map of the Mooi River drainage basin.

of the Vredefort Dome (A-B, Fig. 2). The next element is formed by the outer rim synclinorium, comprised mainly of the Pretoria group (B-C, Fig. 2), while the Hartebeesfontein anticline (C-D, Fig. 2) is the third major structural element. The northward dipping Transvaal sequence meta-sediments and volcanics, dipping towards the Bushveld Complex to the north of the Hartebeesfontein anticline (D-E, Fig. 2), is the last major structural element that plays an important part in the landscape formation of the study area.

The area underlain by the so-called rim synclinorium is in fact far more complex than is shown in Figure 2. In this rim synclinorium, gently dipping domes, basins, an-

488

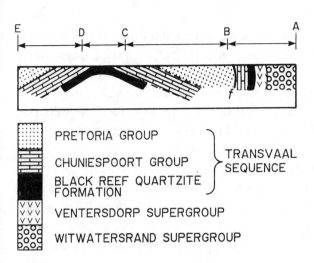

Fig. 2. Cross-section of the Mooi River drainage basin.

Table 1. Percentage area covered by the individual lithological units.

LITHOLOGICAL UNIT	PERCENTAGE AREA COVERED
Dolomite	39.0
Shale	17.0
Quartzitic rocks	14.0
Volcanics	12.0
Alluvium	10.0
Diabase	7.0
Granite	0.5
Ecca sediments	0.5

ticlines are present while the general dip of the strata increases from the north, towards the over-turned collar rocks of the Vredefort Dome in the south. A detailed description of the structure of this portion of the drainage basin can be found in Simpson (1977, 1981).

3. MORPHOLOGY

The effects of these major structural elements, described previously, can be clearly seen in the present day landscape. However, since the time that these major events took place the Mooi River catchment was also subjected to various erosional and depositional cycles. The presence of Dwyka and Ecca sediments, preserved in the paleo-sinkholes in the dolomite, bears evidence to one of these major events during the Karoo times. The sequence of events, resulting in the present day landscape were multicyclic in nature and are reflected in the morphology.

If the morphological map of the area (Fig. 3), originally compiled by Barker (1985), is studied, some idea of the complexity can be formed. However, if this map is compared with the lithological map (Fig. 1), it becomes clear that the underlying geology has indeed played a major role in the formation of the present day landscape. Some of the major correlations between landscape and geology will be discussed in the following paragraphs.

In the area formed by the collar meta-sediments and volcanics the steeply dipping, erosion resistant, quartzites form a series of homoclinal and isoclinal ridges (depending on the angle of dipslope). The undulating hilly terrain to the north of these ridges was formed on the lava of the Ventersdorp Supergroup. The flat terrain developed on the southernmost dolomite exposure effectively divides this area from the area underlain by the rim synclinorium. The southernmost, and a single western exposure of quartzite, form low isoclinal ridges in the rim synclinorium. The rest of the quartzite ridges in this area however, forms homoclinal ridges and cuestas. The areas underlain by shale, lava and diabase usually form flat to slightly undulating terrain. Along the major drainage lines, that are parallel to the strike of the structure, large areas later became covered with alluvium.

The landscapes developed in the extreme north of the basin, as well as those developed on the Hartebeesfontein anticline are very similar to those in the rim synclinorium. The only real differences between these areas are that in the rim synclinorium the general dip of the strata is towards the south, while in the north it is to the north, and depending on the position in relation to the axis of the Hartebeesfontein anticline, it can be either north or south. This has the effect that the free face of the cuestas in the different areas either faces north or south.

The dolomite, in the two major exposures, invariably forms a flat to undulating landscape. Karst-landform development is also evident in these areas. The presence of dolines, sinkholes, caves and karst drainage is well documented.

4. QUANTITATIVE ANALYSIS

A quantitative analysis of drainage basin variables was undertaken with a dual purpose in mind. The first purpose was to try to establish different morphological regions,

MORPHOLOGY

N

CREST LINES
.... BROAD ROUNDED
———— NARROW ROUNDED
———— SHARP
———— CUESTA
███████ ISOCLINAL RIDGE
○ BUTTE
◎ MESA

SLOPES
———— CONVEX
———— CONCAVE
.... STRAIGHT
— — — PROMINENT BREAK IN SLOPE

MINE DUMPS
◢ SLIMES DAM
◖ ROCKWASTE DUMP

0   5   10 km

Fig. 3. Morphological map of the Mooi River drainage basin

491

while the second purpose was to test the validity of the statement that there appears to be a close correlation between the morphology and the underlying geology.

To obtain a large enough sample of basins, underlain by a single lithological unit, it was decided to use only the first order drainage basins of the area. Out of the total number of first order basins (924) only 233 (25.2 %) were homogeneous as far as the lithology is concerned. On the grounds of the previous work on quantitative drainage basin analysis (De Villiers, 1987) two prime variables namely drainage basin size and stream lengths were used in the analysis. The averages for the different basins, for each lithological group, is given in Table 2a. The total number of basins used to calculate these averages is also given in this table. As can be seen from these numbers it is unfortunate that the number of basins for the dolomite, granite and diabase is rather low. This is however unavoidable because they were the only homogeneous basins available.

From the figures contained in Table 2a it can be seen that there appears to be a definite variation in basin size between the different lithological groups. As expected the largest first order basins developed on the dolomite, while the smallest basins developed on the erosion resistant quartzites. As far as the average stream lengths are concerned it is noticeable that the ranking of basin size compared to the ranking of basin lengths are very similar. It is only the shorter streams that are developed on the dolomite that is an exception. This anomaly can however be explained by the normal karst stream development process.

In Table 2b, basin areas and stream lengths are given for the different lithological groups as obtained for the different morphological regions indicated. It needs to be mentioned that the dolomite to the north of the Hartebeesfontein anticline was grouped with the northern Pretoria group sediments and the southern dolomite outcrop was grouped with the Pretoria Group in the rim synclinorium. The southernmost dolomite outcrop has no first order streams and is therefore not taken into consideration.

Only two rocks types, namely quartzite and lava, occur in all four regions. A general increase in both basin size and stream lengths, from the south towards the north, is noticeable. A similar size and length increase is also noticeable for the other rock types. The only exception to this general tendency of increasing stream lengths can be discerned in basins developed on the dolomite.

The reason for the difference in basin size and stream lengths (with the exception of the dolomite) is probably the difference in the general dip of the strata in the different regions. The smallest basins with the shortest streams developed in the collar area with dislopes up to 90 degrees. In the extreme northern area (with dislopes in the order of 3-12°) large basins with long streams have developed. The basins with intermediate drainage areas and stream lengths developed in the central portions of the catchment. In this region there is an average increase in the dip of the strata from

Table 2. Average basin size (Au) and stream length (Lu) for (a) first order basins, and (b) for first order basins in different morphological areas.

LITHOLOGY	NUMBER OF BASINS	Au (sq. km)	Lu (km)
(a) First Order Basins			
Dolomite	16	27.457	2.293
Granite	6	3.629	3.767
Lava	37	3.080	2.419
Shale	71	2.172	1.395
Diabase	19	1.000	0.824
Quartzite	84	0.424	0.372
(b) First Order Basins in Different Morphological Areas			
(i) Collar Area			
Dolomite	-	-	-
Granite	-	-	-
Lava	5	0.460	0.350
Diabase	-	-	-
Shale	-	-	-
Quartzite	7	0.107	0.029
(ii) Rim Synclinorium			
Dolomite	6	15.163	2.416
Granite	-	-	-
Lava	13	2.790	2.019
Diabase	10	1.030	0.845
Shale	44	1.155	1.009
Quartzite	34	0.253	0.162
(iii) Hartbeesfontein Anticline			
Dolomite	-	-	-
Granite	6	3.629	3.767
Lava	11	3.589	3.955
Diabase	-	-	-
Shale	-	-	-
Quartzite	16	0.592	0.853
(iv) Northern Area			
Dolomite	10	34.833	2.220
Granite	-	-	-
Lava	8	4.491	2.256
Diabase	9	0.967	0.800
Shale	27	3.829	2.024
Quartzite	27	0.623	0.411

generally low angles (5-10°) in the north up to 70-80° in the south.

5. CONCLUSIONS

Quantitative basin analysis could be used to define the different morphological regions within the Mooi River Drainage basin. It was further possible to explain these differences on the grounds of differences in lithology and structure.

6. REFERENCES

Barker, C.H. 1985. 'n Geomorfologiese studie van die   Mooirivieropvanggebied. M.Sc.-verhandeling (ongepubl.).   Potchefstroom, P.U. vir C.H.O. 175 p.

De Villiers, A.D. 1987. A multivariate evaluation of a group of drainage basin variables - a South African case study. In: Gardiner, V. (ed.), International Geomorphology 1986 - Part 11. Wiley, London, 21-31.

Simpson, C. 1977. The Structure of the Rim Synclinorium of the Vredefort Dome. Unpublished M.Sc. Dissertation, University of Witwatersrand, Johannesburg, 275 pp.

Simpson, C. 1981. The structure of the rim synclinorium of the Vredefort Dome. Transactions of the Geological Society of South Africa, 81, 115-121.

Stratten, T. 1986. The Dwyka glaciation and its relationship to the Pre-Karoo Surface. Unpublished Ph.D. thesis, Johannesburg, University of Witwatersrand.

Van Zyl, J.A. 1985. 'n Kaart van die landvormstreke van Suid-Afrika. South African Geographer, 13, 105-108.

Wellington, J.H. 1955. Southern Africa - A geographical study. Vol 1. Physical Geography. Cambridge University Press.

THE DIAMONDIFEROUS GRAVEL DEPOSITS OF THE BAM-BOESSPRUIT, SOUTHWESTERN TRANSVAAL, SOUTH AFRICA

TANIA R. MARSHALL

Gold Fields of South Africa, Krugersdorp

1. INTRODUCTION

From 1904 to 1984 the alluvial diamond deposits between Lichtenburg, Ventersdorp, Potchefstroom, and Christiana have produced some 14.4 million carats. The apparent absence of an obvious local primary source for these diamonds has led to the proliferation of hypotheses dealing with the origin of both the gravels and the diamonds. By far the majority of these studies have concentrated on either the dolomite-hosted gravels of the Lichtenburg-Ventersdporp districts or the terrace-type gravels of the lower Vaal River basin (du Toit, 1951; Retief, 1960; Partridge and Brink, 1967; Helgren, 1979).

Few studies have addressed the nature of the alluvial and colluvial deposits of the southwestern Transvaal, between Wolmaransstad, Christiana, and Schweizer-Reneke. Due to the present drought many of the farmers in these districts have opened up large tracts of land in and adjacent to dry tributaries of the Vaal River and are prospecting for diamonds as an alternate source of revenue. This has provided an ideal opportunity to study the gravel deposits that have been hitherto inaccessible. This study has concentrated on the Bamboespruit valley, a tributary of the Vaal River which extends from NNW of Wolmaransstad to SE of Bloemhof (Fig. 1). This affords a total stream-valley study length of approximately 180 km.

The Bamboespruit flows (in wetter years) over the amygdaloidal Ventersdorp lavas of the Makwassie Group. For the most part, these lavas have weathered to a thick fertile soil. Much of the gentle undulating southwestern Transvaal landscape is covered with a layer of red Hutton Sands and yellow-to-red apedal soils (Fitzpatrick et al.,1986).

2. GRAVEL DEPOSITS

There are two fundamentally different types of gravels occurring along the Bam-

Geomorphological Studies in Southern Africa, G.F.Dardis & B.P.Moon (eds)
© 1988 Balkema, Rotterdam. ISBN 90 6191 831 6

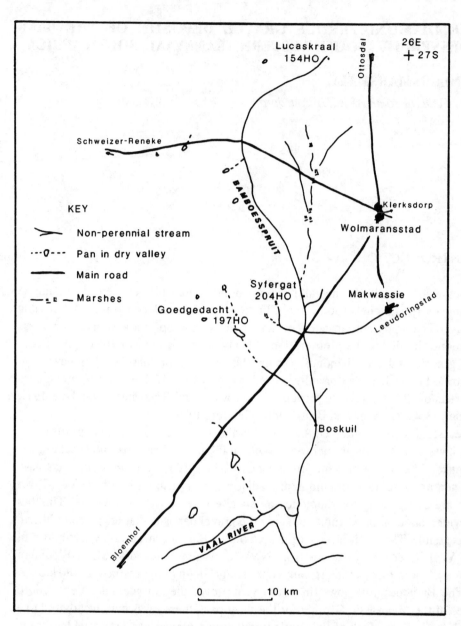

Fig. 1. Location of the Bamboesspruit, southwestern Transvaal, South Africa.

boespruit, namely the derived and the reworked alluvial gravels. The derived gravels occur on hillslopes and interfluves at depths of less than 1m, and the alluvial gravels occur in the present river valleys at depths between 5-8m below surface. The deposits differ in age and, more fundamentally, in terms of the processes which were responsible for their formation.

496

2.1. Derived Rooikoppie Gravels

The Rooikoppie gravels are so named because of their red to reddish-brown colour. They occur in interfluves, the upper section of hillslopes, and on hillcrests. The Rooikoppie gravel profile consists of three components, the overlying red sands, the diamondiferous gravels, and the much altered, uneven basal deposits.

The younger, overlying reddish-red brown sands have been classified as Hutton Sands (Fitzpatrick et al., 1986). A typical Hutton profile has a shallow (0-150 mm) A-horizon consisting of friable, fine sandy-loam and few medium-grained, indurated iron-manganese nodules. The B-horizon is most commonly a red apedal apedal soil and consists of friable, fine sandy-clay-loam, with few medium indurated iron-manganese nodules, fragments from the underlying horizon and is separated from it by a clear undulating boundary. The clay content is usually in the 4-12 % range, but may reach up to 25 %. The sand fraction contains no readily weatherable minerals and is dominated by medium-fine grade, well-sorted and rounded quartz. The red colour is produced by iron-oxide coatings on the sand grains (Helgren, 1979). The Hutton Sands are considered to have been derived from the the ancient floodplains of the Vaal and Harts rivers by deflation processes and have been remobilized several times beween periods of vegetation colonization (Helgren, 1979). Other sources describe the Hutton Sands as having been derived from the Kalahari aeolian dune fields during an arid phase in the Quaternary (Mayer, 1986). In either case, the Hutton Sands are a much younger feature than the underlying gravel deposits.

The layer of diamond-bearing Rooikoppie gravel varies in thickness from a single pebble layer to approximately 40 cm, but can be upwards of 2 m thick in pseudo-karst features in the underlying laterite surface. The clast content of the gravel comprises well-rounded quartz pebbles ranging in size from 4-10 cm. The matrix consists almost entirely of rounded vein quartz with iron-stained rims and a small component   of volcanics (possibly Ventersdorp lavas). Where the gravels have been concentrated in the palaeokarst features of the underlying laterite surface they have delivered extremely high diamond values (Goedgedacht, 197HO, for example, yielded over 14000 carats). The form of the Rooikoppie gravel is determined, to a large extent, by the nature of the surface on which it is deposited. This characteristic highlights the probability that the Rooikoppie is a derived and not a primary gravel.

The surface that the Rooikoppie gravels are deposited on is an ancient one. The original deposits have been so extensively altered by laterization that they are often impossible to identify. On the farm Syfogat, 204HO, the laterite succession consists of a 80 cm thick enriched zone passing progressively through a 50 cm mottled zone into a thick pallid zone (of the order of 2 m) and eventually into fresh unaltered bedrock (Ventersdorp lavas). The surface shows evidence of having undergone extensive leaching. This process has dissolved out many of the soil constituents resulting in a landscape very much akin to karst. Such a landscape has been termed a pseudo-karst landscape. The leaching process has evolved along inherent planes of weakness in the original rocks such as joint planes and results in the development of fissure-like grikes (corrosion opened joints). On the farm Goedgedacht, 197HO, diggings have exposed pseudokarst grikes with dimensions in the order of 1-2 m width, 50-200 m length, a 5-10 m depth. Laterization in these grikes has proceeded to depths of over 1 m, with the pallid zone descending to below the bottom of the opened joint. The original composition of the laterite at this locality appears to have been a small pebble conglomerate. The lateratized conglomerate passes laterally into the interbedded sandstone and shale facies before disappearing beneath the pan sediments of the Goedgedacht pan. Cross-bedding in the sandstones indicates that the palaeocurrent direction was predominantly from south to north. The pseudo-kasrt grike from which the sandstones are exposed is less than 1.5 m and the vertical extent of the profile is not known. It is possible, however, that a gravel horizon could exist at depth (the basal facies of a diamond-bearing palaeo-channel perhaps?).

The Rooikoppie gravels occur as infill in the grikes. The gravels fine inwards from a zone 1.5-3.0 m below the surface (Fig. 2). There are no slump features in the gravels adjacent to the walls indicating that the infilling of the grike occurred contemporaneously with the dissolution or leaching that formed the grike. The in-filling of the grike with gravels is unlikely to have been the result of the river action but, more likely, the result of flash-floods in a semi-arid environment.

Up to now the Rooikoppie gravels have been assumed to be remnants of high level terraces and, thus, referred to as runs. This is because, in many cases, the deposits tend to parallel the modern streams at distances of between 500-1000 m. It is, however, difficult to reconcile these deposits with true alluvial deposits in terms of their clay content (12-25 %) and their morphology; they occur almost parallel to the modern streams on both sides of the valleys, which is difficult to ex-plain as fluvial terraces unless invoking several closely spaced, very wide rivers. The gravels rather have the characteristics of the derived, colluvially reworked gravels of the lower Vaal basin (Partridge and Brink, 1969; Helgren, 1979). As such the Rooikoppie gravels probably represent colluvial and elluvial detritus scattered over a deflation surface that is extensively laterized. The derived gravels in the lower Vaal basin contains Cretaceous age pollen (T.C. Partridge, pers. comm, 1987) and it is possible that the Rooikoppie gravels are of a similar age. If so, then the pseudo-karstic landscape on which the gravels are deposited must represent the African surface which is elsewhere characterized by extensive and intensive weathering and laterization (Partridge and Maud, 1987).

2.2 Younger Terrace Gravels

The terrace gravels occur everywhere along the length of the Bamboespruit, flank-

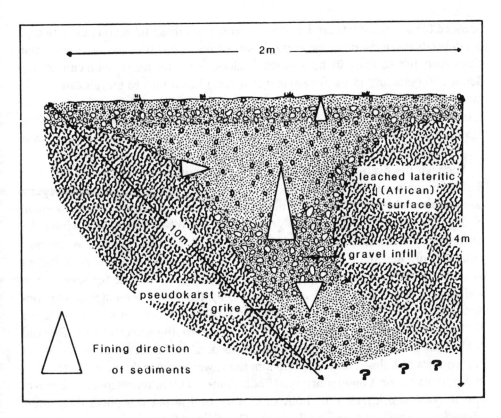

Fig. 2. Pooikoppie gravel infill of laterite psuedokarst grikes, Goegedacht 197H0.

ing the dry and marshy sections of the valley to widths of up to 100 m. In general, the terrace deposits consist of a 5-8 m upward-fining, alluvial sedimentary sequence deposited on an uneven floor of Ventersdorp lavas and capped by 1-2 m of soil (locally known as brown-black turf). The entire sequence is calcretized. The lower portions are either oxidized (brown-orange in colour) or reduced (grey-green) depending on the proximity of the water-table.

Due to the nature of the diggings it is almost impossible to see any sedimentary structures within the gravel profile. The action of the graders and backacters destroys the exposed surface and all but the coarsest sedimentary indicators. From a careful examination of the gravels and overlying calcreted sediments it is apparent that the overall sequence is upward fining. The boulders at the base of the channel are in the range of 30-40 cm (long axis diameter), but can be as large as 3-4 m (long axis diameter). They consist entirely of amygdaloidal Ventersdorp lavas and dolerite. The gravel clasts consists of two populations, (1) a sub-angular population consisting mainly of Ventersdorp lavas and dolerites and (2) a sub-rounded to rounded population of resistant, siliceous rocks, mainly quartz, chert, and quartzite. The latter population represents clasts which have been reworked from an earlier alluvial event (or which have travelled a long distance), and the former population represents locally derived detritus. Since the smooth, sub-

499

rounded to rounded clasts are indistinguishable from those found on the Rooikoppie deposits it is presumed that these deposits have been reworked to release their clasts into the streams to be deposited along with the more abundant locally derived Ventersdorp lavas. The agates that occur in the terrace gravels can also be found, more abundantly, in the older Rooikoppie gravels, and it is, therefore, assumed that this is their immediate (if secondary) source. The original source of the agates is problematic but, has preliminarily been ascribed to the Stormberg lavas (R. Armstrong, Pers. Comm.).

Although the overall form of the gravels is a massive, upward-fining sequence there are numerous sedimentary facies representing changes in hydraulic regime and sediment characteristics along the palaeochannel. An example of this is found on Rooibult, 152HO. The overall gravel sequence is 4 m thick. But within the calcretized gravel sequence, there are numerous fine-grained lenses, which may represent original shale or fine-sand lenses in the gravels. A few tens of metres away, in the same digging the gravel is upward fining at the base for about 2.5 m and then it becomes upward coarsening. Only the lower sequence is diamondiferous. On the Boskuil diggings, on Uitkyk 248HO, gravel lenses and stringers are found within the calcreted sands some 1-2 m above the gravel unit. On Rietkuil 186HO the gravel facies variation is quite considerable. In one instance, the gravel unit consists of a single unit less than 2 m thick, which suddenly increases to 3.5 m in a pothole in the Ventersdorp lava floor. Some 50-60 m downstream the gravel unit thickens considerably to about an average of 4 m but was pervasively interbedded with sandstone and mud lenses. The diamond content of the gravel had also diminished considerably. This leads to the recognition of obvious relationships between hydraulic conditions and deposition of diamonds.

A number of empirical observations were made concerning the depositional sites of the diamonds;

(1). The coarser the gravels the larger the average diamond size.

(2). Diamonds tend to concentrate in pockets in potholes in the floor of the channel; in front of (and in some cases behind) obstructions in the channel; in traps formed by the presence of large (3-4 m) boulders in the channels; downstream of constrictions in the channels; and in the point-bars of meanders in the palaeochannel (Fig. 3).

(3). The terrace gravels tend to be richer where the nearby Rooikoppie gravels had been (or are) rich and, similarly, where the Rooikoppie gravels are not so rich the terrace gravels also tended to be relatively poor.

For the most parts, the alluvial gravels occur at depths of between 4-8 m. Both the oxidized and the reduced varieties are usually at the same elevation. There is only one recorded occasion where the two gravel terraces are at two different elevations. On the farm Rooibult 152HO reduced gravels developed below 6-7m of calcrete and turf. Approximately 100 m away (upslope) from these gravels are distinctly shallower oxidized gravels at 3-4m below surface. According to the digger-farmer the reduced gravels extend 50-60 m from almost the present stream into the mielielands at roughly the same elevation (not depth below surface, due to the ef-

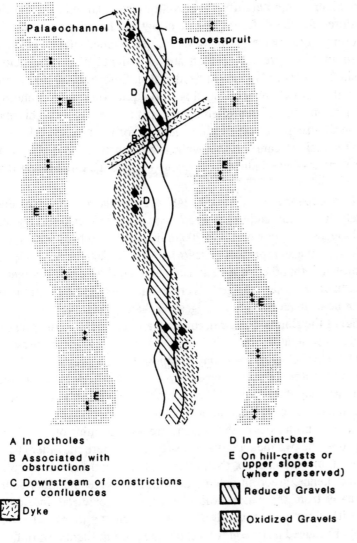

Fig. 3. Schematic plan diagram of the interference pattern set up by the meandering of the palaeochannel about the present dry Bamboesspruit. The stippled pattern represents the possible original extent of the Rooikoppie (derived colluvial) gravels and the diamond shapes represent economic concentrations of diamonds.

fects of topography). There is then a gap of some 10-20 m, where the calcreted sediments lie directly on top of Ventersdorp lavas, until the oxidized gravels are encountered at 3-4 m below the surface. The oxidized terrace gravels at this locality are not sufficiently different from those developed elsewhere along the Bamboespruit to warrant the assumption that they represent an earlier terrace than either oxidized gravels in the area. That the oxidized gravel is older than the reduced gravel on Rooibult 152 is undisputable, but it does not suggest that all such oxidized gravels are older than, or at different elevations from, all reduced gravels. The explanation offered for the different terrace elevations here is as follows; At the time of downcutting by the palaeo-Bamboespruit, the channel was in a position somewhat to the south of its present position and at a depth of +- 2m below present, and deposited what is today the oxidized gravel. Before the period of incision was complete the river channel changed position quite substantially and re-established itself further north and deposited the present reduced gravel. This may have been the result of stream cut off of a meander to form an oxbow lake at a higher elevation to the main stream. Alternatively, a local obstruction in the channel up- or downstream may have induced the channel to change direction at that point.

In two localities (Roodepan, 163 and Uitkyk, 248) the gravels occur in a channel that is approximately 30 m wide and 6-7 m deep. But in other localities diggings and refilled dumps suggest that gravels up to 200 m in width have been excavated. It is most unlikely that these represent palaeochannels of 200 m wide rivers. Rather, it is more probable that they represent channel migration of palaeostreams less than 30 m wide. In places where the gravels are mined over widths greater than 50 m there is little or no channel definition, suggesting channel migration.

The palaeochannel of the Bamboespruit occupies the same valley as the present stream. The meander wavelength, however, is different from that of the present. This is as a result of a wider channel (+- 30m as opposed to the present +- 10m) and a higher volume of water that must have flowed through it. As a result, the palaeochannel meanders about the present Bamboespruit. The gravels of the older channel, therefore, sometimes occur on one or other side of the present stream, in the present channel, or on both sides of the present stream depending on the interference pattern set up by the two meander patterns (Fig. 3). Where the older channel sediments occur in, or very near to, the present channel and the water table, they are reduced and, hence, green in colour. Conversely, where the older sediments are further away from the water table they are more oxidized and, thus, more yellow-orange in colour. Consequently, there is no regular pattern to the occurrence of oxidized or reduced terrace gravels about the present Bamboespruit.

3. DISCUSSION

3.1. Correlation with Vaal River stratigraphy

The stratigraphy of the lower Vaal River gravels has been subdivided into the Older and the Younger gravels (Partridge and Brink, 1969; Helgren, 1979). Two distinct types of Older Gravels have been described, namely the Primary Alluvial

Gravels and the Derived Gravels. Distinctive similarities exist between the Rooikoppie Gravels and the Derived Gravels. Both deposits consist almost entirely of chemical lithologies mixed with a small proportion of local Ventersdorp lavas and exhibit colluvial redistribution features. They contain no record of the time, place, or character of their initial, presumably alluvial aggradation, but are simply a weathered lag derived from an earlier deposit.

Correlations of the Primary Alluvial Gravels (the initial, presumably alluvial deposits) have not yet been encountered in the southwestern Transvaal. It is the opinion of the present author that such alluvial gravels may be found beneath the Rooikoppie Gravels where these lie on a calcrete bottom. Where no Rooikoppie Gravels overlie a lateratized pseudo-karst surface, these are likely the only vestiges of a colluvially reworked slope deposit.

The Younger Gravels of the lower Vaal River basin have been divided into the Rietputs (A-D) and Riveton (I-V) formations, representing numerous episodes of cut and fill under varying climatic regimes (Helgren, 1979). The Terrace gravels of the southwestern Transvaal may be tentatively correlated with the Rietputs-A and -B formations. The basal fines and the gravels themselves may be attenuated correlations of the Rietputs-A formation (representing humid aggradation). The calcreted sediments overlying the gravels are possibly time equivalents of the Rietputs-B formation. In both cases the deposits are calcreted. The calcretization is not pedogenic but the result of precipitation from throughflow and surface waters draining the upslope areas.

3.2. Environment of Deposition

From an analysis of the Rooikoppie and Terrace gravels and a comparison with the Older and Younger gravels of the lower Vaal Basin the following sequence is proposed for their deposition and subsequent evolution;

(1). During the long period of stillstand associated with the African erosion surface primary alluvial gravels were reworked predominantly by slope processes and redistributed over an intensively lateratized, pseudo-karst surface. Patterns found in similar Derived Gravel deposits in the lower Vaal Basin indicate that they may be Cretaceous in age (T.C. Partridge, pers. comm.).

(2). Following this extended period of stillstand incision of the Vaal River and its tributaries occurred. This was followed by aggradation of the Rietputs-A basal fines and gravel formation under relatively humid conditions. The incision is probably associated with uplift along the Griqualand-Transvaal axis (Marshall, 1986).

(3). During this period of aggradation the climate appears to have become significantly drier, with the deposition of the Rietputs-B sediments and subsequent calcretization of the entire sequence. The period of aridification was finally climaxed by the deposition of the Hutton Sands - mixed alluvial and aeolian sands of apparently Kalahari origin.

(4). Subsequent phases of incision and aggradation took place along the main Vaal River and extended along many of its tributaries (the Bamboespruit, for example). These deposits, which occur in the present channel of the Bamboespruit, have reworked the diamondiferous gravels in places and may themselves be diamond-bearing.

4. CONCLUSIONS

Field evidence indicate that there are two fundamental types of gravel deposits in the southwestern Transvaal, namely the Rooikoppie (Derived) Gravels and the Terrace Gravels. The Rooikoppie gravels usually occur as a thin veneer on hillslopes and as infill of pseudokarst grikes in the underlying lateratized surface. These colluvially reworked and redistributed gravels represent the remains of an ancient, presumably alluvial deposit. The Terrace Gravels which range in colour from green through grey to brown, occur in the valley of the Bamboespruit. The variation in colour is the result of the degree of calcretization and oxidation-reduction that has occurred in the gravels. These gravels are a result of incision and subsequent aggradation of the Vaal River and its tributaries, most likely during the Tertiary warping of the central interior.

The present distribution of the gravels can be explained in terms of colluvial and alluvial reworking of earlier deposits. The diamond concentration within the gravels is controlled by hydraulic conditions present in the channel. Economic concentrations of the diamonds occur in potholes in the channel floor, associated with obstructions (such as dykes) in the channel, downstream of constrictions and confluences, and in the meander point-bars of the palaeostream.

The question regarding the ultimate source of the gravels and diamonds still remains unanswered. If the Rooikoppie Gravels are the colluvial remains of the original alluvial deposit then these have to be sought by other means outside of the Bamboespruit valley. A possible relationship exists between palaeodrainage relating to this early epoch and small pans adjacent to some Rooikoppie deposits. This line of investigation is being further pursued by the author.

5. SUMMARY

The field evidence indicates that there are two fundamentally   different types of diamondiferous gravels associated with the Bamboespruit. The oldest (rooikoppie) gravels occur as a 1-2 m thick unsorted, matrix-supported, generally upward-fining unit that has been completely lateratized. These gravels usually occur in hillcrests or on the upper sections of hillslopes, and are generally overlain by a thin (0.5-1.0 m) soil overburden. Younger terrace-type deposits (the oxidized and reduced gravels) occur on the lower slopes of the present tributary valleys. These gravels average 1-4 m thick units, developed below 8-10 m of sand, mud, calcrete overburden. The gravels closer to the present stream channels have a greenish clay matrix which becomes more oxidized (and, hence, brownish) away from the channels. Deposits of economic concentrations of diamonds in these gravels are found in potholes in the lavas of the channel floors, behind obstructions in the channels, associated with large boulders, and in meander point-bars. The youngest (spruit gravel) deposits occur in the present channels of the Vaal River tributaries and represent recent reworking of all the older deposits.

6. REFERENCES

du Toit, A.L. 1951. The diamondiferous gravels of Lichtenburg. Geological Survey of South Africa, Memoir 44, 58pp.

Fitzpatrick, R.W., Hahne, H.C.H. and Terreblanche, S.P. 1986. Soil Mineralogy. In: MacVicar (ed.), Land Types of the Maps SE 27/20 Witdraai, 2720 Normieput, 2722 Kuruman, 2724 Christiana, 2820 Upington, 2822 Postmasburg, Memoir of the Department of Agriculture and Natural Resources of South Afrrica No. 3.

Helgren, D.M. 1979. River of Diamonds: An Alluvial History of the Lower Vaal Basin, South Africa. University of Chicago, Department of Geography Research paper No. 185, 309pp.

Marshall, T.R. 1986. The Alluvial Diamond Fields of the western Transvaal. Information Circular No. 188, Economic Geology Research Unit. University of the Witwatersrand, Johannesburg, 13pp.

Mayer, J.J. 1986. Differential erosion of possible Kalahari aeolian deposits along the Vaal-Orange drainage basin and the upper reaches of the Harts River. Transactions of the Geological Society of South Africa, 89, 401-408.

Partridge, T.C. and Brink, A.B.A. 1969. Gravels and terraces of the lower Vaal basin. South African Geographical Journal, 49, 21-38.

Partridge, T.C. and Maud, R.R. 1987. Geomorphic evolution of Southern Africa since the Mesozoic. South African Journal of Geology, 90 (2), 179-208.

Retief, E.A. 1960. The diamondiferous gravels in the Lichtenburg-Ventersdorp area. Geological Survey Report No. 1960-0004, Pretoria, 68pp.

LIST OF CONTRIBUTORS

V.R. Baker, Department of Geosciences, University of Arizona, Tucson, Arizona 85621, U.S.A.

H.R. Beckedahl, Department of Geography, University of Transkei, Private Bag X1, Unitra, Umtata, Transkei, Southern Africa

T.A.S. Bowyer-Bower, School of Geography, University of Oxford, Mansfield Road, Oxford, OX1 3TB, United Kingdom

G.F. Dardis, Department of Geography, University of Transkei, Private Bag X1, Unitra, Umtata, Transkei, Southern Africa

M. De Dapper, Laboratory for Physical Geography, Geological Institute of the State University, Krigslaan 281, B-9000 Gent, Belgium

A.B. De Villiers, Department of Geography, Potchefstroom University for C.H.E., P.O. Potchefstroom 2520, South Africa

F.J. Dewey, South African Sugar Association Experimental Station, P.B. X02, Mount Edgecombe 4300, South Africa

M.C.J. De Wit, Geology Department, De Beers Consolidated Mines Ltd., P.O. Box 47, Kimberley 8300, South Africa

K.J. Hall, Department of Geography, University of Natal, P.O. Box 375, Pietermaritzburg 3200, South Africa

P.M. Hanvey, Department of Geography, University of Transkei, Private Bag X1, Unitra, Umtata, Transkei, Southern Africa

J.T. Harmse, Department of Geography, Rand Afrikaans University, P.O. Box 524, Johannesburg 2000, South Africa

J. Hövermann, Geographisches Institut, University of Göttingen, FR Germany

W. Illenberger, Department of Geology, P.O. Box 1600, University of Port Elizabeth, Port Elizabeth 6000, South Africa

E.O. Jacobs, South African Forestry Research Institute, Saasveld Forestry Research Centre, Private Bag X6525, George 6530, South Africa

V. Klein, 41000 Zagreb, Yugoslavia

N. Lancaster, Department of Geology, Arizona State University, Tempe, Arizona, U.S.A.

C.A. Lewis, Department of Geography, University of Zululand, Private Bag X1001, Kwa-Dlangezwa 3886, South Africa

Geomorphological Studies in Southern Africa, G.F.Dardis & B.P.Moon (eds)
© 1988 Balkema, Rotterdam. ISBN 90 6191 831 6

M.E. Marker, Department of Geography, University of Fort Hare, Private Bag X1314, Alice, Ciskei, Southern Africa

T.R. Marshall, Luipaardsvlei Geological Centre, Gold Fields of South Africa, P.O. Box 53 Krugersdorp 1740, South Africa

R.R. Maud, P.O. Box 4122, Durban 4000, South Africa

M.E. Meadows, Department of Environmental and Geographical Sciences, University of Cape Town, Private Bag, Rondebosch 7700, South Africa

B.P. Moon, Department of Geography and Environmental Studies, University of the Witwatersrand, 1 Jan Smuts Avenue, Johannesburg, South Africa

G.J. Mulder, Department of Geography, University of Zululand, Private Bag X1001, KwaDlangezwa 3886, South Africa

T.C. Partridge, Department of Geology, University of the Witwatersrand, 1 Jan Smuts Avenue, Johannesburg, South Africa

E. Plumstead, Department of Zoology, University of Transkei, Private Bag X1, Unitra, Transkei, Southern Africa

G. Prasad, Institute for Southern African Studies, National University of Lesotho, P.O. Roma 180, Lesotho

J. Rogers, Marine Geoscience Unit, University of Cape Town, Private Bag, Rondebosch 7700, South Africa

K.M. Rowntree, Department of Geography, Rhodes University, P.O. Box 94, Grahamstown 6140, South Africa

N.W. Rutter, Department of Geology, University of Alberta, Edmonton, Alberta, Canada

I. Rybak, Aquatic Research and Ecological Consultants, Mississauga, Ontario, Canada

M. Rybak, Aquatic Research and Ecological Consultants, Mississauga, Ontario, Canada

H. Sänger, Institute for Geography, University of Hamburg, FR Germany

R. Sarracino, Department of Physics, National University of Lesotho, P.O. Roma 180, Lesotho

J.M. Sugden, Department of Environmental and Geographical Sciences, University of Cape Town, Private Bag, Rondebosch 7700, South Africa

J.T. Teller, Department of Geological Sciences, University of Manitoba, Winnipeg, Manitoba, Canada

D.S.G. Thomas, Department of Geography, University of Sheffield, Sheffield S10 2TN, United Kingdom

R.N. Thwaites, School of Earth Sciences, Macquarie University, North Ryde, N.S.W. 2109, Australia

C.R. Twidale, Department of Geography, University of Adelaide, Adelaide, Australia

P.W. van Rheede van Oudtshoorn, Department of Geography, Vista University, Private Bag X03, Tsiawelo 1818, South Africa

T.H. van Rooyen, Department of Geography, University of South Africa, P.O. Box 392,

0001 Pretoria, South Africa

E. Verster, Department of Geography, University of South Africa, P.O. Box 392, 0001 Pretoria, South Africa

J.D. Ward, Geological Survey of South West Africa, Windhoek, Namibia

H.K. Watson, Department of Geography, University of Durban-Westville, Private Bag X54001, Durban 4000, South Africa.

A. Van B. Weaver, Department of Geography, Rhodes University, P.O. Box 94, Grahamstown 6140, South Africa

M.J. Wilkinson, Department of Geography and Environmental Studies, University of the Witwatersrand, 1 Jan Smuts Avenue, Johannesburg 2050, South Africa